T0342531

A Framework of Human Systems Engineering

A Framework of Human Systems Engineering

Applications and Case Studies

Edited by

Holly A. H. Handley
Andreas Tolk

WILEY

Library of Congress Cataloging-in-Publication Data:

Names: Handley, Holly A. H., editor. | Tolk, Andreas, editor.
Title: A framework of human systems engineering : applications and case
  studies / edited by Holly A. H. Handley, Andreas Tolk.
Description: Hoboken, New Jersey : Wiley-IEEE Press, [2021] | Includes
  bibliographical references and index.
Identifiers: LCCN 2020038951 (print) | LCCN 2020038952 (ebook) | ISBN
  9781119698753 (cloth) | ISBN 9781119698777 (adobe pdf) | ISBN
  9781119698760 (epub)
Subjects: LCSH: Systems engineering. | Human engineering.
Classification: LCC TA168 .F736 2021 (print) | LCC TA168 (ebook) | DDC
  620.8/2–dc23
LC record available at https://lccn.loc.gov/2020038951
LC ebook record available at https://lccn.loc.gov/2020038952

Cover Design: Wiley
Cover Images: Power robotic arm© Phanie/Alamy Stock Photo, Young woman with wearing smart glass
© Tim Robberts/Getty Images, Self driving vehicle © metamorworks/Getty Images, Abstract background
© MR.Cole_Photographer/ Getty Images

10 9 8 7 6 5 4 3 2 1

To Mark
Quarantined together for 86 days and still married.
HAHH

To my father, who taught me the love for numbers and precision,
And to my mother, who taught me the love for the human mind and inclusion.
Andreas Tolk

Contents

Section 3　Focus on Training and Skill Sets　*129*

8　Building a Socio-cognitive Evaluation Framework to Develop Enhanced Aviation Training Concepts for Gen Y and Gen Z Pilot Trainees　*131*
Alliya Anderson, Samuel F. Feng, Fabrizio Interlandi, Michael Melkonian, Vladimir Parezanović, M. Lynn Woolsey, Claudine Habak, and Nelson King

9　Improving Enterprise Resilience by Evaluating Training System Architecture: Method Selection for Australian Defense　*143*
Victoria Jnitova, Mahmoud Efatmaneshnik, Keith F. Joiner, and Elizabeth Chang

Editor Biographies

Holly A. H. Handley is an associate professor in the Engineering Management and System Engineering Department of Old Dominion University (ODU). Her research focuses on developing models and methodologies to better represent the human component during the architecting and design of sociotechnical systems. She received her PhD from George Mason University in 1999 and is a licensed professional engineer. Her education includes a BS in Electrical Engineering from Clarkson College (1984), a MS in Electrical Engineering from the University of California at Berkeley (1987), and an MBA from the University of Hawaii (1995). Prior to joining ODU, Dr. Handley worked as a design engineer for Raytheon Company (1984–1993) and as a senior engineer for Pacific Science & Engineering Group (2002–2010). Dr. Handley is a member of the Institute of Electrical and Electronic Engineers (IEEE) Senior Grade, the International Council on Systems Engineering (INCOSE), and the Human Factors and Ergonomics Society. She is currently the chair of the IEEE Systems Council Human Systems Integration Technical Committee and was recently named an HFES Science Policy Fellow.

Andreas Tolk is a senior divisional staff member at The MITRE Corporation in Charlottesville, VA, and adjunct full professor at Old Dominion University in Norfolk, VA. He holds a PhD and MSc in Computer Science from the University of the Federal Armed Forces of Germany. His research interests include computational and epistemological foundations and constraints of model-based solutions in computational sciences and their application in support of model-based systems engineering, including the integration of simulation methods and tools into the systems engineering (SE) education and best practices. He published more than 250 peer-reviewed journal articles, book chapters, and conference papers and edited 12 textbooks and compendia on SE and modeling and simulation topics. He is a fellow of the Society for Modeling (SCS) and Simulation and senior member of IEEE and the Association for Computing Machinery (ACM) and received multiple awards, including professional distinguished contribution awards from SCS and ACM.

Contributors List

Hussein A. Abbass
School of Engineering and Information
Technology
University of New South Wales
Canberra, Canberra, ACT, Australia

Alliya Anderson
Khalifa University of Science and Technology
Abu Dhabi, United Arab Emirates

Philip S. Barry
George Mason University
Fairfax, VA, USA

Marius Becherer
University of New South Wales at Australian
Defence Force Academy
Canberra, ACT, Australia

Jean Bogais
University of Sydney
Sydney, NSW, Australia

Amy E. Bolton
Office of Naval Research
Arlington, VA, USA

Guy André Boy
CentraleSupélec, Paris Saclay University
Gif-sur-Yvette, France; ESTIA Institute of
Technology, Bidart, France

Thien Bui-Nguyen
University of New South Wales at Australian
Defence Force Academy
Canberra, ACT, Australia

A. Peter Campbell
SMART Infrastructure Facility
University of Wollongong
Wollongong, NSW
Australia

Elizabeth Chang
University of New South Wales at
Australian Defence Force Academy
Canberra, ACT, Australia

Hugh David
Chartered Institute of Ergonomics
and Human Factors
Birmingham, UK

Steve Doskey
The MITRE Corporation
McLean, VA, USA

Mahmoud Efatmaneshnik
Defence Systems Engineering at the
University of South Australia (UNISA)
Adelaide, SA, Australia

Samuel F. Feng
Khalifa University of Science and
Technology
Abu Dhabi, United Arab Emirates

Florian Gottwalt
University of New South Wales at
Australian Defence Force Academy
Canberra, ACT
Australia

Stuart Green
University of New South Wales at Australian
Defence Force Academy
Canberra, ACT, Australia

Claudine Habak
Emirates College for Advanced Education
Abu Dhabi, United Arab Emirates

Holly A. H. Handley
Old Dominion University
Norfolk, VA, USA

Fabrizio Interlandi
Etihad Aviation Training
Abu Dhabi, United Arab Emirates

Victoria Jnitova
School of Engineering and Information
Technology, University of New South Wales at
Australian Defence Force Academy, Canberra,
ACT, Australia

Keith F. Joiner
Capability Systems Center, School of
Engineering and Information Technology,
University of New South Wales at Australian
Defence Force Academy, Canberra, ACT,
Australia

Michael Joy
IDEMIA National Security Solutions
New York, NY, USA

Grace A. L. Kennedy
SMART Infrastructure Facility
University of Wollongong
Wollongong, NSW, Australia

Nelson King
Khalifa University of Science and Technology
Abu Dhabi
United Arab Emirates

Pravir Malik
First Order Technologies, LLC
Berkeley, CA, USA

Angus L. M. T. McLean
Collins Aerospace
Cedar Rapids, IA, USA

Michael Melkonian
Emirates College for Advanced Education
Abu Dhabi, United Arab Emirates

Kelly J. Neville
The MITRE Corporation
Orlando, FL, USA

Vladimir Parezanović
Khalifa University of Science and Technology
Abu Dhabi, United Arab Emirates

Maria Natalia Russi-Vigoya
KBR, Houston, TX, USA

George Salazar
Johnson Space Center, NASA,
Houston, TX, USA

Christian G. W. Schnedler
CISSP®, CSEP®, PMP®, and PSP®; IDEMIA
National Security Solutions, New York,
NY, USA

William R. Scott
SMART Infrastructure Facility
University of Wollongong, Wollongong
NSW, Australia

Sarah M. Sherwood
Naval Medical Research Unit Dayton
Wright-Patterson AFB, OH, USA

Farid Shirvani
SMART Infrastructure Facility, University
of Wollongong, Wollongong, NSW,
Australia

Andreas Tolk
The MITRE Corporation,
Charlottesville, VA, USA

Melissa M. Walwanis
Naval Air Warfare Center Training
Systems Division
Orlando, FL, USA

M. Lynn Woolsey
Emirates College for Advanced Education
Abu Dhabi
United Arab Emirates

Kate J. Yaxley
School of Engineering and Information
Technology, University of New South Wales,
Canberra, Canberra, ACT, Australia

Michael Zipperle
University of New South Wales at Australian
Defence Force Academy, Canberra, ACT,
Australia

Foreword

No one would question that we are today living in the age of connectivity. Global communications, global commerce, and global pandemics epitomize current affairs.

From a system of systems perspective rarely do we ever design and employ a system in isolation. Systems are developed and used in new and innovative ways as needs change, often working with other systems in ways not considered when the systems themselves were conceived. Complex supply chains integral to the modern economy include connections and dependencies that go beyond common understanding. With unknown billions of nodes in the Internet, connectivity between systems and people is a bedrock of contemporary society.

People are connected to their workplace, retailers, and their friends and family electronically. The majority of Americans possess "smart phones" that connect them into a growing network of cyber–physical systems – the Internet of Things – where they are part of a complex collaborative exchange with people and systems. People have moved beyond being "users" of systems to become an integral part of systems of systems both in the small and large sense. People no longer simply consume services of systems, but they and their actions are a core part of the dynamics of the larger system of systems. Their actions can affect the systems of systems in ways often not well understood, and changes in human behavior can have considerable ripple effects on large complex societal capabilities.

All of this has profound implications for human systems engineering. While a premium continues to be placed on human-centered design focusing on the direct relationship between systems and their users, human systems considerations have expanded in this age of connectivity putting new demands on systems engineers as they factor human systems considerations into engineering endeavors.

We as systems engineers are no longer just expected to ensure that our systems are usable by an individual, but we are also expected to integrate users into complex distributed systems of systems where the users are part of the systems of systems and their behavior is part of the larger system of systems dynamics.

Systems engineers are no longer just expected to design systems, so they have value for the users but increasingly are asked to build systems that also bring value to the system owners through generation of data to support other aspects of the enterprise or to influence people's economic, political, or social behavior.

Particularly in safety critical situations, it is no longer enough for systems engineers to design systems that enable people to operate systems to meet their immediate needs, but as these systems are part of a larger dynamic environment, a growing need exists to provide sufficient situational awareness to understand the impacts individual actions may have on other systems and people in the larger systems of systems.

Finally, as systems take on functions that had in the past been done by people, there is an increased emphasis on developing approaches to human systems teaming, a challenge heightened by the increased use of machine learning, where the balance between human and systems may shift over time based on experience.

These changes make this book both timely and important. With the framework provided by Handley in the opening chapter to the research agenda by Tolk at the close, the papers here explore numerous dimensions of human systems engineering, providing a window on experiences today and challenges for the future.

Judith Dahmann
MITRE Corporation Technical Fellow
INCOSE Fellow
Alexandria, Virginia

Preface

The International Council on Systems Engineering (INCOSE) defines Systems Engineering (SE) as an interdisciplinary approach and means to enable the realization of successful systems. SE focuses on defining customer needs by documenting requirements and then proceeds with functional analysis, design synthesis, and system validation. Throughout this process the complete system life cycle is considered: operations, performance, test, manufacturing, cost and schedule, training and support, and disposal.

SE promotes a team effort integrating various disciplines and specialty groups into a structured development process that considers both the business and the technical needs of all customers with the goal of providing a quality product that meets the users' needs. It is therefore considered a methodical, disciplined approach for the design, realization, technical management, operations, and retirement of a system. In all these aspects, humans play a vital role. They define, use, maintain, and, as operators and decision makers, are part of the system. Since a system is only as strong as its weakest component, human potentials, capabilities, constraints, and limitations are pivotal for the successful engineering of systems.

The Human Systems Integration (HSI) Technical Committee (TC) of the IEEE Systems Council was formed in order to increase awareness of the user during SE processes. It focuses on identifying and improving methods to integrate human concerns into the conceptualization and design of systems. It encourages early understanding of human roles and responsibilities, along with limitations and constraints that may impact system design. This consideration of human concerns from the system design perspective is termed human systems engineering (HSE). HSE describes the engineering efforts conducted as part of the system design and analysis processes to evaluate the appropriateness and feasibility of system functions and roles allocated to operators. The importance of this topic is apparent from notable design errors, i.e. the placement of the iPhone 4 antenna resulting in poor performance when holding the phone, to design successes, for example, the Xbox Kinect that allowed users to interact with the game system without a handheld interface.

One of the goals of the HSI TC is to improve communication between the HSI and SE communities to provide better integration of human and systems to expedite resolution of issues. The HSI TC members promote this collaboration through conference presentations and workshops, as well as cooperation with other societies through joint events. Our members serve as technical reviewers and society liaisons to promote the role of human factors in engineering. This volume is a continuation of our technical committee outreach efforts.

This book was written for both systems engineers and HSI practitioners who are designing and evaluating different types of sociotechnical systems across various domains. Many engineers have heard of HSE but don't understand its importance in system development. This book presents a series of HSE applications on a range of topics, such as interface design, training requirements, personnel capabilities and limitations, and human task allocation. Each chapter represents a case study of the application of HSE from different dimensions of sociotechnical systems. The examples are organized using a sociotechnical system framework to reference the applications across multiple system types and domains. These case studies serve to illustrate the value of applying HSE to the broader engineering community and provide real-world examples. The goal is to provide reference examples in a variety of domains and applications to educate engineers; the integration of the human user is listed as one of the enablers of SE in the Systems Engineering Body of Knowledge (SEBoK).

As IEEE is primarily concerned with the engineering of electrical technologies, our goal is to include the perspective of design engineers who may be removed from the end user and unaware of potential concerns. The book chapters represent specific projects from the HSI TC members; the result is a set of stories that show the value of HSE through the development of human interfaces, improvement of human performance, effective use of human resources, and the design of safe and usable systems. The examples cross traditional SE sectors and identify a diverse set of HSE practices. Our contributed book is a source of information for engineers on current HSE applications.

Holly A. H. Handley, PhD, PE and Andreas Tolk, PhD

Section 1

Sociotechnical System Types

1

Introduction to the Human Systems Engineering Framework

Holly A. H. Handley

Old Dominion University, Norfolk, VA, USA

1.1 Introduction

Many human-centered disciplines exist that focus on the integration of humans and systems. These disciplines, such as human factors (HF), human systems integration (HSI), and human factors engineering (HFE), are often used interchangeable but have distinct meanings. This introductory chapter identifies these varied disciplines and then defines the domain of human systems engineering (HSE). HSE implies that human has been "engineered" into the design, in contrast to "integrating" the user into the system at later stages of design.

The use of HSE for increasing complex and varied sociotechnical systems requires a more context-specific suite of tools and processes to address the combination of human and system components. More often a wider range of system stakeholders, including design and development engineers, are becoming involved in, and are vested in, the success of both HSE- and HSI-related efforts. To assist these efforts, a framework was developed based on the dimensions of sociotechnical system and domain types, with relationships to specific HSI and SE concerns. The development of this framework and its dimensions is also described in the chapter.

Finally, the framework is used to organize a wide range of case studies across a variety of system types and domains to provide examples of current work in the field. These case studies focus on both the systems engineering (SE) applications and the HSE successes. Linking the cases to the framework identifies the contextual variables, based on both sociotechnical system and domain characteristics, and links them to specific human system concerns. Our goal with this volume is to emphasize the role of systems engineers in the development of successful sociotechnical systems.

1.2 Human-Centered Disciplines

HF is a broad scientific and applied discipline. As a body of knowledge, HF is a collection of data and principles about human characteristics, capabilities, and limitations. This knowledge base is derived from empirical evidence from many fields and is used to help minimize the risk of systems by incorporating the diversity of human characteristics (England 2017). Ergonomics is the

A Framework of Human Systems Engineering: Applications and Case Studies, First Edition.
Edited by Holly A. H. Handley and Andreas Tolk.

scientific discipline concerned with the understanding of interactions among humans and other elements of a system and the profession that applies theory, principles, data, and methods to design in order to optimize human well-being and overall system performance (IEA 2018). The term "human factors" is generally considered synonymous with the term "ergonomics." HF engineers or ergonomics practitioners apply the body of knowledge of HF to the design of systems to make them compatible with the abilities and limitations of the human user.

HF has always employed a systems approach; however, in large complex systems, it was recognized that the role of the human must be considered from multiple perspectives (Smillie 2019). HSI is the interdisciplinary technical process for integrating multiple human considerations into SE practice (DOA 2015). Seven HSI areas of concerns have been identified – manpower, personnel, training, HFE, health and safety, habitability, and survivability – all of which need to be addressed in an interconnected approach. The emphasis of the HSI effort is on the trade-offs within and across these domains in order to evaluate all options in terms of overall system performance, risk, and personnel-related ownership cost (SAE6906 2019). HSI provides a comprehensive snapshot of how human systems interaction has been addressed throughout the system development process by evaluating each of these domains as the system design progresses through different stages. It identifies what issues remain to be resolved, including their level of risk, and suggests potential mitigations.

Human factors integration (HFI) is a systematic process for identifying, tracking, and resolving human-related issues ensuring a balanced development of both technological and human aspects of a system (Defence Standard 00-251 2015). HFI is the term used in the United Kingdom equivalent to HSI. Similar to HSI, HFI draws on the breadth of the HF disciplines and emphasizes the need to facilitate HFI management activities of concern across seven similar domains: manpower, personnel, training, HFE, system safety, health hazard assessment, and social and organizational (England 2017). The methods and processes available for HFI can be broken down into both technical activities and management activities; HFI has a well-defined process and can draw on many methods, tools, standards, and data in order to prevent operational and development risks (Bruseberg 2009).

1.3 Human Systems Engineering

The HSI discipline was established with the primary objective to enhance the success of the Department of Defense (DoD) systems by placing humans on more equal footing with design elements such as hardware and software (SAE6906 2019). SE is an interdisciplinary field of engineering and engineering management that focuses on how to design and manage complex systems over the system life cycle. While HSI is considered an enabler to SE practice, systems engineers need to be actively engaged to continuously consider the human as part of the total system throughout the design and development stages. HSE is the application of human principles, models, and techniques to system design with the goal of optimizing system performance by taking human capabilities and limitations into consideration (DOD 1988). HSE approaches the human system design from the perspective of the systems engineer and views the human component as a system resource. Human-focused analyses that occur as part of the HSE evaluations determine the required interactions between users and technology and are essential to insure efficient processes and data exchange between the technology elements and the human users (Handley 2019a). In the United Kingdom, human-centric systems engineering (HCSE) seeks better ways to address HF within mainstream SE while building on and optimizing the coherence of existing best practice.

Similar to HSE, HCSE approaches HF from an SE viewpoint and aims to develop core SE practices that help engineering organizations adopt the best HF processes for their needs (England 2017).

HSE applies what is known about the human to the design of systems. It focuses on the tasks that need to be performed, the allocation of specific tasks to human roles, the interactions required among the human operators, and the constraints imposed by human capabilities and limitations. A key focus of HSE is on the determination of the human role strategy; this allocation determines the implications for manning, training, and ultimately cost (ONR 1998). The human elements of the system possess knowledge, skills, and abilities that must be accounted for in system design, along with their physical characteristics and constraints, similar to other technical elements of the system. The goal of HSE is to augment the system descriptions with human-centered models and analysis; these purposeful models inform trade-off analyses between system design, program costs, schedule, and overall performance (Handley 2019a). As part of the SE process, HSE incorporates the human-related specifications into the system description to improve overall system performance through human performance analysis throughout the system design process.

1.4 Development of the HSE Framework

The HSE framework was developed for the SE community to provide a basis for categorizing and understanding applications of HSE for different types of sociotechnical systems. It was developed by cross-referencing and aligning different aspects of domains, system types, and design stages with applicable HSE and HSI tools and methods. The goal was to categorize projects in such a way that systems engineers and HSI practitioners could leverage tools, processes, and lessons learned across projects (Handley 2019b).

The original framework was developed by a team of Army HSI practitioners and subject matter experts (SMEs). The HSE framework was part of a larger project designed to mitigate human performance shortfalls and maximize system effectiveness by integrating well-defined HSE (and where applicable HSI) processes and activities into the acquisition life cycle and to make these analyses explicit to stakeholders to increase "buy-in" early in the design process (Taylor 2016). The resulting ontology could be expanded as needed to provide a common framework to identify elements and relationships important to the application of HSE, including classifying different stakeholders, system types, acquisition timelines, and user needs. This would allow HSI practitioners, systems engineers, and program managers to determine appropriate tools, methodologies, and information. The overall goal was to provide an overall organizing structure for HSE processes and products relevant to the SE effort that could be linked to a comprehensive repository of information and concurrent and past projects (Taylor 2016).

The original HSE framework is shown in Figure 1.1; it is a subset of the envisioned comprehensive ontology. This framework was used successfully to categorize different projects that involved the intersection of SE and HSI, including the Army's transition to cross-functional teams (Handley 2018). The framework represents the initial effort to provide a consistent taxonomy to determine appropriate tools and methodologies to address sociotechnical system concerns by offering an organizing structure to identify similar efforts.

The dimensions and descriptions of the original framework are as follows:

A) Sociotechnical system type – This dimension represents the different ways that users interact with systems. From the "users are the system," which represents organizations and teams, to the other extreme "no direct system," which represents autonomous systems, the intermediary points suggest different interaction points between users and systems.

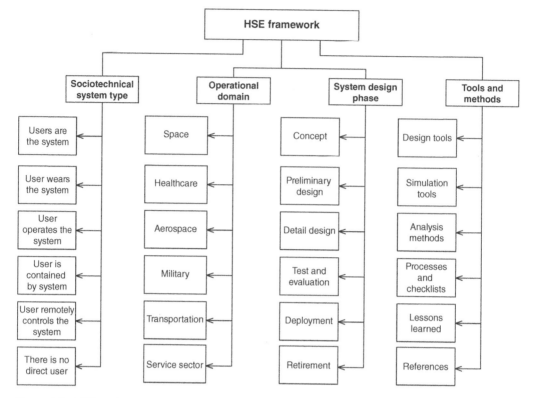

Figure 1.1 HSE original framework.

B) Domains – This dimension represents the different contexts of use for systems, as different domains can induce different considerations and restrictions. Domain-induced constraints include environmental variables, operator state, organizational factors, and personnel characteristics. While the framework was developed specifically for military systems, it can be extended and adapted across various domains such as space, transportation, and aerospace.

C) System design phases – The intent of the original framework was to capture the impact of different tools and methods at different phases of system design, i.e. concept, preliminary design, detailed design, test and evaluation, deployment, and retirement. This approach emphasized the benefits of applying human-centered analyses early in the system development.

D) Tools and methods – By mapping the three previous dimensions to available tools and methods, the intent of the framework was that it could be used to suggest tool sets for different human-centered analyses depending on the system type, domain, and stage of system development.

The framework acts as an index to identify essential information and previously validated findings. It can be used to suggest tools, methods, processes, data, standards, and expertise across similar systems and/or domains. The intent in developing the framework was that the dimensions could be expanded or modified as needed to capture evolving elements in sociotechnical systems and provide the metadata to classify the required HSE efforts.

1.5 HSE Applications

The original framework has been repurposed here to classify the case studies that compose this volume. The original dimensions have been slightly modified to better provide an index to the cases presented. This revised framework maintains the sociotechnical and domain dimensions; however, the second two dimensions were modified slightly to represent both HSE and SE concerns as shown in Figure 1.2. Note that for simplicity, both the HSE and SE concerns dimensions were limited to those that appear in the case studies. The modified framework presents a better categorization of the cases provided and facilitates easy identification of cases that best match the readers' interest.

Additionally, the rendering of the framework has changed from the original tree structure to a multi-axis plot. Each axis represents one of the framework dimensions, and the hash marks identify the subcategories. This visualization allows the cases to be "plotted" as an intersection of two (or more) dimensions. While the original framework identified the categories for each domain, the new rendering allows these categories to be used as a classification system, easily identifying the key content of each case study. As the applications in this volume are quite varied, the framework provides a logical way to organize and connect the case studies.

As shown in Figure 1.2, each chapter has been located on the framework to show its intersection among the dimensions. The first section of the book contains applications that describe different sociotechnical system types and their relationships with the human user. For example, Chapter 2

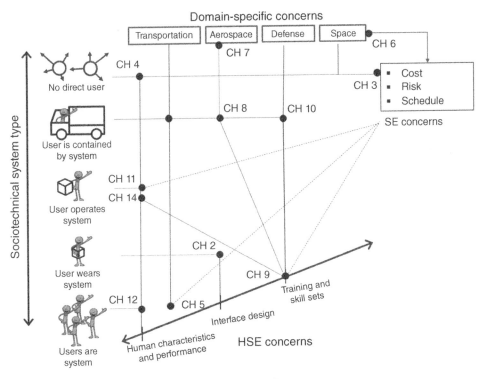

Figure 1.2 HSE framework as an index for the case studies.

describes human considerations for domain awareness and focuses on human interface design. The authors make a comprehensive analysis of situational awareness platforms for public safety and stress the importance of traditional training methods coupled with cutting-edge technology. Chapter 3 defines the sociotechnical factors shown to improve success using artificial intelligence in a system development. With the integration of artificial intelligence into every system domain, the authors employ a quantitative model of the sociotechnical space to identify the discrepancies between not considering the stakeholder and high risks in complex agile projects. Chapter 4 considers both technology readiness and autotomy level to determine meaningful human control based on trust in human-autonomy teaming. The authors use an example of herding sheep with airborne drones to provide a validation scenario for the proposed concept and process.

The second section of the book provides a "deep dive" focus on specific domains. These chapters provide examples of HSE impacts in specific contexts. For example, Chapter 5 looks at the Australian heavy rail industry and the use of sociotechnical modeling. The authors describe how integrating HF models with SE can be used to introduce new capabilities from an integrated organizational standpoint. Chapter 6 focuses on the engineering life cycle for space exploration systems and the use of human-centered programs to mitigate risk. The authors describe how HSE can play an important role throughout the SE phases to optimize total system performance. Chapter 7 reviews the evolution of cockpit design based on the impact of evolving technologies the aerospace domain. The author describes how traditional human–computer interaction practices have given way to user experience "UX" and interaction design methodologies.

The next section, section three, focuses on training and skill sets with cross-references to different domains. Chapter 8 discusses the impact of generational differences of users on the design of training programs. The authors describe a socio-cognitive framework that combines the social aspects, i.e. generational differences, with the cognitive aspects, such as neuropsychology, that allows researchers to assess the effectiveness of gamified learning interventions. Chapter 9 investigates how training resiliency impacts readiness in the military domain. The authors identify basic workforce resilience measures that can be used to guide SE efforts to migrate to new training systems. Finally, Chapter 10 describes research that evaluates the introduction of virtual and constructive technology into live air combat training systems. The authors use qualitative methods, influenced by cognitive engineering and action research, to iteratively identify, assess, and mitigate risks stemming from the change of training techniques.

Section four presents two chapters that focus on the intersection of the socio-component and human characteristics. Chapter 11 presents an approach to build trustworthy blockchain applications for large complex enterprises based on HSE principles. The methodology develops a human data integration and interaction methodology through establishing trust and security links. The authors illustrate their approach through an operational risk management example. Chapter 12 offers a unique took at the impact of light technologies on organizational information. The author describes the association between the implicit properties of light on the four organizational principles of presence, power, knowledge, and harmony.

Finally, section five offers some observations "from the field." Chapter 13 provides a lighter note, offering an unedited account of some observations and suggestions for real-time control room future designs. Chapter 14 concludes the volume with a selection of research topics challenges compiled into several categories. The chapter author hopes that members of the scholastic community will contribute to the improvement of this first topology of challenges as well as the framework for HSE itself.

1.6 Conclusion

While many systems engineers understand that the human operator and maintainer are part of the system, they often lack the expertise or information needed to fully specify and incorporate human capabilities into the system design (INCOSE 2011). Human systems engineers are actively involved in the development of the system and ensure human-centered principles are incorporated into design decisions. HSE provides methods for integrating human considerations with and across system elements to optimize human system performance and minimize total ownership costs.

The case studies in this volume provide insights into HSE efforts across different sociotechnical system types across a variety of domains. Currently, most of the existing sociotechnical system case studies are from the HSI perspective, i.e. working with users to improve the system usability and interfaces in deployed systems. The focus of this book, however, is from the SE viewpoint, encouraging early consideration of the human in the system design. While some of the chapters will overlap with the traditional HSI approaches, the goal of the book is to encourage systems engineers to think about the human component earlier in the system development. The chapters are organized and indexed by the framework; the book can be read in order to follow the progression across the framework, or Figure 1.2 can be used to identify specific chapters of interest to the reader based on any one of the four dimensions. The goal of this book is to serve as a reference volume for HSE.

References

Bruseberg, A. (2009). *The Human View Handbook for MODAF (Pt. 2, Technical Description)*. Somerset, UK: Human Factors Integration Defence Technology Centre.

DOA (2015). *Soldier-Materiel Systems Human Systems Integration in the System Acquisition Process*. Department of the Army Regulation 602-2. Washington, DC: DOA.

DOD (1988). *Manpower, Personnel, Training, and Safety (MPTS) in the Defense System Acquisition Process*. DoD Directive 5000.53. Washington, DC: DOD.

England, R. (2017). *Human Factors for SE*. INCOSE UK, Z12, Issue 1.0 (March 2017). http://incoseonline.org.uk/Groups/Human_Centric_Systems_Engineering_WG/Main.aspx (accessed 16 March 2020).

Handley, H. (2018). *CFT by System Type and HSI Domain, Deliverable to Human Systems Integration (HSI) Tool Gap Analysis Report for Deputy Director*. US Army Human Systems Integration.

Handley, H. (2019a). Human system engineering. In: *The Human Viewpoint for System Architectures*. Springer.

Handley, H. (2019b). A socio-technical architecture. In: *The Human Viewpoint for System Architectures*. Springer.

IEA (2018). *What Is Ergonomics?* International Ergonomics Association. https://iea.cc/what-is-ergonomics (accessed 16 March 2020).

INCOSE (2011). *Systems Engineering Handbook: A Guide for System Life Cycle Processes and Activities*, 3.2e (ed. H. Cecilia). San Diego, CA: INCOSE.

ONR (1998). *Human Engineering Process*. Technical Report, SC-21 S&T Manning Affordability Initiative. Washington, DC: Office of Naval Research.

SAE6906 (2019). Standard Practice for Human System Integration, SAW6906, 2019-02-08.

Smillie, R. (2019). Introduction to the human viewpoint. In: *The Human Viewpoint for System Architectures* (ed. H. Handley). Springer.

Taylor, A. (2016). *The Human Systems Integration Workbench*. White Paper PJF-18-425. US Army Materiel Command (AMC).

UK Defence Standardization (2015). *Def Stan 00-251 Human Factors Integration for Defence Systems*, Public Comment Draft, Issue 1, Version 1.0 (September 2015).

2

Human Interface Considerations for Situational Awareness

Christian G. W. Schnedler[1] and Michael Joy[2]

[1] CISSP®, CSEP®, PMP®, and PSP®, IDEMIA National Security Solutions, New York, NY, USA
[2] IDEMIA National Security Solutions, New York, NY, USA

2.1 Introduction

The field of situational awareness (SA) arguably embodies the most urgent demand for human systems integration (HSI) as it encompasses the real-time application of (increasingly machine-assisted) human decision making in all-too-often life and death circumstances. Birthed in the maritime and military domains, SA concepts are now applied to fields as diverse as public safety and first responders, facility and border security, autonomous vehicles, and digital marketing. Common across these domains is the need to understand relevance within vast amounts of disparate data and present this information to human operators in an intuitive, timely, and conspicuous manner. To achieve these objectives, SA systems must disambiguate the definition of "relevant" by understanding the rules governing an operator's potential range of actions and the specific context of the operator receiving the information.

Emerging developments in the technology platforms of sensors, data, artificial intelligence (AI), computer vision, and mobile devices are enabling advancements in the SA platforms that provide real-time decision-making opportunities in both structured and unstructured space. These developments challenge the traditional ways that information has been collected, aggregated, collated, analyzed, disseminated, and provide opportunities to empower operators and citizens to gain greater awareness of their surroundings in order to make better informed and more meaningful decisions. Inherent challenges with the volume, variety, velocity, and veracity of this information demand novel approaches to HSI across multiple, concurrent operational theaters.

This chapter summarizes major considerations given to SA platforms and illustrates these through their application to the public safety domain. The authors draw on their decades-long experience designing and implementing SA systems in municipal and federal public safety organizations in regions as diverse as the United States, Middle East, and Africa. Due consideration is

---

This chapter contains the personal views of the authors based on their experience designing, implementing, and operating a variety of mission-critical situational awareness platforms. The opinions contained herein do not reflect the official position of the government agencies or private sector companies with which the authors served. The reader should remember that while technology can provide significant advancements to public safety, officer safety, and transparency, there is no substitute for systematic training and adhering to a culture of relentless improvement.

A Framework of Human Systems Engineering: Applications and Case Studies, First Edition.
Edited by Holly A. H. Handley and Andreas Tolk.

given to the growing concerns around privacy in Western nations and the apparent paradox around the need to promote transparency within public safety organizations without empowering terrorists, criminals, and others' intent on disrupting the lives and liberties of those engaged in democratic societies.

2.2 Situational Awareness: A Global Challenge

Situational awareness is a concept, a system, and a solution. There are well-established SA definitions and related organizations for the maritime domain, the space domain, and the Arctic. In her seminal *Designing for Situation Awareness* (Endsley 2011), Dr. Mica Endsley summarizes SA as "being aware of what is happening around you and understanding what that information means to you now and in the future." Elsewhere, Dr. Endsley has defined SA as "the perception of the elements in the environment within a volume of time and space, the comprehension of their meaning, and the projection of their status in the near future."[1] It is the internal mental model of the dynamic environment, which when combined with more static system and procedural knowledge allows decision makers in these domains to function effectively.

In the wake of the 9/11 attacks, the New York Police Department (NYPD) led a public–private partnership (PPP) effort to create what became the Domain Awareness System (DAS) to counter future terrorist attempts and to improve public safety.[2] This initial DAS effort by the NYPD provided a subsequent technology framework for the development of real-time SA solutions to address a broad range of public and private use cases, from high value facility security and border management to conflict zone and environmental protection, to healthcare, to opioid crisis response, and to the recovery of persons at risk from human traffickers. In each of these use cases, development was led by industry in partnership with government.

The Chinese central government has led PPP development of its "Sharp Eyes" surveillance system.[3] By intertwining digital commerce with public safety, China has created an unprecedented surveillance apparatus with near limitless opportunities for machine learning and analytics to process, categorize, and contextualize information for human operators. This surveillance system model now exported around the world as "Safe City" solutions challenges the Western notion of privacy and human rights when employed against targeted population groups like the Muslim Uighurs in western China.

In light of this range of applications, the definition of "situational awareness" remains somewhat ambiguous and nonpractical. For the purpose of this paper, SA refers back to the foundational definition espoused by Dr. Endsley and refers to the real-time presentation of pertinent information to a human operator to inform subsequent action. The geographic domain is relevant only in so much as its relevance to the human operator in question. Similarly, historic information and trends are relevant only in as much as they apply to the real-time context of the operator. Multiple operators may be involved in a single event, and the SA platform must consider the perspective and context of each in order to achieve its intended purpose.

---

1 https://www.researchgate.net/publication/285745823_A_model_of_inter_and_intra_team_situation_awareness_Implications_for_design_training_and_measurement_New_trends_in_cooperative_activities_Understanding_system_dynamics_in_complex_environments.
2 https://www1.nyc.gov/site/nypd/about/about-nypd/equipment-tech/technology.page.
3 https://asia.nikkei.com/Business/China-tech/China-s-sharp-eyes-offer-chance-to-take-surveillance-industry-global.

2.3 Putting Situational Awareness in Context: First Responders

Although much literature has been written on SA concepts in the aerospace, military, and maritime domains, the proliferation of Internet of Things (IoT) devices and advancements in machine vision and AI have enabled the democratization of SA capabilities. Under the banner of "smart cities," municipalities have begun implementing static surveillance capabilities and outfitting first responders with mobile and body-worn devices that act as both a sensor and a means of improving SA. By some estimates, the market for surveillance equipment will reach $77B by 2023.[4] This explosion in sensors has led to increased public safety expectations, as well as greater scrutiny over the actions taken by first responders.

To meet these expectations, law enforcement agencies in particular employ a variety of surveillance tools to achieve awareness of events occurring in the geographic domain under their authority. These tools include closed-circuit television (CCTV) cameras; license plate readers; and chemical, biological, radiation, nuclear, and explosive (CBRNE) sensors. Historically compartmented information warehouses containing criminal histories, emergency calls, use of force logs, and similar are increasingly being fused and made available for real-time search. Moreover, noncriminal information ranging from social media and other open-source datasets to credit histories and other quasi-public records are increasingly accessible to provide context to an event. The use of such noncriminal records to assist law enforcement is often vigorously contested and will be addressed later in this chapter, but regardless of a particular agency's implementation, today's challenge remains a big data problem. In other words, identifying the particular set of information relevant to an event is paramount; with few exceptions, the requisite data points to improve an officer's SA are available.

Complicating the analysis and dissemination of pertinent information to SA are the layers of information security policies applied to the first responder community. For example, law enforcement agencies in the United States must adhere to the Criminal Justice Information Standards established by the Federal Bureau of Investigation.[5] These standards mandate, among other requirements, that anyone accessing law enforcement data first authenticate themselves as a qualified operator and further establish a need to know the information requested. These regulations are often further restricted by agency-specific policies, such as preventing the disclosure of information pertaining to active cases to anyone not specifically associated with the case in question. Such policies and regulations were generally enacted and expanded in the wake of inadvertent (or deliberate) misuse of information over many decades. Few contemplated the ramifications on nonhuman actors, such as the potential of AI, and fewer still considered how persistent access to such information may contribute to real-time SA platforms charged with improving the safety and effectiveness of modern-day first responders.

It is in this context that the demands on first responders to employ SA platforms for decision support are being placed. With this comes a myriad of HSI concerns, ranging from the physical real estate available to first responders to interact with SA platforms to the means by which this complex set of information can be presented. Underpinning all considerations is the paramount importance of officer safety and the need to understand the operator's context in order to establish information relevance and right to know.

---

4 https://www.prnewswire.com/news-releases/the-global-video-surveillance-market-is-expected-to-grow-over-77-21-billion-by-2023-808999313.html.
5 https://www.fbi.gov/file-repository/cjis-security-policy-v5_6_20170605.pdf.

2.4 Deep Dive on Human Interface Considerations

With the advent of IoT sensors and significant increases in capabilities for both connectivity and storage, big data has become the prime dependency for many new technologies and solutions, especially SA. In public safety, and more particularly with first responders, the sheer breadth of information available is overwhelming. Designing human system interfaces that can retrieve, parse, and organize relevant data based on real-time activities and events, as well as present it in a meaningful, concise, and unintrusive (yet attentive) way, is a defining challenge.

At its core, public safety focused SA is predicated on alerting to noteworthy events in real time while increasing the knowledge and expanding the experience of responding personnel by drawing upon all pertinent historical, concurrent, and predictive information available to the agency. With a primary focus on officer safety, users of this system only have a few minutes upon being notified of the event to ingest the relevant data, make a determination on tactics, and adjust their response accordingly. This is all while they are also driving, communicating with dispatch, and coordinating with colleagues and supervisors. As such, the intelligence generated and presented must offer substantive benefits as rapidly and concisely as possible. The immediate goal of all first responders is to protect life, and much of the data available to police departments can support key areas such as subject identification, threat assessment, and response tactics, all of which greatly enhance SA and help to keep everyone safe.

Machine-assisted data retrieval, organization, and presentation not only improve the safety of all those involved, but it supports officer decision making by informing them of supplementary details and historical activities and actions. These characteristics are unique to every call for service, and a better understanding of them within the context of the current interaction is invaluable. However, the same mechanisms that collate the appropriate information must also exclude the rest. Considering the highly mobile nature of first responders and the inherent limitations of portable hardware in a public safety setting, it is not practical to expose all associated data, even if it could potentially be relevant in some ancillary contexts. Conversely, ignoring that information has its own tangible detriments, most notably, indicating an incorrect narrative to responding personnel that causes them to make poor judgments that have lasting impacts.

Computers have a unique ability to project truth, regardless of the quality and completeness of the underlying data. This "machine heuristic"[6] easily combines with algorithmic bias, which can corrupt the decision-making process for first responders with little apparency. As an example, domestic violence incidents are some of the most volatile and dangerous in policing and one in which historical context can greatly sharpen the officer's picture of the situation. Here, the prioritization of arrest history of the involved individuals would seem prudent, since it can assist with the identification of the primary aggressor. However, if that lessens the visibility, or excludes completely, non-arrest situations where the incident was resolved without enforcement action, it can skew the opinion of the officer prior to the interaction. In a public safety context, there is no single right answer. Extreme care must be given to ensure that relevant data sources are identified through a careful review by experienced personnel, that the presentation and prioritization is tailored to the specific situation, and that the assumptions exhibited are revisited frequently as part of a never ending cycle of improvement.

---

6 https://doi.org/10.1145/3290605.3300768.

2.5 Putting Human Interface Considerations in Context: Safe Cities

Public safety is a team effort. First responders do not work in a silo, but rather, they communicate regularly with dispatch, supervisors, operation centers, and a multitude of others. Additionally, support organizations may receive nonemergency tasks and follow-ups related to the work of those responding personnel. While the first responder in the field, that same data being leveraged for officer safety and rapid decision making is being reorganized and expanded to support the medium and long-term goals of the agency. Specifically, frontline supervisors will monitor the duration and amount of resources allocated to ensure continued coverage for this and other potential incidents; commanders will want to immediately recognize and respond to trends; operation centers will coordinate the response of supplemental agencies when required. All of these moving parts will leverage the same pool of data in dramatically different ways. In addition, many of these partners will have a broader picture of an incident, as they have the time and space to ingest more information and make more meaningful connections among the disparate sources. As such, human interfaces built around public safety data must support staff in many positions and at all levels of an organization.

All of the dependencies on data mean that there are real and significant impacts to public safety when that information is wrong. Historically, government has been very good about ensuring there is documentation of incidents, actions, and outcomes; however, the quality of that documentation varies wildly. Data can be incomplete, poorly structured, badly transcribed, or any combination thereof. Any of these deficiencies will inherently flow down to SA and data democratization platforms, where their inaccuracies will distract and delay the efforts of public safety professionals at best or compromise them completely at worst. Separately, with the explosion of IoT sensor platforms, such as gunfire detection, license plate recognition, video analytics, etc., the accuracy and pertinence of real-time alerts are just as important. For example, acoustic gunfire detection draws attention to shootings faster than any witness phone call. However, if those alerts are frequently inaccurate, the data itself becomes meaningless. False alarms become background noise, and the system is ignored. Worse, responding personnel become distrusting of the data, increasing the risk to their personal safety through complacency. Sensor limitations and bad data will likely slow the advancement of automation and enforce human independence for the foreseeable future.

While legacy data may contain inaccuracies and some sensors themselves may have a high false positive rate, human system interfaces are increasingly becoming smarter gatekeepers. The failings of the underlying technology and information are being counterbalanced through sheer volume. Individual data points may be important within the context of a single event, but it is the aggregation of these elements that build complex trend and pattern of life analyses. Here, individual errors are drowned out, and modern visualization solutions present this intelligence in a human-readable and actionable format. What started as a method to present real-time information to a user in order to address a specific incident has grown into an endless parade of data that can be stored indefinitely.

This boundless repository of information brings with it a host of security, policy, and legal concerns. Law enforcement is increasingly becoming augmented through technology, which often evolves faster than the necessary companion legislative changes, privacy guidelines, and security enhancements. Beyond the straightforward need to minimize the exposure of this data outside of its intended distribution, there must be controls within the organization as well, maintaining the concept of "need to know." Access itself should be routinely audited, reviewed, and revised, with secure mechanisms for data distribution and sharing. Just as important as the technical solutions, policy and procedures set a sturdy foundation for an organization's data security. In addition, having a written policy that has been properly vetted, reviewed, and socialized outside of the organization will express to the general public that the security of what is inherently their data is taken seriously. It also provides a roadmap for how to handle data breaches and inadvertent disclosures. Notably, with such large datasets, bad actors no longer require personally identifiable information (PII), like names or social security numbers. Rather, they can cause significant harm through anonymized sources with techniques such as data reidentification. As such, all data can be vulnerable to targeting and must be treated with as much care as traditional PII.

2.6   Human Interface Considerations for Privacy-Aware SA

The ability for HSI systems to understand the complex data used to achieve SA, as well as the context of the human operator consuming the information, will reach its apex in the form of minimum viable data presentation. Demands for better protection of PII have evolved over the past decade to include right to PII data ownership and demands on disclosure of PII use and storage. These rights have been codified through legislation including the European Union's General Data Protection Regulation (GDPR),[7] California Consumer Privacy Act (CCPA),[8] and Illinois Biometric Information Privacy Act (BIPA).[9] Though most legislation currently allows for lawful government use, the expectation is that government will adhere to the underlying principles in the performance of its duty.

To illustrate the impact such privacy-aware requirements have on SA platforms, consider the example of electronic identification (eID) mechanisms such as the mobile driver's license (mDL). Visitors to secure facilities, individuals involved in routine traffic stops, and even patrons to bars have historically been required to present a government-issued document verifying their name, date of birth, and other PII. Such credentials often violate the principle of minimum viable data as only portions of this information or, more precisely, derivatives of this information are needed to affect the requisite vetting. The root question of the individual accessing the secure facility is whether he or she is authorized to access the site, just as the bouncer is fundamentally concerned with whether or not the individual is of legal age to enter the bar. All other PII contained on the credential is incidental to this core concern and represents an elevated risk to the SA operator – providing them the opportunity to act on information irrelevant to the interaction – and to the information provider by disclosing more PII than necessary to complete the transaction. To mitigate these risks, the Secure Technology Alliance has advocated the widespread adoption of eIDs such as mDLs.[10] Such solutions provide the operator the pertinent derivative information – whether or not the person is of legal age – without disclosing the underlying PII (name, date of birth, address, etc.). Such solutions not only have the potential to improve the efficiency of SA operations but also mitigate the risk of excess PII disclosure.

Another manifestation of the requirement of SA platforms to incorporate privacy-aware concepts in their design and operation comes in the form of demands for greater transparency and

---

7 https://eur-lex.europa.eu/eli/reg/2016/679/oj.
8 https://leginfo.legislature.ca.gov/faces/billTextClient.xhtml?bill_id=201720180AB375.
9 http://www.ilga.gov/legislation/ilcs/ilcs3.asp?ActID=3004&ChapterID=57.
10 https://www.securetechalliance.org/mobile-drivers-license-initiative.

auditability. Regulators and privacy advocates are increasingly interested in the means by which public safety organizations achieve SA, concerning themselves not only with the underlying data employed but also with the tools used to analyze and correlate aggregate data sources. Interest in this tooling is only increasing with the proliferation of AI as the determination of relevance and scoring of risk shifts from human operators to machines. The debate around facial recognition epitomizes this concern, as highlighted by Georgetown Law's "The Perpetual Lineup" report.[11]

The impact such privacy-aware demands will have on SA platforms is manifold. As minimal viable data presentation capabilities progress, the data available to SA platforms will shift from primary source data to derivative data, and the need for contextualization and inference will increase. To overcome these and other challenges, SA platforms will increasingly incorporate AI to interpret information and ascertain relevance to the operator. As AI becomes increasingly pervasive, SA platforms must develop the capacity to explain nonhuman decision-making processes and accommodate transparency and independent auditing.

The impact these changes will have on HSI concerns is equally complex. On one hand, the ability to develop and interpret derivative information will improve the efficiency of man–machine interfaces by focusing operators on the most pertinent information available. On the other, the use of AI and reliance on derivative datasets will increase the demand on auditors tasked with ensuring the SA platform is making reasoned decisions, without bias, and that traceability is maintained to primary data where appropriate.

Reference

Endsley, M.R. (2011). *Designing for Situation Awareness*. Boca Raton, FL: CRC Press, Inc.

---

11 www.perpetuallineup.org.

3

Utilizing Artificial Intelligence to Make Systems Engineering More Human*

Philip S. Barry[1] and Steve Doskey[2]

[1] George Mason University, Fairfax, VA, USA
[2] The MITRE Corporation, McLean, VA, USA

3.1 Introduction

Systems engineering (SE) can be broken into several major phases since its inception during World War II. These major phases can be binned into four epochs, where each ensuing epoch builds on the knowledge and insights of the previous epochs as shown in Figure 3.1.

Epoch 1 began with the origins of SE being driven by the advent of large systems being developed such as the telephone system and operations research concepts employed during World War II. Epoch 2 picked up in the mid-1940s as Bell Labs (Fagen, 1978), DoD, and universities begin to formalize engineering development processes. While great strides were made, SE remained a methodology to maintain control and enforce stability in large programs. Epoch 3 changed SE by introducing technology as a force multiplier allowing industry to build ever more powerful SE tools that extended and leveraged the traditional processes developed in Epoch 2.

SE is now entering Epoch 4 that will integrate artificial intelligence (AI) and sociotechnical integration into the development and deployment of systems and systems of systems. AI has permeated our society in such diverse areas: improving mobile phone reception, spam filters, ride-sharing apps, autopilot systems, and fighting fraud. Epoch 4 will integrate AI, not only in the deployed systems, but change how we engineer these systems as part of human systems engineering (HSE). Furthermore, AI coupled with advances in sociotechnical system design will change SE tools and methods used to develop systems.

There is increasing recognition of the importance of considering stakeholders as part of the system development ecosystem. Epoch 4 explicitly recognizes the "human" as a part of the system and requires development environments to take into consideration human behavior and humans' proclivity for making choices not aligned with traditional statistical optimization. As the development of system capabilities evolve, it is apparent that a static interpretation of human integration into the system is insufficient. AI will enable real-time interpretation of continuous human integration and will be interdependent, collaborative, and cooperate between people and systems

---

*Approved for Public Release; Release Unlimited. Public Release Case Number 20-1208.

A Framework of Human Systems Engineering: Applications and Case Studies, First Edition.
Edited by Holly A. H. Handley and Andreas Tolk.
© 2021 The Institute of Electrical and Electronics Engineers, Inc. Published 2021 by John Wiley & Sons, Inc.

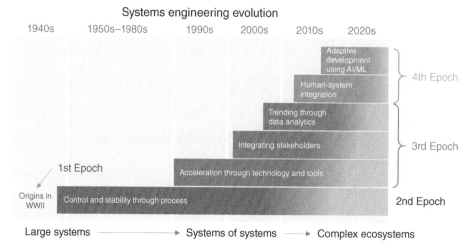

Figure 3.1 Systems engineering evolution.

(Kevin Reilly, 2020). To both identify existing risks and project into the future, AI-based models mimicking human behavior and learning will be necessary, as part of the development ecosystem. As HSE matures, these models will be a common component of HSE to design and build highly adaptive and resilient systems needed to keep pace with the velocity of change in business. This chapter presents a framework for defining and identifying the elements for the next evolution of SE, Epoch 4.

3.2 Changing Business Needs Drive Changes in Systems Engineering

SE, and more recently HSE, seeks to deliver successful systems that realize programs' targeted outcomes and the value derived from realizing those outcomes. Early in the evolution of SE, complexity drove the need for stability and control in engineering practices in order to reduce development risks and improve quality in operation that led to the definition of a normative set of processes, frameworks, and guidelines to increase stability and control as well as address risk in system development (Jamshidi, 2008).

Over the next 50 years, the continual drive for greater speed, agility, and innovation as well as the tremendous increase and integration of information technology (IT) into systems necessitated an evolution – if not a revolution – of SE. By the early 2000s, advances in computing power and SE tools allowed engineers to move the field from a process-driven, traceability-focused discipline to a discipline based on speed to value. This shift in addressing business needs was also driven by the realization that systems rarely existed in isolation but were part of a larger ecosystem of systems of systems, where systems of systems engineering codified the modeling of this interconnectedness into an evolving set of principles (Jamshidi, 2008).

Concurrently, systems engineers also recognized that the sociotechnical aspects of system development are key to success. Various methodologies (Scaled Agile Framework [SAFe] [Alexander, 2019; Leffingwell et al., 2018], Agile [Guru99, n.d.], SCRUM, etc.) were developed to increase the amount of communication between stakeholders to continually align perspectives as well as shorten the time to delivered value. Agile methodologies have migrated from domains that were primarily for software development to being widely used in the broader SE development environment. Model-based approaches and digital engineering provide for high-fidelity concept exploration, rapid integration, and continual verification during all phases of system development through the integration of digital engineering (Baldwin, 2019; Zimmerman and Gilbert, 2017), model-based systems engineering (MBSE) (Ramos et al., 2012), modeling and simulation, and DevOps constructs, where DevOps is defined as a set of practices combining development practices and IT operations with the goal of shorten the system development life cycle. This digital foundation has expanded the development of the ecosystem (Tolk and Hughes, 2014; Tolk et al., 2017) to integrate stakeholder perspectives, further reinforcing the value of exploration and experimentation of the sociotechnical solution space.

There is an emerging realization that the success or failure of a system development effort is often driven by the sociotechnical aspects of the development environment. By explicit quantitative modeling of sociotechnical factors, with a focus on delivering capabilities, stakeholder alignment and coordinated action can be engineered. Sociotechnical factors can describe the quality of stakeholder alignment in a complex, multi-stakeholder environment. These factors assess the degree and strength of alignment between stakeholders and are combined with traditional SE evaluation criteria to create a complete picture of the project environment, allowing for a holistic risk analysis, early warning of anti-pattern presence, and success determination. AI makes it feasible to deal with the complexity and dynamic nature of adding temporal sociotechnical factors across large numbers of stakeholders. Critical aspects of the projects often have significant, dynamic uncertainties that challenge integrating evidence from stakeholder behaviors with traditional project management techniques. Innovative use of AI to assess the implication of new evidence can more fully identify risks and emergent trends and suggest corrective action or where continued actions are unwarranted.

As business owners realize the benefits of HSE and AI, the fourth epoch will integrate HSE and AI into a system development ecosystem that can effectively address the development and release of incremental value for complex, human-technology integrated systems while promoting innovation and effectively managing risks.

3.3 Epoch 4: Delivering Capabilities in the Sociotechnical Ecosystem

As discussed above, the serial approach to SE from phase 1 has morphed into an almost continual delivery of value, particularly in IT. For example, the increased popularity of DevOps has provided a basis for continual improvement in enterprise IT systems as exemplified by Netflix (Netflix Technology Blog, 2017). DevOps is "a set of practices intended to reduce the time between committing a change to a system and the change being placed into normal production, while ensuring high quality" (Mersino, 2018), which can be viewed as accelerating continual improvement, but perhaps at the cost of thoroughly investigating the implications of changes across the entire stakeholder base. Epoch 4 will:

- Provide the ability to explore the implications of changes across the stakeholders, as well as experimentation to better understand the effects of the integration of possible enhancements and modifications.
- Augment the continuous development cycle of DevOps and the rapid delivery of value in SAFe with persistent quantitative sociotechnical validation through AI. It can be viewed as

accelerating continual improvement while concurrently investigating the implications of changes across the entire stakeholder base.

- Provide the ability to explore the implications of changes across the stakeholders as well as experiment to better understand the effects of the integration of possible enhancements and modifications.
- Augment the continuous development cycle of DevOps and the rapid delivery of value in SAFe with persistent quantitative sociotechnical validation.

3.3.1 A Conceptual Architecture for Epoch 4

Epoch 4 builds on the prior epochs, particularly in the areas of Agile methodologies, DevOps, and MBSE and adds the sociotechnical components as shown in Figure 3.2. For purposes of this discussion, we have highlighted four areas of interest for Epoch 4:

- *Temporal sociotechnical measures*: Measures over time of the complex organizational interactions between people and technology in workplaces.
- *SE frameworks*: Incorporates a variety of SE methods and techniques that work together as a cohesive set for development activity.
- *Sociotechnical models*: Illuminate the multidimensional interconnections between humans and technology.
- *Digital twins*: A digital replica of an entity, human, or system.

3.3.2 Temporal Sociotechnical Measures

System development has been based on measures for decades. Historically, measures could be traced to the "iron triangle" of cost, schedule, and performance used in traditional project performance management. If the system development effort was expending resources in accordance with a predetermined plan, value was realized when expected, and the functionality of the emerging system was consistent with requirements and constraints (e.g. earned value in line with investment and time constraints).

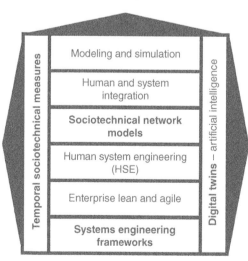

Practitioners of SE have realized that in addition to tracking the development effort along the iron triangle, alignment between stakeholders is critical toward achieving successful delivery of capabilities, as well as the early identification of risks. This is evidenced by modern development processes using frequent iterations of nominal group techniques (e.g. scrum) to ensure consistent communication and strive to identify disconnects and errors early. Epoch 4 takes this one step further and identifies specific measures to ensure alignment between stakeholders as a quantitative addition to the existing alignment processes. Previous work by the authors (Barry and Doskey, 2019; Doskey and Barry, 2018) posited a measure based on both strength of

Figure 3.2 A conceptual architecture for Epoch 4.

belief and alignment between sets of stakeholders on key issues in the system development environment. Epoch 4 frequently and quantitatively assesses misalignment between stakeholders as well as determining risk based on stakeholder belief (or lack thereof) that key elements of the development effort will succeed.

3.3.3 Systems Engineering Frameworks

Existing and emerging SE frameworks can be viewed as a set of rules and best practices, nominal group techniques, and documentation standards. Epoch 4 adds additional techniques to drive quantitative measures of sociotechnical alignment such as iterative surveys, natural language processing of communications, and explorative modeling to proactively identify and quantify sociotechnical risk.

Historically, SE frameworks have been shown to produce, on average, better development results than efforts lacking discipline. This is not to say that viable products cannot be developed outside of an established framework, it merely suggests that well-defined frameworks reduce risk and provide a basis for communicating intent. Thus, by expanding the framework to explicitly examine sociotechnical measures and calculate additional areas of risk, Epoch 4 provides the foundation for addressing these risks and raising the likelihood of a successful development effort.

The importance of defining risk in sociotechnical systems has been previously noted by other researchers. For example, Greenwood and Sommerville (2011) demonstrated an analysis approach for the identification for sociotechnical risks associated with coalitions of systems, and Johansen and Rausand (2014) posited a framework for defining complexity for risk assessment. Epoch 4 integrates this body of work into the main development environment.

3.3.3.1 Sociotechnical Network Models

Higher-fidelity models of system development environment of the hardware and software typically ignore the necessity for the explicit recognition and modeling of the sociotechnical network. Luna-Reyes et al. (2005) noted that many information system development activities fail to deliver because of social and organizational factors. A concurrent modeling and analysis environment that recognizes the effects of stakeholder interactions would address this shortcoming.

While modeling of sociotechnical networks is not new (Hu et al., 2010), Epoch 4 fully integrates both the modeling and subsequent exploration and experimentation of the sociotechnical network into the development cycle and uses tools such as systemigrams (Mehler et al., 2010), agent-based models, and system dynamics models to visualize and model the influence between stakeholders. This development of a representative sociotechnical network model provides the ability to monitor the trends and alignment between stakeholders and potentially identify existent and emerging risks.

This concept of modeling of the sociotechnical network during system development has been demonstrated by the authors who used changes in information entropy (IE) to assess whether the stakeholders were moving closer or diverging in their alignment of key belief areas (El Saddik, 2018). The temporal belief and alignment measures discussed in Section 3.3.1 indicated risk to system development. Epoch 4 explicitly recognizes both the source of sociotechnical risks and the importance and necessity to ameliorate them.

3.3.3.2 Digital Twins

Epoch 4 recognizes that the acceptance of any development effort is dependent upon successfully meeting the preferences of myriad stakeholders. System development environments that advocate frequent stakeholder interactions attempt to create an environment where there are few surprises and disagreements are worked out in near real time. In most cases this represents an improvement from the staged formal interactions that may occur asynchronously (such as the V model) and are document based. However, frequent interactions with stakeholders can be costly and, in some cases, may not be feasible. By augmenting the development environment with digital twins, Epoch 4 provides a mechanism to reduce the burden on the full set of stakeholders as well as allowing real-time experimentation during all phases of development.

There are a number of definitions of digital twins, but the most useful for the discussion here is from El Saddik who defines a digital twin as: "A digital replica of a living or non-living physical entity. By bridging the physical and the virtual world, data is transmitted seamlessly, allowing the virtual entity to exist simultaneously with the physical entity" (Rosen et al., 2015). Developing digital twins is used extensively in industry (Tao et al., 2018) and in healthcare (Baillargeon, 2014; Bruynseels et al., 2018). While there has been work in modeling stakeholder in various domains with an array of preferences (Le Pira et al., 2015; Tracy et al., 2018), quantitatively modeling the preferences of stakeholders in the development space per the aforementioned set of sociotechnical measures is not nearly as prevalent.

Epoch 4 advocates the creation of digital twins to model the preferences of the stakeholders as they relate to the development activity and the associated effects that the implementation of the development may have. The models of the digital twins can be straightforward. For example, Bayesian networks have been used to model stakeholders. The key is that a digital model of each of the agents is created with the interactions between them quantitatively described and how the introduction of new evidence affects preferences.

3.4 The Artificial Intelligence Opportunity for Building Sociotechnical Systems

Consider the definition of AI offered by Poole et al. (1998): "The study of intelligent agents: any device that perceives its environment and takes actions that maximizes its chance of successfully achieving its goals." As the fourth epoch of SE matures, there is significant opportunity to employ various AI techniques to enhance quality and reduce risk during capability development. AI can be seen as computational agents that work with the development team throughout the development life cycle, essentially human–machine teaming from a sociotechnical perspective.

Figure 3.3 provides a high-level taxonomy of AI. For the application discussed here, techniques discussed above can be viewed as approaches to enable the identification of risk events in the sociotechnical space and a forward projection of what will happen if the risks are ignored or mitigation actions are specifically taken. For example, in the context of risk identification, it is

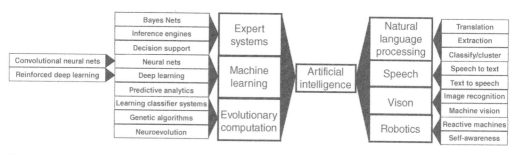

Figure 3.3 A taxonomy of artificial intelligence.

straightforward to examine specific applications of AI. Conditional decision trees can be developed to identify risks based upon temporal sociotechnical measures. Natural language processing is useful to assess the tone, and possible misalignment, by examining electronic communications. Bayesian reasoning can be employed to predict the possible consequences of the misalignment. Artificial neural networks can be used to learn from previous cases as well as dynamic data.

Initially, it is anticipated that AI will be used primarily for the development of complex sociotechnical systems that will be long lived and are likely to evolve over their lifespan. These systems have diverse stakeholders, and the stakeholder population will evolve over time. Understanding the perspectives of the stakeholders who may not be available or even exist during development is a necessary requirement to field a system that meets the requirements of the diverse stakeholder population. Further, the evolution of the deployed system as well as the sociotechnical ecosystem provides additional challenges. Modeling stakeholders' behaviors and evolution with AI provides an approach for prospective analysis of risks.

3.5 Using AI to Track and Interpret Temporal Sociotechnical Measures

For truly effective HSE in system development, it is essential to measure alignment between stakeholders and their belief that the project will succeed along numerous axes to rapidly identify existing and emerging risks. In the previous work cited by the authors (Barry and Doskey, 2019; Doskey and Barry, 2018), the approach taken to ascertain this information was via survey or forensic assessment. While this can work effectively on small development efforts, this approach does not scale when the number of stakeholders begins to increase into the tens or more.

Using AI there is an opportunity to conduct real-time and near-real-time analysis of sociotechnical measures using natural language processing and sentiment analysis. Sentiment analysis is well established in social media (Ortigosa et al., 2014), and APIs exist, such as Google Natural Language API (Putra, 2019), which allow individuals to quickly create sentiment analyzers. By employing agents that comb through communications like e-mail, text messages, and information on collaboration sites (e.g. Kanban boards), discontinuities and misalignments can be identified across the sociotechnical ecosystem. This approach has the potential to recognize existing and emerging sociotechnical risk. By enabling communication between various risk tracking intelligent agents, an assessment can be made as to whether there is enterprise or localized risk.

3.6 AI in Systems Engineering Frameworks

Modern SE frameworks have recognized the importance of frequent communications between stakeholders. Further, the popularization of agile methods has provided a backdrop for these communications and the associated record of success. AI bots (Wünderlich and Paluch, 2017) and other applications can act as an unbiased risk assessor across the enterprise as the system is developed.

Figure 3.4 provides a taxonomy of representative systemic risks that the AI can identify and assess:

- Structural risk assessments are conducted to determine if there are inherent weaknesses in the sociotechnical ecosystems and relation risk looks at groups of stakeholders.
- Relation risk is based on sociotechnical measures.

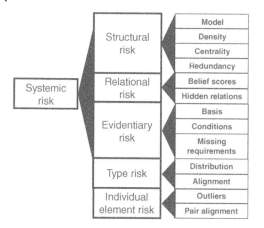

Figure 3.4 Risk taxonomy.

- Evidentiary risk is an assessment of the progress of the system development (e.g. measurements to determine if value streams being delivered regularly and as scheduled).
- Type risks aggregate groups of specific elements risks to identify systemic risk across the enterprise (e.g. widespread disagreement on technical approach).
- Individual element risks focus on the specific aspects of the sociotechnical space (e.g. assessing if the objectives of the stakeholders aligned across the enterprise).

In Epoch 4, AI will bring persistent, holistic risk analysis across the enterprise. This will be combined with a historical assessment of progress using data from analogous systems and from the progress on the current development effort. AI will provide recommendations for addressing material weakness in the sociotechnical ecosystem and track progress along these lines.

It is important to realize that during system development activity, risks evolve. Thus, when constructing a holistic risk analyzer, various measures and interpretation mechanisms are necessary. There is a need to identify situations that are rapidly emerging and indicate an immediate risk (e.g. replacement of a key stakeholder). These immediate sociotechnical risks can be quantified via persistent analysis of sociotechnical measures and network analysis.

Perhaps more insidious than detectable risk is the scenario of a slowly growing risk, where cumulative deviations add up and/or there is a drift in beliefs from one perspective to another. Often, stakeholders have some idea that things are amiss but have no consistent methodology to identify the cause. A persistent AI that monitors, examines, and assesses the sociotemporal measures can infer the presence of these emerging risks, either from a structural perspective (e.g. lower than expected communication between stakeholders) or disparate indicators that when taken together can integrate into a full-blown risk scenario.

3.7 AI in Sociotechnical Network Models

There is a multifaceted opportunity for AI with respect to sociotechnical models. Consider the composition of the sociotechnical model. Typically, the development of the sociotechnical network is largely manual. The composition exercised is designed to identify the major components as well as their interactions. Defining the interactions consists of first describing and then modeling the interaction in a quantitative fashion. The interactions between disparate components require both an understanding of how the interactions occur and how the interaction affects the components. AI can assist in the definition and development of the networks by analyzing data in the ecosystem. As an example, it is well-proven technology to develop and use link analysis (Oliveira and Gama, 2012) to create and explore networks. This process can also be complicated with the situation of an evolving social network as discussed by Cordeiro et al. (2018). AI can be used with evolving networks as well to continually assess flow relations, identify important relations, affinity relations, and alignment.

Typically, an SE ecosystem is rife with unstructured and semistructured data. Typically, large amounts of e-mail are exchanged along with texts messages. Using natural language processing and uncertain reasoning, a base social network can be constructed that can then be adjusted as evidence is introduced. This network can then be employed to focus the measurement of socio-technical measures. Without such analysis, the number of possible connections in a medium-sized network can become very large, i.e. $n(n-1)/2$. Moreover, as some of the connections are likely not significant, without a network analysis, nonrepresentative emergent effects may be inadvertently modeled.

3.8 AI-Based Digital Twins

Perhaps the most intriguing use of AI in the development of sociotechnical systems is creating digital twins to represent the interests and proclivities of the various stakeholders. Representative stakeholder digital twins provide a mechanism to better understand how events in the development of the sociotechnical system will be interpreted. For example, if some stakeholders have significant interest in a rapid delivery of specific value, the effect of delays in that delivery can be modeled.

The mechanisms for building representative models can be simple. For example, a simple version of this would be the development of a Bayesian network that processes the introduction of evidence of successful or unsuccessful progress with an assessment of the likelihood of successful capability delivery (Misirli and Bener, 2014). One can posit a digital mapping of the stakeholders' preferences that can describe influence propagation across the stakeholders that will either increase or decrease alignment and belief in the potential success of the capability development.

Digital twins that model stakeholder beliefs provide an opportunity to do exploratory experiments to examine how to best manage risks that may result from emergent effects of influence propagation across the sociotechnical network. There has been interesting work in this area, for example, agent-based models have been used to model bidirectional influence propagation (Li et al., 2016).

3.9 Discussion

Practitioners can apply the methods and tools, as described in the previous sections to create models of the sociotechnical environments where risks are actively modeled, monitored, and predicted; alternative courses of action can be simulated and assessed to provide practitioners with insight on how to lower risk and maximize value delivered. Using AI-based approaches to develop explicit models of stakeholders, how they relate and how their beliefs directly affect the chances for successful capability development is foundational to the constructs employed in the fourth epoch. These "proxy stakeholders" provide a powerful mechanism to conduct explorations into possible futures and allows proactive risk mitigation as the scope and scale of development efforts increases. The opportunity for the development effort to potentially get off track due to the multitude of stakeholders and the rapid pace of development necessitates holistic active monitoring and continual risk assessment that is infeasible for manual human-intensive approaches.

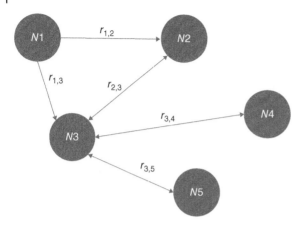

Figure 3.5 Example project social network.

To illustrate the necessity for computational aids, consider the example of a system development effort with five stakeholders. The connections between the stakeholders form a network, albeit not a fully connected network, as shown in Figure 3.5, where stakeholders are represented as nodes in the directed graph and the interactions between the stakeholders are shown as arcs. Thus, the set of stakeholders $N = \{N1, N2, N3, N4, N5\}$ and the relationships between them are described as $r_{x,y}$ where x and y are stakeholders X and Y.

The relationship between the stakeholders $r_{x,y}$ can be represented as $r_{x,y}(d, s, a)$ where d, s, and a are defined as follows:

- d represents directionality: In Figure 3.5, $r_{1,3}$ is shown as stakeholder 1 influencing stakeholder 3, but not vice versa. Compare this to $r_{3,4}$ where this influence is bidirectional. In long-lived capability development environments, a given relationships $r_{x,y}$ may not persist over time, or new relationships may emerge. Failure to recognize the network structure between relationships can result in unintended consequences that likely inject adverse effects into the development cycle or operations of the ensuing system.

- S indicates strength of influence: Strength of influence is defined as the degree to which the change in one node affects another. The strength of influence can be positive, negative, or neutral. Upon a change in a stakeholder, positive influence will increase the value of the temporal sociotechnical measures, negative influence will decrease them, and neutral influence indicates that a change in one node will have no effect on the other. For example, in Figure 3.5, if $N1$ was the leader of the organization, it is reasonable to assume that the $r_{1,3}(s)$ would be strongly positive. The diagram above also indicates that the leader is not significantly influenced by stakeholder $N3$ as arc is unidirectional; in other words, the leader is not listening. Similarly, the discussion with respect to relationships and the strength of influence of a relationship may change over time.

- a is the alignment between stakeholders: Alignment (a) is defined as the difference between the beliefs in both project execution and the underlying project ecosystem between stakeholders. Large differences in beliefs may portend risk as tactical measures may be taken that are not congruent with the success metrics of the parties and larger strategic measures of success may be out of alignment. The alignment is explicitly assessed using the temporal sociotechnical measures, such as the previously discussed belief approach. Whereas the relationship and strength of influence form the underlying substrata for sociotechnical network, risk is directly assessed by alignment (or lack thereof) of the belief structures of the stakeholders.

As a system development effort progresses, evidence is presented to the stakeholders that affects their beliefs. AI models of stakeholders can be used to model how beliefs can morph and drive how the stakeholder processes evidence and chooses actions. Understanding and modeling how misaligned belief structures can result in actions that are counter to overall goals of the development effort is key to proactively managing emerging risks. Frameworks such as belief, desire, and intention (Kim et al., 2012) have been used successfully in developing high-fidelity agent-based models and can be used here.

Accurate AI modeling of stakeholders using digital twin concepts provides a solid representation of the stakeholders and mechanism to track the evolution of the preferences. These AI-based risk assessors can look at atomic measures, group measures, and holistic sociotemporal measures to assess risks as shown in Figure 3.5. Structural risk can be assessed using appropriate interpretations of the sociotechnical network, consistently looking for over-connectiveness as well as sparsity. Structural risks are identified when measures exceed the tolerance of network metrics within a degree of error.

The canonical definition of risk can be described as a tuple represented as *risk{event, likelihood, consequence}*. Using this definition, AI-based models can spot misalignment across the sociotechnical measures and have the added capability of identifying localized risk, enterprise level risks, and emergent risks as misalignment grows over time. Further, with the introduction of evidence, and stakeholders' interpretation, may result in the identification of hidden risks. For example, a significant reduction in electronic communication between certain stakeholders can be a signal if impending misalignment and the advent of a relational risk. Relational risk occurs when a risk assessor looks at the relationship between two stakeholders and identifies significant incongruence between the sociotemporal measures leading to a lack of alignment. Here again the application of AI can model incidental and emerging risks as the scale of a large enterprise and the rapid pace of change make typical knowledge acquisition efforts across the stakeholders unrealistic. AI-based models can assist in proactive decision support when coupled with SE to identify measures, establish metrics, and define acceptable deviation limits for the measure.

A large enterprise may have tens or even hundreds of relevant stakeholders, all of whom are seeing different aspects of the development effort. As evidence is introduced to stakeholders in these complex sociotechnical networks, it is improbable that manual, or mental, methods can adequately track and assess the impact of the evidence. AI-based models can easily track and learn from these events and provide the practitioner with insight on how best to lower risk and improve value delivery.

Traditional risks in typical system development projects and their evolution are well documented in literature (Warkentin et al., 2009), but analytical analysis of risk particularly in the sociotemporal space is largely a manual process. An AI-based risk assessor can monitor trends in the sociotechnical measures to examine if there are emerging risks from a growing misalignment between groups of stakeholders. An AI risk assessor can develop forecasts of emerging risks that can be modeled by deliberate introduction of possible evidence that may occur in the future. Modeling in this fashion provides the capability to proactively assess system risk areas and take preventive steps.

In addition to localized risk identification, a holistic view of systemic risk across the sociotechnical ecosystem becomes more feasible using AI-based models. While localized gross individual misalignment may not manifest itself, a general trend toward misalignment across the enterprise can be discovered and identified. Whereas the values for sociotemporal measures typically have been discovered through interviews, AI offers the opportunity to make assessments based upon noninvasive approaches such as analyzing e-mails and text messages. Applications like the hedonometer (Dodds et al., 2011) have been used for many years; it is suggested here that AI can tap into the results of the hedonometer for both localized and global assessments of risk.

The second element in the risk tuple is the likelihood that the risk will occur. Misalignment on the sociotechnical measures provides an indication of an impending risk event. Traditional uncertain reasoning approaches, e.g. using a Bayesian network, can be employed to determine the likelihood of the risk (Feng et al., 2008; Hui and Liu, 2004). If the misalignment is observed, there is a certitude that the risk event has now become an issue, at which point the AI can be used to make an estimate of the consequences should the issue not be resolved in a timely fashion.

With an estimate of the likelihood of the risk event, the last step is to assess the consequence(s) if the risk event happens and becomes an issue as mentioned above. This is again an opportunity for the AI to employ uncertain reasoning: taking a risk event and the likelihood that it will happen, what are the anticipated consequences? The assessment can be based upon historical data or generalized rules that categorize the risk and its associated consequences and iterate multiple outcomes based on actions taken to determine enterprise-level success vice local optimization. Generally, the consequences will manifest themselves as a negative effect along the traditional lines for system development, impact on cost, an extension of the schedule, or a decrease in the capability or value delivered with the system increment. This mapping to the traditional measures of system progress facilitates the movement toward the use of AI to risk amelioration.

A conceptually straightforward approach for amelioration is to use AI to calculate the expected value of the impact of a given risk if it comes to fruition and then compare the costs of various amelioration activities with the expected value of the courses of action that could be taken. In general, taking steps to reduce risks that are more expensive than the expected value of the payoff is a poor decision. The other aspect of choosing a risk reducer is to consider the global effect that may have an aggregate cost that is unacceptable even though the local risk calculation may indicate a local improvement. Thus, having AI make a global assessment as to the possible future state to best calculate the expected cost is an important capability that will differentiate the fourth epoch of SE. Essentially, AI will conduct localized sociotechnical risk assessments and heuristic risk assessments and suggest corrective actions based upon a quantitative risk calculation and projections of the effects of the actions.

3.10 Case Study

The concepts described above were used to analyze a large governmental IT development effort. The project was designed to modernize an antiquated system that provided benefits to hundreds of thousands of recipients. Initially, a traditional development effort was undertaken that focused almost solely on the technical aspects of the effort. While the technical engineering was adequate, the project failed in large part due to sociotemporal factors and inherent structural risks. After the project was halted, a new system develop approach was implemented. The project was put back on track and is successfully delivered its primary capability suite. Further development of ensuing capability is on track for successful deployment on time, at cost, and with the requisite quality. For the remainder of this discussion, the first effort will be called Project One, and the second will be called Project Two.

Project One started in a precarious position, but for all intents and purposes, the stakeholders were unaware. There was significant structural risk in that lines of communication were not fully open; the network had insufficient communication channels between the stakeholders, resulting in discontinuities in beliefs; discontinuities in beliefs resulted in disparate actions that can act at cross-purposes. There was significant relational risk resulting from an apparent unwillingness of the stakeholders to change beliefs even after evidence was presented, which further contributed to the project inability to successfully deliver on time.

To further compound the situation, Project One suffered from type risk that resulted in an inability to recognize systemic risks and trends across the enterprise. In part, there was no mechanism in place to measure the trends, particularly in the area of sociotechnical risks. Further, there was individual cognitive anchoring as described above that further contributed to a lack of alignment. These risks compounded into a situation where the project began to fail, and the failure was not able to be corrected.

Project Two actively implemented changes specifically designed to address the reasons why Project One failed. First, recognizing that the misalignment of sociotechnical factors was a major cause of the failure for Project One; Project Two moved to a SAFe development paradigm as previously discussed. The frequent interactions of various components of the stakeholders, from developers to business owners, provided numerous opportunities to identify and address misalignment in beliefs and prevent actions that are counter to achieving the project objectives. Second, active measurement of sociotechnical factors allowed identification of hidden and emerging problems. This data-driven approach provided a mechanism to flag biases as well as individual stakeholder risks. The net result was project stability that placed Project Two on a clear road to success.

3.11 Systems Engineering Sociotechnical Modeling Approach

The case study in Section 4 was used to model the development effort in Section 5. Figure 3.6 illustrates a taxonomy of terms that were used to more fully characterize the project and the associated environment and how it directly influences the necessity and frequency for information updates: measurability, situational awareness, information availability, novelty, and complexity. Considerations such as uncertainty, novelty, and complexity also describe the inherent volatility and drive information requirements.

Achieving project objectives is directly dependent upon the quality of the project plans as well as the project sensors, where sensors are humans or project tools that collect the data over time. As the project is executed, the project sensors inform the comparison with the project plan. After the initiation of these actions in concert with or opposition to the plan, the progress is assessed and then repeats until the plan is successfully executed or the project is terminated.

To characterize the project environment, an information ecosystem classifier was developed to characterize the stability of the project environment and the alignment of the project team's beliefs

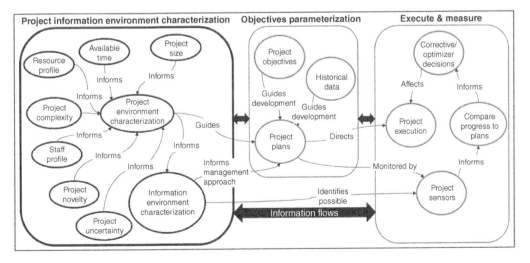

Figure 3.6 Project information ecosystem.

with what the beliefs of the organization requesting the project. Descriptive questions or phrases were used to provide a better and more consistent means of collecting perceptions versus using single word attributes. To fill out the classifier, each of the following 10 criteria was scored on a 0–10 Likert scale, where 0 means "fully disagree" and 10 means "fully agree." The project team assessed their beliefs and then provided complementary ratings for the perception of the stakeholder procuring the project. The environmental analysis was developed using a cross section of common attributes identified in literature such as *Relationships Between Leadership* and *Success in Diverse Types of Project Complexities* (Muller et al., 2012). The criteria were as follows:

1) All stakeholders are committed to attaining the project objectives.
2) Stakeholders are willing to adapt/change.
3) Project objectives are known and measurable.
4) Timely information is available from stakeholders.
5) Sufficient time is provided to complete the project on schedule.
6) Budget is sufficient for the project.

The project classifier offered a translation to quantitatively assess each aspect of the project ecosystem in terms of where information and communication will be critical to the success of the project. Each aspect considered was assessed for both symmetry and stability/maturity. For example, the project team is wholly committed to the achieving the project outcome, but the stakeholder's commitment is perceived as lacking, thereby creating an imbalance and likely area of risk. Similar assessments can be done for each of the criteria used in the project classifier.

The difference and the size for each element in the classifier was used to derive a coefficient that provided an indication of risk in achieving that element. Equation 3.1 describes the *belief* coefficient, a tunable coefficient derived from heuristics and project experience:

$$Belief = e^{(-1*((f^d+1)/(s^g+1)))}$$ (3.1)

where

- f is the difference for a given criterion
- s is the value for stability/maturity
- d and g are real values used to emphasize the contribution of size over maturity or vice versa
- *Belief* values run from 0 to 1, where lower values indicate significant risk for project success for a given criterion

The *belief* coefficient equation was developed for large-scale, contract-based, custom IT projects in which a relationship with the stakeholder is integral to the delivery of the project. In essence, alignment with the stakeholder has been found to be equal to or greater in priority than maturity or stability of the environment as determined by the responses to the project classifier questions.

To assess the overall information risk profile of the project, the authors computed a single value that aggregated the *belief* coefficients across the project. The authors calculated the geometric mean of the *belief* values as shown in Equation 3.2. The geometric mean is commonly used to find a single value when trying to aggregate criteria with different properties:

$$\left(\prod_{i=1}^{n} x_i\right)^{\frac{1}{n}} = \sqrt[n]{x_1 x_2 \cdots x_n}$$ (3.2)

where

- x's are individual *belief* coefficients for each criterion in the project classifier

The intent of using the geometric mean of the belief coefficients is to provide a singular aggregate risk for a project in terms of where information and communication are critical to project success, where *belief* values approaching 1 presume high alignment and ecosystem stability for a program and values approaching 0 denote areas of concern that require attention. Additionally, as *belief* coefficients approach 0, the velocity of decision making should be expected to increase to account for the abundance of uncertainty.

As stated, the goal of this research was to identify sources of hidden risk in the programmatic underpinnings of system development efforts. To do it, the authors used the following logic to translate the beliefs of the stakeholders and the evidence that drives these beliefs into a Bayesian belief network (BBN). Expert elicitation of individuals involved in the system development projects and along with a review of published artifacts was used to develop prior probabilities, such as success and failure rates for project based on the Standish Group's Chaos Report for 2015 (The Standish Group, International, Inc., n.d.). Figure 3.7 is provided as an exemplar of BBN construction.

The scalar measures were then translated into a logistics distribution curve, i.e. the *belief* coefficients. The *belief* coefficients indicate areas of misalignment in beliefs between stakeholders, where the beliefs are not objective measures of reality but instead the best understanding of the beliefs of other stakeholders' expectations, if direct collection of beliefs is unavailable. The *belief* coefficients were categorized based on their position of the logistics distribution curve. This categorization gives practitioners an indication of the material weaknesses within the system development environment and profiles where risk may be present yet unidentified. Note that these categorizations will vary for different types of system development project genres.

3.11.1 Modeling the Project

There are two major components of designing the social system and technical system in tandem in order for them to work together so that the interaction of social and technical factors creates the conditions for successful organizational performance. IE was used as a measure of environment stability, and belief coefficients were used as a measure of alignment of sociotechnical attributes.

IE was used as a means of gauging confidence in the conditions for success in a multi-stakeholder environment. IE calculations provided a measure of project environment stability. Changes in entropy provided a mechanism to examine the project over time and to understand the patterns that model environments for success or failure.

The analysis began by modeling the project stakeholders and their relationships as a graph, where edges are assumed to be bidirectional (see Figure 3.8). The edges represent the relationships between the stakeholders, and the strength of the relationships as well as the alignment is defined by the integration of the *belief* coefficients scores for the *belief* questions.

As can be seen, the project is composed of stakeholders that must work together to deliver the project. For expediency, calculations were made for the edges connected to the program team/PIO.

In addition to looking at the overall enterprise modeled as a graph structure, it is important to look at the evolution of the alignment coefficients and the overall alignment coefficient over time. Specifically, it is desirable to determine whether the analysis can assess whether the project will succeed based on current and past behavior. For example, random behavior between measuring periods will probably indicate poor predictability of future states. Correspondingly, reasonably stable behavior is an indicator of future performance. Information theory provides a ready measure of the randomness in the system with IE. IE is defined as

$$(x) = -\Sigma_{k=1} \, p\left(x_k\right) \log_k p\left(x_k\right) \tag{3.3}$$

where k refers to the number of distinct states the system may be in.

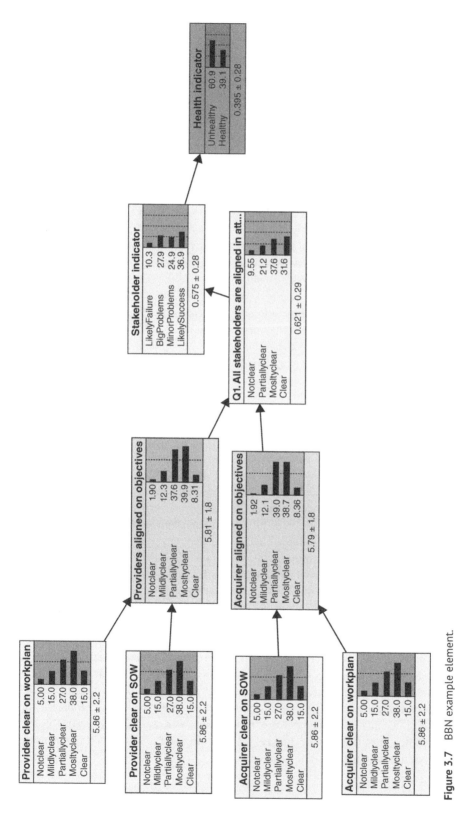

Figure 3.7 BBN example element.

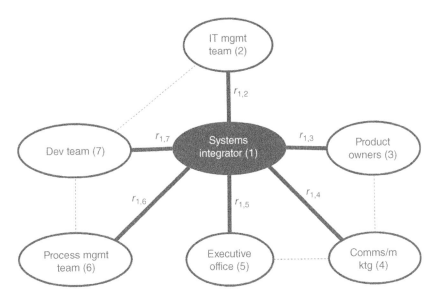

Figure 3.8 Case study project social network.

As the IE rises for a relation (edge) or the system, the randomness rises, and the stability of the project environment falls as does the ability of the model to predict future behavior. Correspondingly, as the IE falls, the stability of the project increases, and the ability of the model to predict future behavior rises. To apply this discrete formula for assessing stakeholder alignment, the continuous distribution of stakeholder alignment was broken down into several distinct states.

IE was chosen as a measure that is at a maximum when the probability of all possible outcomes is equivalent; in essence there is no information that would allow a more educated prediction. Experimental testing indicates the values of both edges, and the system at large generally will cluster around several peak values. If variance (σ) was chosen as a measure, it would obscure this information as it is not sensitive to multiple modes. In fact, it is possible that (x) might increase, while σ could decrease. Consequently, using variance alone would indicate less disorder and more confidence in the model than would be warranted.

This methodology is designed to identify areas of risk to successful project execution. Risk is defined here as the belief that the project will be challenged or fail given the alignment coefficient and the IE. Figure 3.2 illustrates the possible outcomes of the IE.

For this discussion, stability S is defined as follows:

$$S = 1 - IE \tag{3.4}$$

S is used with the alignment coefficient to calculate confidence. If the project has a high alignment coefficient score across the relationships and S is large, it can be plotted in Zone 1, and there is good confidence that the project will succeed. A project that has high alignment coefficient, but low S (Zone 2), is a project of concern with low confidence of success and an analysis warranted to identify why the stability of the environment of the project is low, potentially indicating hidden factors. Zone 4 indicates a project that has both a low alignment coefficient score and low S, which would indicate there should be serious lack of confidence that the project will succeed. Zone 3 has high S but low alignment coefficient; this is an area of concern that may be recoverable, and deeper analysis is warranted to understand why the alignment coefficient scores are low. Figure 3.9 plots

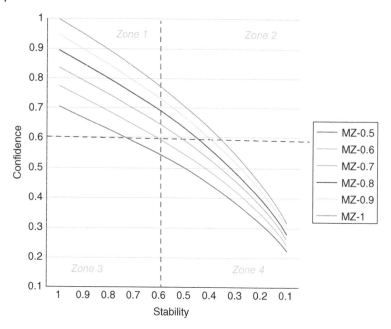

Figure 3.9 Confidence vs. stability.

Project 1		Project 2	
1.	Kick-off	1.	Kick-off
2.	Preliminary design review	2.	Project increment planning
3.	Development	3.	Sprints
4.	Demo	4.	Solution demo
5.	Test and fix	5.	Inspect and adapt
6.	User testing	6.	UAT

Figure 3.10 Case study project events.

belief alignment against environment stability; for the purposes of this figure, confidence is defined as

$$\text{Confidence} = \left(S^2 + \text{Alignment Coefficient}^2\right)^{1/2} \tag{3.5}$$

For both projects, a retrospective study was employed in which the concept of digital twins was used to construct the model using the concepts described above. Six events were identified in which information from sensors was collected and reviewed. Figure 3.10 shows the project events used for the model. Different, yet equivalent, events were used because Project One followed a traditional SE and project management approach and Project Two employed SAFe, agile, and scrum methodologies to manage the project.

3.12 Results

The case study comparison of Project One and Project Two demonstrates the effects of employing sociotechnical constructs, AI, and HSE into the development environment. Comparatively, the results from the model are plotted showing belief alignment and stability (see Figure 3.11). As

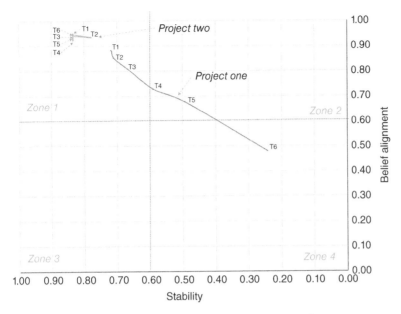

Figure 3.11 Belief alignment and stability results for case study.

noted, belief alignment is generated by calculating the geomean of belief coefficients derived from a series of questions, noted previously. Stability is calculated using 1 minus IE ($S = 1 - $ IE). Four zones were defined to aid in analysis:

- **Zone 1** (belief > 0.6; stability > 0.6): Projects scoring in Zone 1 generally have a high probability of success.
- **Zone 2** (belief > 0.6; stability < 0.6): Projects scoring in Zone 2 demonstrate a high level of confidence in the alignment and cohesiveness of the stakeholders but present challenges in that the environment itself and/or the belief alignment is unstable due to exogenous influences.
- **Zone 3** (belief < 0.6; stability > 0.6): Projects scoring in Zone 3 demonstrate a low level of confidence in the alignment and cohesiveness of the stakeholders, while the environment itself and/or the belief alignment is relatively stable with minimal expected change or exogenous influences. Note that this stable structure does not mean it is good – only that it is deterministic.
- **Zone 4** (belief < 0.6; stability < 0.6): Projects scoring in Zone 4 exhibit one or more behaviors and associated risks that are commonly identified root causes of failed projects.

Analysis was performed using the conceptual architecture for Epoch 4 as a means of comparing and determining value of evolving SE toward Epoch 4 (see Table 3.1).

Overall, for Project One, eroding belief alignment was evident over the life of the program as the stakeholders did not learn from previous activities and outcomes and instead hardened their own beliefs. Once these patterns were set, the culture of the project made it hard to change the patterns of behavior and project structure. As can be seen, the project had a consistent trajectory leading from initial confidence to the reality of a poorly run project. Conversely, the culture of Project Two employed open communications across all stakeholders and maintained the idea of being a learning organization that actively sought to improve throughout the project life cycle. Continuous alignment and high stability in the project are evidenced by the low variation in measurements for Project Two over time.

While this is an initial study, indications are that the constructs of Epoch 4 will improve success in quality project deployment. As more projects follow these constructs, it is hoped that future studies will prove the value of employing HSE and AI into the development environment.

Table 3.1 Case study comparative analysis.

	Project One	Project Two
Systems engineering frameworks	Employed a generic and linear systems engineering model where each activity was independent and completed its full activity before moving to the next phase, e.g. all requirements gathered prior to development	The project team used SAFe, agile SW development, and scrum techniques to continuously evolve the solution and deliver incremental value where feedback was incorporated into future design considerations and development environment structure
Sociotechnical network models	Measures of belief were ascertained after events occurred and misalignment had already degraded relationships. In post-project analysis, it was clear that stakeholders were unaware of belief misalignment or were anchored in their beliefs and unwilling to change	The project team developed a structure network of stakeholders identifying roles and responsibilities and used the network to ensure communications challenge were actively being used to ensure alignment between stakeholders
Temporal sociotechnical measures	Traditional project measures were used to measure progress – cost, schedule, and quality. No quantitative measures of stakeholder alignment nor stability were calculated. Small groups aired concerns among themselves created an environment where factions were actively working against project progress in order to protect their personal interests	Stakeholders used active measuring and a cultural shift toward seeking out and resolving misalignment of beliefs. Open communications were the norm where identifying and sharing risks, discrepancies, and differing opinions were seen as a positive behavior
Digital twin/AI	No modeling or use of AI to model the ecosystem transpired. Nor were there efforts to understand effects of events prior to their occurrence	The project team used basic model-based engineering techniques and rudimentary AI techniques to assess alternative courses of action throughout the project life cycle

3.13 Summary

As SE enters Epoch 4, the complexity and scale of projects have increased dramatically. While the advent of MBSE has brought a level of rigor and repeatability to the practice, formal methods that recognize the importance of the sociotechnical aspects of system development are lagging. However, quantitative real-time analysis of the sociotechnical space has yet to be widely employed. Historically, the difficulties in using sociotechnical models could be traced to a lack of data and computational power. With the proliferation of digital communications, natural language processing, and methodologies, such as SAFe, these fundamental barriers are falling. Yet even with agile approaches, over half of projects fail to be delivered within the time, cost and performance constraints, or fail outright (Mersino, 2018).

This chapter explored how a quantitative model of the sociotechnical space can be used to identify incongruities between stakeholders and highlight risks. While SAFe provides a methodology for continuous communication and data exchange, in most implementations, the data is informal or at best captured in meeting notes. With improvements in natural language processing, text processing, and uncertain reasoning as well as machine learning, there is an opportunity for both model stakeholder belief systems to predict their actions and train the models by analyzing the correlation between text communications, e-mail, and other sources.

The chapter also illustrated how the conceptual architecture described in Section 3.3.1 can be readily instantiated as a tool to identify incidental and emerging risk. Results indicated promise that sociotechnical risks can be identified early on in the development effort and dealt with. The case study presented illustrated the possible positive effects of employing this approach.

The quest to find more effective methodologies for system development efforts continues to be a research challenge across numerous industries. While improving digital representation can result in faster, more accurate system development, it still does not address the most frequent reason for project failures, the misalignment between stakeholders' beliefs and expectations. In addition to the obvious challenge of mismatched expectations, misalignment can have a more insidious risk. When belief structures are misaligned, noncomplementary actions, and even actions that are cross-purposes, can be taken. Quantitative sociotechnical modeling and risk analysis have the potential to significantly improve successful development by explicitly monitoring and analyzing belief structures and making proactive predictions of emerging risks. Mature sociotechnical modeling as part of the SE capability development is the realization of the promise of Epoch 4.

The authors' affiliation with The MITRE Corporation is provided for identification purposes only and is not intended to convey or imply MITRE's concurrence with, or support for, the positions, opinions, or viewpoints expressed by the author. This paper has been approved for Public Release; Distribution Unlimited; Case Number 20-1208.

References

Alexander, M. (2019). What is SAFe? The Scaled Agile Framework explained. *CIO Magazine* (30 August).

Baillargeon, K. (2014). The Living Heart Project: a robust and integrative simulator for human heart function. *European Journal of Mechanics – A/Solids* 48: 38–47.

Baldwin, K. (2019). Journal of Defense Modeling and Simulation (JDMS) special issue: transforming the engineering enterprise – applications of Digital Engineering and Modular Open Systems Approach. *The Journal of Defense Modeling and Simulation: Applications, Methodology and Technology* 16 (4): 323–324.

Barry, P. and Doskey, S. (2019). Identifying hidden risks in multi-stakeholder, dynamic system development environments. *2019 IEEE International Systems Conference (SysCon)*, Orlando, FL, USA, pp. 1–7.

Bruynseels, K., Santoni de Sio, F., and van den Hoven, J. (2018). Digital twins in health care: ethical implications of an emerging engineering paradigm. *Frontiers in Genetics* 9: 31.

Cordeiro, M., Sarmento, R.P., Brazdil, P., and Gama, J. (2018). Evolving networks and social network analysis methods and techniques. In: *Social Media and Journalism – Trends, Connections, Implications*. IntechOpen https://doi.org/10.5772/intechopen.79041.

Dodds, P.S., Harris, K.D., Kloumann, I.M. et al. (2011). Temporal patterns of happiness and information in a global social network: Hedonometrics and Twitter. *PLoS One* 6 (12): e26752. https://doi.org/10.1371/journal.pone.0026752.

Doskey, S. and Barry, P. (2018). A systems engineering perspective of timing information for optimized decision making. *2018 Annual IEEE International Systems Conference (SysCon)*, Vancouver, BC, pp. 1–8.

El Saddik, A. (2018). Digital twins: the convergence of multimedia technologies. *IEEE Multimedia* 25 (2): 87–92.

Fagen, M.D. (ed.) (1978). *A History of Engineering and Science in the Bell System: National Service in War and Peace (1925–1975)*. Bell Telephone Laboratories.

Feng, N., Li, M., Xie, J., and Fang, D. (2008). A data-driven model for software development risk analysis using Bayesian networks. *2008 IEEE Symposium on Advanced Management of Information for Globalized Enterprises (AMIGE)*, Tianjin, pp. 1–5.

Greenwood, D. and Sommerville, I. (2011). Responsibility modeling for the sociotechnical risk analysis of coalitions of systems. Computing Research Repository – CORR. doi: https://doi.org/10.1109/ICSMC.2011.6083832.

Guru99 (n.d.). Agile vs. scrum: know the difference. https://www.guru99.com/agile-vs-scrum.html (accessed 16 January 2020).

Hu, F., Mostashari, A., and Xie, J. (eds.) (2010). *Socio-technical Networks: Science and Engineering Design*. CRC Press.

Hui, K.T. and Liu, D.B. (2004). A Bayesian belief network model and tool to evaluate risk and impact in software development projects. *Annual Symposium Reliability and Maintainability, 2004 – RAMS*, Los Angeles, CA, USA, pp. 297–301.

Jamshidi, M. (ed.) (2008). *Systems of Systems Engineering: Principles and Applications*, 1e. CRC Press.

Johansen, I.L. and Rausand, M. (2014). Defining complexity for risk assessment of sociotechnical systems: a conceptual framework. *Proceeding of the Institution of Mechanical Engineers, Part O: Journal of Risk and Reliability* 228 (3): 272–290.

Kevin Reilly, C.W. (2020). Quantum computing, climate change, and interdependent AI: academics and execs predict how tech will revolutionize the next decade (29 January 2020). https://www.businessinsider.com/davos-microsoft-tech-henry-blodget-panel-2020-1.

Kim, S., Xi, H., Mungle, S., and Son, Y.-J. (2012). Modeling human interactions with learning under the extended belief – desire – intention framework using agent-based simulation. *62nd IIE Annual Conference and Expo 2012*.

Le Pira, M., Inturri, G., Ignaccolo, M. et al. (2015). Simulating opinion dynamics on stakeholders' networks through agent-based modeling for collective transport decisions. *Procedia Computer Science* 52: 884–889.

Leffingwell, D., Knaster, R., Oren, I., and Jemilo, D. (2018). *SAFe® 4.5 Reference Guide: Scaled Agile Framework® for Lean Enterprises*. Scaled Agile, Incorporated.

Li, W., Bai, Q., Zhang, M. et al. (2016). Agent-based influence propagation in social networks. *2016 IEEE International Conference on Agents (ICA)*, pp. 51–56.

Luna-Reyes, L.F., Zhang, J., Gil-García, J.R., and Cresswell, A.M. (2005). Information systems development as emergent socio-technical change: a practice approach. *European Journal of Information Systems* 14 (1): 93–105. Special Issue: from Technical to Socio-technical Change: Tackling the Human and Organisational Aspects of Systems Development Projects.

Mehler, J., McGee, S., and Edson, R. (2010). Leveraging systemigrams for conceptual analysis of complex systems: application to the U.S. National Security System. *8th Conference on Systems Engineering Research*, Hoboken, NJ (17–19 March 2010).

Mersino, A. (2018). Agile project success rates are 2X higher than traditional projects (1 April [2019 update]). https://vitalitychicago.com/blog/agile-projects-are-more-successful-traditional-projects (accessed 18 December 2019).

Misirli, T. and Bener, A.B. (2014). Bayesian networks for evidence-based decision-making in software engineering. *IEEE Transactions on Software Engineering* 40 (6): 533–554.

Muller, R., Geraldi, J., and Turner, J.R. (2012). Relationships between leadership and success in different types of project complexities. *IEEE Transactions on Engineering Management* 59 (1): 77–90.

Netflix Technology Blog (2017). How we build code at Netflix (19 April). https://netflixtechblog.com/how-we-build-code-at-netflix-c5d9bd727f15 (accessed 19 December 2019).

Oliveira, M.D.B. and Gama, J. (2012). An overview of social network analysis. *Wiley Interdisciplinary Reviews: Data Mining and Knowledge Discovery* 2 (2): 99–115.

Ortigosa, A., Martín, J.M., and Carro, R.M. (2014). Sentiment analysis in Facebook and its application to e-learning. *Computers in Human Behavior*: 527–541.

Poole, D., Mackworth, A., and Goebel, R. (1998). *Computational Intelligence: A Logical Approach*. New York: Oxford University Press.

Putra, D.W. (2019). How to make your own sentiment analyzer using Python and Google's Natural Language API (12 February). https://www.freecodecamp.org/news/how-to-make-your-own-sentiment-analyzer-using-python-and-googles-natural-language-api-9e91e1c493e (accessed 20 February 2020).

Ramos, A.L., Ferreira, J.V., and Barceló, J. (2012). Model-based systems engineering: an emerging approach for modern systems. *IEEE Transactions on Systems, Man, and Cybernetics – Part C: Applications and Reviews* 42 (1): 101–111.

Rosen, R., von Wichert, G., Lo, G., and Bettenhausen, K.D. (2015). About the importance of autonomy and digital twins for the future of manufacturing. *IFAC-PapersOnLine* 48 (3): 567–572.

Tao, F., Cheng, J., Qi, Q. et al. (2018). Digital twin-driven product design, manufacturing and service with big data. *The International Journal of Advanced Manufacturing Technology* 94 (9–12): 3563–3576.

The Standish Group, International, Inc. (n.d.). *The Chaos Report 2015*. https://www.standishgroup.com/sample_research_files/CHAOSReport2015-Final.pdf (accessed 19 December).

Tolk, A. and Hughes, T.K. (2014). Systems engineering, architecture, and simulation. In: *Modeling and Simulation-Based Systems Engineering Handbook* (eds. D. Gianni, A. D'Ambrogio and A. Tolk), 11–41. CRC Press.

Tolk, A., Glazner, C.G., and Pitsko, R. (2017). Simulation-based systems engineering. In: *Guide to Simulation-Based Disciplines – Advancing Our Computational Future*, Simulation Foundations, Methods and Applications, 75–102. Switzerland: Springer International Publishing AG.

Tracy, M., Cerdá, M., and Keyes, K.M. (2018). Agent-based modeling in public health: current applications and future directions. *The Annual Review of Public Health* 39: 77–94.

Warkentin, M., Moore, R., Bekkering, E., and Johnston, A. (2009). Analysis of systems development project risks: an integrative framework. *Database* 40: 8–27. https://doi.org/10.1145/1531817.1531821.

Wünderlich, N.V. and Paluch, S. (2017). A nice and friendly chat with a bot: user perceptions of AI-based service agents. ICIS.

Zimmerman, P. and Gilbert, T. (2017). Digital engineering transformation across the Department of Defense. *The Journal of Defense Modeling and Simulation: Applications, Methodology and Technology* 16 (4): 325–338.

4

Life Learning of Smart Autonomous Systems for Meaningful Human-Autonomy Teaming

Kate J. Yaxley[1], Keith F. Joiner[2], Jean Bogais[3], and Hussein A. Abbass[1]

[1] *School of Engineering and Information Technology, University of New South Wales, Canberra, Canberra, ACT, Australia*
[2] *Capability Systems Center, School of Engineering and Information Technology, University of New South Wales at Australian Defence Force Academy, Canberra, ACT, Australia*
[3] *University of Sydney, Sydney, NSW, Australia*

4.1 Introduction

Cyber–physical systems augment all aspects of society (Zanero 2017), with many industries embracing the benefits available (Brennan et al. 2019a; Muller 2017). While the increase of sensors, information sharing, and computer-aided mechanization has improved industrial capabilities, it has also introduced a level of complexity that means significant analysis is required for performance assurance before decision making (Joiner and Tutty 2018; US Government Accountability Office 2018). Take, for example, a pilot operating an aircraft representing a classic cyber–physical system. The aircraft is equipped with computers to enable easier mechanical operation, sensors to interpret data on the environment, under varying operating conditions robustly and an ability to share information with other operators through networked connectivity. However, without the cognitive ability provided by the pilot to operate the aircraft and interpret information, that is, containing the user in the system, the cyber–physical aircraft is not useful. Together, the pilot and cyber–physical aircraft create a cognitive cyber–physical (C2P) system that can contribute to fulfilling missions' objectives meaningfully.

An extension of this example of C2P is the concept of a smart autonomous system (SAS), where the cognitive agent controlling the system is artificial. Artificial intelligence (AI) already forms part of many cyber–physical systems (Abbass et al. 2016b; Endsley 1997; Klemas and Chan 2018), performing task-specific functions to assist a cyber–physical system's performance. The limitation of current AI is that while we understand how it has been programmed, with algorithms to allow for machine learning, what is not yet understood is how a lifetime of learning impacts the SAS. Hence, through-life SAS is likely to partner with humans who rely on them or put another way; SAS systems need governing. The partnership between humans and SAS can be termed human-autonomy teaming (HAT) (Lyons et al. 2018; Strybel et al. 2018), which will be subject to all the complex systems traits of an organismic model such as those documented by Keating and Bradley (2015) like emergence, uncertainty, interdependence, complexity, and ambiguity. Hence, any through-life HAT should conform to complex systems governance (CSG) such as the foundational CSG model by Keating and Bradley (2015), supported where necessary by the pathological

A Framework of Human Systems Engineering: Applications and Case Studies, First Edition.
Edited by Holly A. H. Handley and Andreas Tolk.
© 2021 The Institute of Electrical and Electronics Engineers, Inc. Published 2021 by John Wiley & Sons, Inc.

methods of Katina (2016). The CSG model can be considered "organismic" because it does more than operate; it can monitor and develop to sustain policy and identity – put simply, to evolve. This organismic approach to HAT would see the evolution of the sociotechnical system type from *user contained by the system* to *users are the system*, acknowledging that users may be human or human-trusted artificial agents. By implementing CSG methodologies and modeling (Katina et al. 2021: accepted), in conjunction with our SAS competency matrix, organizations will be able to confidently answer the question: *do we understand systems laws and their impact on design and performance* (Keating and Katina 2020)? What is required by engineers, educators, and evaluators is governed agility using a meta-system approach to support system development, lessons and training, and the ability to assure system serviceability with minimal risk.

To illustrate, consider that as with the pilot earlier, a SAS would also undergo an iterative through-life learning regime to both develop skills and forge relationships with other cognitive agents. As with any system, an iterative process between system operations and system development can ensure improved SAS functionality. However, due to its cognitive nature, the SAS will also undergo learning throughout, creating risks of emergent behavior, uncertainty, and so forth, which in turn require clear communication, policy, and identity governance. This iterative learning will be introduced from development and continue to end of life, with interactions that involve both artificial and human cognitive agents.

Successful through-life interactions should be meaningful and contribute to the success of the end goal of the system, with users able to identify whether lessons have had the desired impact of meaningful learning. The use of "meaningful" here is loaded for any SAS, in part due to the widespread military use of such future systems. For example, at a United Nations Convention on Certain Conventional Weapons Meeting on Lethal Autonomous Weapons Systems (Roff and Moyes 2016), a call for more *meaningful human control (MHC)* of SAS whereby there is "a threshold of human control that is considered necessary" and protocols such as those in Table 4.1 propose to set that threshold (Roff and Moyes 2016). There are also new engineering standards to design SAS with "fail-safe mechanisms in a robust, transparent, and accountable manner" (IEEE 2017), where there are similar new protocols for what is considered an acceptable manner. The strict application of ethical guidelines concerning these protocols, such as those documented by Jobin et al. (2019), is necessary to further the development of symbiotic relationships that have significant implications for human society. By applying a human systems engineering (HSE) approach, specifically by engineering both the cognitive agent and agreed ethics, into the lifelong development, education, and evaluation of SAS, organizations will be able to ensure human-centered principles are upheld, balanced with the characteristic properties of intelligence-based systems (Tolk et al. 2011).

Table 4.1 Key elements of meaningful human control (MHC).

Code	Element descriptor
K1	Predictable, reliable, and transparent technology
K2	Accurate information for the user of the outcome sought, operation and function of technology, and the context of the use
K3	Timely human action and potential for a timely intervention
K4	Accountability to a certain standard

Source: Based on Roff and Moyes (2016). © 2016 United Nations.

Finally, in organizations that employ cognitive agents to produce outcomes, such as the military, teams are an integral part of executing missions (DeCostanza et al. 2015; Stearns 2018). For such teams to be successful, both trust (Douglas and Strutton 2009) and good leadership (Quigley 2013) are essential. While such trust and leadership are crucial in human teaming, the extension of these value-laden concepts to, and with governance over, meaningful human-autonomy teaming (M-HAT) is what concerns this research.

We start by defining trust and the role of trust in successful teaming, allowing for the formation of M-HAT. We then present concepts of HAT to present our definition of M-HAT. We then present a systematic taxonomy and a method to characterize the maturity, or meaningfulness, of HAT in the development and through-life operations of SAS. Such a taxonomy is quintessential to the many interactive meta-functions of governing complex systems (Keating and Bradley 2015) in ways that meet the essential AI principles (Jobin et al. 2019). Following our systematic taxonomy, we discuss concepts to ensure successful SAS. Finally, we illustrate aspects of M-HAT using our Australian research program into smart autonomous sky shepherding.

4.2 Trust in Successful Teaming

Trust is an essential key to the performance of an organization (Berens 2007), with teams that exhibit trust between agents, at all levels, producing positive outcomes and displaying harmonious relationships, relishing the payoff of building trust (Reina and Reina 1999). However, what is trust? Trust is a complex array of feelings, emotions, attitudes, and beliefs combined with elements of control, risk, and power (Petraki and Abbass 2014), which is essential for interdependent teams (Mayer et al. 1995). Trust requires an understanding of values as the objects of moral conscience experienced by all individuals and communities. Values do not exist in an inaccessible metaphysical realm but exist when they function in cognitive thought and action. They are the objects of intentions (Kelly 2011). Because trust impacts on the actions and behavior of an agent, it becomes a needed characteristic for an agent to be considered fully autonomous (Abbass et al. 2016a). For a team to be successful, each agent must be capable of both understanding and displaying trust – in this case, both the humans and SAS that form a human-autonomy team.

When considering the complexity inherent in C2P, trust is a root cause for the emergence of successful teaming. First, consider two agents and the trust relationship between each other. One agent (i.e. *truster*) must be willing to be vulnerable to the other (i.e. *trustee*), with the reward being successful completion of the request (Abbass et al. 2016a). With this established trust, the agents, within their group, may later form part of a team. The formation of trust must then extend beyond two agents and to that of multiple. Here *learned trust* (Abbass et al. 2016a) may assist in propagating the trust beyond the two original agents, through transparency and communication, as well as repeated reliability (Abbass et al. 2016a). Hence, trust can be carefully and deliberately built through *team-building* exercises that reinforce reliability, transparency, communication, and alignment of objectives and intents, thereby allowing a trust relationship to follow. With an established team, freely demonstrating trust, overall performance will improve (Berens 2007; Mayer et al. 1995), allowing the trust relationship, and by proxy, the sociotechnical system evolution, to be measured. While task completion is one metric to measure the existence of a trust relationship within a team, additionally, Costa and Anderson (2011) recommend measuring individual agent performance (*perceived trustworthiness* and *propensity to trust*) and collective performance (*cooperative behaviors* and *monitoring behaviors*), which includes quality of achievement and the ability of the team to accept commands outside of the team network. By applying such broad metrics, testers and developers will be able to meet the MHC elements presented in Table 4.1. Specifically, task completion relates to K1 – *predictable, reliable, and transparent technology*. While task completion may not be a complete indication of transparency, combining this with K2, *accurate information for the user of the outcome sought*, with a correlation between precise information and transparent performance between cognitive agents, forms an enabler for task completion. The metric of individual agent performance applies to both human and artificial cognitive agents. Successful agent performance requires disclosing *accurate information for the outcome sought and the context of use* (K2) while also remaining *accountable* for actions (K4). The cognitive agent, who demonstrates trust through transparency, will understand the importance of being accountable for tasks completed, which may assist with the reinforcing of transparency in relationships. The level of transparency will contribute to the development of a trust relationship, linking back to individual performance, quality of achievement, and team cohesion. While working in a team, it is often necessary to accept commands from agents external to the group. The *timeliness* in receiving and *acting* on this external information (K3) is a measure of both individual performance and team cohesion.

An insight into how to structure the evaluation of HAT is evident in the precepts of *usability testing* for human–computer interaction where there is a significant emphasis on appropriately diverse metrics and enough iteration (Wickens et al. 2014). Perhaps the most significant extension of these usability testing precepts necessary to cater for SAS is the increased dependency on continued through-life testing[1] governance with meaningful control being "as frequent as necessary." Research in the governance of cyber–physical systems that are under constant emergent and adaptive threats has found there is a significant overhead in test infrastructure and skills from such through-life governance (Joiner et al. 2018), which likely will be the case for M-HAT, unless the risk is mitigated through properly designed cognitive agents specializing in lifelong testing of SAS, in adequately prepared scenarios for meaningful behavior and decision-making assessment (Tolk et al. 2009).

4.3 Meaningful Human-Autonomy Teaming

HAT is the shared operation of systems by task-specific autonomous systems and humans. Matessa et al. (2018) present one example of this form of HAT, where automation enhances the planning system for emergency landing that allows operator interaction, a form of transparency, and human-autonomy communication. The task-specific autonomous system, in this case, an Autonomous Constrained Flight Planner (ACFP), reduces human workload by selecting the optimal flight path for various emergencies. The user operates the ACFP by influencing weightings for flight path optimization. The ACFP provides the operator with not only the apparent optimal path but correspondingly alternative paths, therefore achieving transparency and reinforcing a degree of trust both through justifying the recommendations and the human-autonomy communication.

While the use of the HAT by Matessa et al. (2018) is simple, it seeks to provide an exemplar foundation for M-HAT, that is, one that seeks to form a meaningful relationship. However, the ACFP does not learn from suggestions it provides the pilot, nor whether the pilot accepts its input. Further, there is no reward for the pilot choosing to trust the ACFP; therefore, the success of this system becoming a team built on trust in the future is, in our opinion, considered low. Further, the ability of such a system to evolve as a sociotechnical system is low and can therefore remain a system that is engineered for *user operates system* only (Chapter 1).

---

1 A system is defined to have a "life cycle" as per IEEE (2015). Through-life testing refers to testing during the life cycle, from *conception* to retirement.

As presented earlier, fostering trust between any two agents, or more, is complex, requiring an understanding of beliefs, attitudes, and expectations (Petraki and Abbass 2014; Strybel et al. 2018), but, together with values, must also be measurable when developing autonomous systems designed for integration within teams. Lyons et al. (2018) identified that the initial introduction of systems where humans are required to trust autonomous agents is comparable with teams with members who have not worked together before. This incremental approach may work if it is accretional; however, few developmental acquisitions of SAS are likely to be ideally incremental since the underlying architecture and interface changes will mean significant project investment. If a major information communication technology (ICT) project implemented M-HAT as simple addition, that is, without consideration of the interaction between systems within a family of systems (Clark 2008), there are substantial risks that the traits of such a complex system could undermine trust during development and certainly before operational implementation. Introducing HAT, without the influence of meaningful relationships throughout development, risks creating a bias by humans preferring human-human teaming, thereby inefficiently avoiding the autonomy or of humans deferring to the SAS and providing the type of nonmeaningful control implied by Roff and Moyes (2016). Such relationship would result in a sociotechnical system that remains *user operates systems*, instead of evolving to *users are system* that includes human-SAS users (Chapter 1).

Trust in automation requires the considerations of reliability, performance, error type, transparency, shared awareness, and shared intent (Abbass et al. 2016a; Lyons et al. 2018) just as it does in human teams, commonly used in performance appraisal (Denisi and Pritchad 2006). To foster a meaningful relationship, the development of the autonomous system should be integrated with human teams from the creation of the project, in such a manner that both the autonomous system(s) and the human team members must be able to each provide and accept feedback (Abbass et al. 2016a). By ensuring transparency, team-building interactions, and human-autonomy communication, the systems, engineered through consideration of human-centered principles, will build trust in a meaningful and measurable way. Human teams that operate in a manner promoting trust demonstrate greater collaboration and performance, with views to improve productivity and overall team behavior (Cummings 1983; Favero et al. 2016): hypothetically human-autonomy teams can be expected to do the same. Irrespective of the efficacy of such careful team development and maintenance, a systematic approach should promote transparency and reward accuracy and task completion for meeting the principle of *predictable, reliable, and transparent technology* outlined earlier (Jobin et al. 2019; Roff and Moyes 2016) – tenets of MHC.

4.4 Systematic Taxonomy for Iterative Through-Life Learning of SAS

As discussed, establishing M-HAT from the beginning of a teaming system life cycle allows for the development of meaningful relationships between the cognitive agent teams. Given the complexity of SAS, project managers, human systems engineers, researchers, and developers will need to alter their view of testing and operational introduction of SAS to one of continual learning and development of cognitive agents. In research presented by Klemas and Chan (2018), agencies are seeking to promote cyber resilience by introducing machine learning, data analytics, and automated testing; however, there is no defined way to appropriately assess the competency of such technology. Promoting the cyber resilience of ICT systems through testing has been presented (Joiner et al. 2018), and augmenting this testing with competency measures will better communicate system readiness and security to users and managers alike.

Currently, the development of a system includes conducting a technology readiness assessment, using a technology readiness level (TRL) as a metric (U.S. Department of Defense 2016b), shown in Table 4.2.

While conducting a technology readiness assessment for SAS will remain relevant, the metric will need to consider not only the TRL but also the ability of the SAS to learn and demonstrate concepts. That is, what lessons are required for the SAS, not only to be technologically ready but just like a human apprentice, cognitively prepared for the task. Consequently, we propose SAS assessments must consider not only TRL but also the SAS cognitive competence level to produce an overall SAS readiness level (SRL). We define SAS cognitive competence level as the ability of the SAS to articulate its understanding of a task to another cognitive agent, as given in Table 4.3, based on similar human concepts reviewed by Krathwohl (2002). Identifying the SAS cognitive competence level is vital to ensure M-HAT principles of interaction, human-autonomy communication, and transparency. Human team members will better accept a learning SAS, or through-life a relearning SAS if it is part of the normal progression of the team's development. Importantly, managers of programs, simulationists, engineers, and trainers can then systematically allocate the schedule, time, and funding to constructively develop, introduce, and refine SAS in the M-HAT construct. For example, Ryan (2018) presents that the future workforce of militaries will include both soldiers

Table 4.2 Technology readiness level (TRL).

1	Basic principles observed and reported	Survey and review
2	Technology concept or application formulated	Analytic study
3	Analytic and experimental critical function or characteristic proof of concept	Proof of concept
4	Component or breadboard validation in a laboratory environment	Laboratory (lab) integration
5	Component or breadboard validation in a relevant environment	Early integration
6	System and subsystem model or prototype demonstration in a relevant environment	Lab or simulated (sim) environment (env) demonstration (demo)
7	System prototype demonstration in an operational environment	Demo in operational (op) env
8	Actual system is completed and qualified by test and demonstration	Accepted
9	Actual system has been proven in successful mission operations	Operational

Source: Modifed from U.S. Department of Defense (2016b).

Table 4.3 Proposed SAS cognitive competence levels.

Level	Definition
A. Remembering	Recall gameplay, recite an instruction
B. Understanding	Explain process using own words, explain movements
C. Applying	Create new moves, create a procedure for others to follow
D. Analyzing	Troubleshoot, identify tasks for training, conduct a test
E. Evaluating	Select the most suitable player, justify a process
F. Creating	Integrate information from multiple sources to create a solution, network with others

Source: Adapted from Krathwohl (2002).

and SAS. SAS will be part of the team and structure, equating to both following and leading a team of cognitive agents. To fulfill this role, SAS must be able to not only learn but also demonstrate concepts with others. To ensure success, organizations will need to continue to employ the iterative development of teams.

We illustrate how to assess SAS using SRL by considering *AlphaGo* (Silver et al. 2016, 2017). *AlphaGo* learned to play a strategic game through simulated gameplay and actual gameplay with humans. While learning, *AlphaGo* used information from historical gameplay, as well as reinforcement learning during simulated gameplay. *AlphaGo* then applied this knowledge in a game against Fan Hui, winning five games to zero. While *AlphaGo* used its knowledge and demonstrated gameplay closer to human play, it was unable to communicate the methodology. To understand how *AlphaGo* played, researchers analyzed gameplay in isolation and not in conjunction with *AlphaGo*. Using the proposed SAS cognitive competence levels (Table 4.3), *AlphaGo* demonstrated Level A and aspects of Levels B and C; however, these are insufficient as they are to classify the readiness of the SAS adequately and therefore require further refinement. Combining the technological readiness level (Table 4.2) and SAS cognitive competence level (Table 4.3), we propose that the two-dimensional SRL be defined for use, as shown in Table 4.4.

In the *AlphaGo* example, we assign an SRL nomenclature of "*7.Abc*" as follows:

- Number "7" indicates that *AlphaGo* was demonstrated in an operational environment.
- Uppercase "A" indicates *AlphaGo* has competency in *remembering*.
- The lowercase letters "b" and "c" indicate *AlphaGo* has potential, yet need for improvement, in *understanding* and *applying*.

As per our M-HAT definition, to ensure meaningful interactions, effective human-autonomous communication, and transparency, SAS will need to hold a minimum of SRL "*4.A.B.C.d.e.f.*" SAS with a competency of "*4.A.B.C.d.e.f*" will be able to operate with a human or operator, in a simulated environment, and be able to demonstrate meaningful behavior to foster trust, collaborate on ideas, and learn through interactions with either self, human(s), or another SAS. By compiling a SAS competency matrix, modular components making up the SAS may be tracked, allowing

Table 4.4 Proposed SAS readiness levels.

| | | | | SAS cognitive competency level | | | | | |
| | | | | A | B | C | D | E | F |
				Remember	Understand	Apply	Analyze	Evaluate	Create
Technological Readiness Level	Basic research	1	Survey and review						
		2	Analytic study						
	Applied research	3	Proof of concept						
		4	Lab integration						
		5	Early integration			SAS competency matrix			
		6	Lab or sim env demo						
	Early acquisition	7	Demo in op env						
		8	Accepted						
	Final acquisition	9	Operational						

flexible design and good project management. Further, simulationists and testers will be able to assess M-HAT teams in terms of task completion (K1 and K2), individual agent performance (K2 and K4), and collective performance (K3). This minimum standard would also allow testers to assess whether outcomes achieved are a result of proper function, or chance, as the system will be able to explain what it has sought to achieve. An assessment of the autonomous agent as displaying operational or functional morality (Sharkey 2017) is also possible, or at least "collaborative intelligence" as described by Goldberg (2019), where humans collaborate with SAS to mutually complement each other. An autonomous agent exhibiting *operational morality* (value based) would show lower levels of ethical sensitivity and would be unsuitable for any safety-critical system applications beyond full human control. Such systems would remain tools to assist or augment humans, whereby the user operates the system to support system outcomes, yet not collaborate at an intellectual level. However, an autonomous agent exhibiting *functional morality* (values/virtues based) would also be ethically sensitive and could integrate with a team of cognitive agents, utilizing MHC. SAS would be capable of collaborating with humans intellectually, and by assessing actions, assessors will be better able to query the agent on both the outcome sought and the purpose behind such a result. In this manner, agents will be able to articulate decisions, measurable in terms of both competence and ability to express responsibility for actions (Laukyte 2017), evolving to a system of users, where users are both human and SAS. Responsibility for actions may align to normative outcomes, with an autonomous agent capable of making meaningful decisions when faced with a choice of good/bad or right/wrong (B), taking these normative conditions into account (E) and exercising a degree of control in making decisions on that basis (ABCDEF). Should the explanation be incorrect, there is likely a need to introduce further learning curriculum to ensure correct operation, human-centered principles, and function as well as to assign the agent to a team to enable success correctly. Using the SRL in Table 4.4 to assess SAS, it is possible to instill a multi-level interdisciplinary research, development, assessment, and operation process. Such a process introduces what Brey (2000) termed *disclosive computer ethics*, focusing on justice, autonomy, democracy, and privacy and involved disciplines ranging from philosophers to computer scientists.

Robbins and Steffen (2018) also propose that completing testing of SAS in phases of simulation, hardware-in-the-loop, and safe-level operational environments. By leveraging the results from previous phases to shape the test plan of future phases, a continuous assessment of readiness is possible. Further, Robbins and Steffen (2018) present that to adequately test the decision-making logic, the SAS should be exposed to highly stressful situations before being placed in an operational environment. It is not impossible to stress an agent in a simulated environment in safety-critical domains, such as by using a flight simulator or in gameplay (gaming for soldiers). However, it is important to caution against only assessing the decision-making logic of SAS in levels of stress as it is likely to reduce the ability to measure the cognitive competence level of the SAS effectively. It is important to simultaneously evaluate the errors a SAS makes and the cognitive capacity of the SAS, which represents a SAS capacity to learn and overcome the mistakes. A system with a medium level of errors but with no capacity to be educated could deem unviable, especially compared with a system that initially has a higher level of errors, but a significant capacity to be educated and reach a lower level of errors.

As such, we recommend developing SRL testing to identify the optimum performance in both general tasks and high-stress task environments caused by congestion and contestation while simultaneously considering the system capacity for further education and learning. By conducting SRL evolution tests using proper and thorough methodologies, transparently reporting outcomes during testing and evaluation, and collaborating beyond bespoke disciplines, the risk of unexplainable emergent behavior is reduced. The success of through-life development will be dependent not only on the ability of a SAS to demonstrate competence but also on the ability of the entire team to uphold code of practices[2] and transparently report outcomes to stakeholders, including society.

With the systematic technological phases of simulation, hardware-in-the-loop, and safe-level operational environments, the testing process will need to be structured to ensure a successful assessment of the SRL throughout. Using regular phases, underpinned with contemporary test and evaluation (Joiner et al. 2019), will evolve the SAS to promote self-awareness, trust, and, therefore, M-HAT. The systematic progression proposed would substantially help meet the conventions of Roff and Moyes (2016) (i.e. Table 4.1) and the new IEEE standard for SAS (IEEE 2017). The next section of the chapter focuses on the specialist teaming expected to govern the SAS through education, assessment, and testing.

4.5 Ensuring Successful SAS

To reliably and quickly identify emergent behavior, ambiguity, or uncertainty in decision-making logic during development and through-life of SAS, it is likely to require more than just the teamed humans and systematic readiness testing. Early concepts of identifying emergent behavior have been proposed using the conceptual terms *sidecar* or *cyber sidecar* (U.S. Department of Defense 2016a, pp. 28–30; Brennan et al. 2019b). What we propose here is the alternative conceptual term *watchdog AI agents* (Abbass et al. 2018), as it conveys the role as being less passive, more intelligent, and more analogous to augmenting human attention than a *sidecar*. Chan (2018) outlines an AI system for cybersecurity monitoring that provides oversight against threats and deals with uncertainty and ambiguity in a manner that aids timely human decision making. Such uses for *watchdog AI* are likely to be led by cybersecurity applications seeking to regain the initiative from the prevalence of malicious threats, albeit in ways to avoid accelerating such threat sophistication (Klemas and Chan 2018).

For a *watchdog AI* to be valuable requires it to be developed independently of, but concurrent with, the SAS utilized for M-HAT. *Watchdog AI agents* would independently monitor the SAS during testing and operational implementation to continually assess the readiness level progress yet also mitigate risks that may be present by continuously iterating education of the SAS (Abbass et al. 2018). Specifically, by defining and assessing risky behaviors, or directly interrogating SAS to understand actions, education of SAS will still have bounds on the meta-cognitive level. By both assessing and interrogating, *timely intervention* is more ensured (Jobin et al. 2019; Roff and Moyes 2016). As with any system exposed to the cyber environment, monitoring performance is key to identify whether vulnerabilities have developed or a change in interactive behavior has resulted through a meaningful interaction shift assessment. Using *watchdog AI agents* potentially alleviates complexities in assessing the nature and speed of information processing of SAS that cannot be achieved by humans alone (Abbass et al. 2018). The oversight of the HAT SAS through a *watchdog AI* and supervising human agents is itself a governing layer of HAT, belonging to the strategic monitoring component of policy and identity in the CSG model. Emergent behavior detected by the teaming of *watchdog AI* and governing humans can then trigger adjustments to system operations and system development with the necessary information and communications

---

2 Many disciplines, such as research (The Australian Code for the Responsible Conduct of Research, 2018), engineering (IEEE Code of Ethics 2020), and simulation (Tolk 2017), include a code of ethics to govern professionals.

Table 4.5 Human–robot interaction roles.

Role	Responsibility	Definition
Supervisor	Testing	Monitor and control the overall situation
Operator	Training	Modify internal software of models when behavior not acceptable
Mechanic	Training	Physical intervention, including hardware changes to ensure correct behavior
Peer	M-HAT	M-HAT team members, working together to achieve the outcome
Bystander	M-HAT	An external observer who may be called upon to interact with SAS
Educator[a]	Testing	Responsible for assessing SRL and developing curriculum to assess achievements and enable lifelong learning

Source: Based on Scholtz (2003).
[a] The role of educator is proposed in this chapter and not in Scholtz (2003).

to regain trust and promote human-centered principles. Such governance aligns with the "timely human action" and "accountability" *to a certain standard* of Roff and Moyes (2016).

Scholtz (2003) presents five roles for human–robot interactions, shown in Table 4.5 for completeness. We propose two changes to these roles for HAT. The first change is to add a SAS educator or clarify who among the supervisor, operator, or mechanic fulfills that secondary role. The second change is to relabel the bystander as human team members.

The human supervisor alone will be unable to adequately assess the success of SAS promptly, as the SAS cognitive processes occur at a much faster rate (Chan 2018; Klemas and Chan 2018). Using *watchdog AI* and meeting SRL milestones, it is possible to assess whether intervention is required. Intervention from a supervisor would trigger an evaluation of the SAS to reassess the SAS readiness against a known state. This assessment should be undertaken by a SAS educator, who complements both the operator and mechanic roles presented by Scholtz (2003). Using the educator may identify the need to modify either software or hardware components of the SAS with specific knowledge for the SAS of goals, intentions, actions, a priori curriculum, and lifelong experiences. Specifically, the educator would be responsible for monitoring, educating, and assessing. Establishing a hybrid team of agents to ensure the successful integration, operation, and education of SAS is an example of extended agency and, therefore, hybrid responsibility (Gunkel 2017), enhanced with the critical elements of MHC.

As presented by Scholtz (2003), the different roles may assist in assessing how well the human–robot team interacts. Concerning fostering trust between agents and successful M-HAT, this is essential. Given the SAS readiness assessments are to determine and later monitor whether the SAS is ready to be part of the team, there must also be a way to identify whether interactions are meaningful. As mentioned, one aspect is the achievement of outcomes. However, this may not determine whether the team is genuinely cohesive. Therefore, a valuable addition proposed is "situational awareness" assessments that promote meaningful interactions that are a result of perceptions from SAS and peers. Situational awareness is the processing of incoming information from many systems, including the environment and other team members, to better enable decision making (Endsley 2013). These proposed *situational awareness* assessments are akin to the team-building exercises mentioned earlier but adapted to account for SAS being on the team, with a greater focus on information flows around critical decisions and with less intuition than a purely human exercise (i.e. where the monitoring and information may be subliminal or in some other way not explicit).

Table 4.6   Situation awareness levels.

Level	Definition
1	Perception of the elements in the environment
2	Comprehension of the current situation
3	Projection of the future status

Source: Based on Endsley (2013).

Given that situational awareness is influenced by several factors, including abilities, experience, and training (Endsley 1995), iterative learning (SAS education and training) will be essential in promoting M-HAT. Endsley (2013) identifies the three levels of situational awareness in Table 4.6 that, when applied to HAT, reaffirms a level of educational understanding of self, situation, and task for the partnered SAS. Through-life education can focus on developing HAT through regular education to improve situational awareness in different environments and adapted roles, including adaptive threats. The team will involve varying aspects of human-autonomy interaction, catered to the unique situation of the education environment. Crucially the governing supervisor, *watchdog AI,* and SAS educator will need to be involved throughout, focused on read-across of readiness levels to the adapted roles and threats. As such, the proposed situational awareness assessments are an extension to the first analogy of a pilot, where the pilot must undergo reeducation for adapted roles and threats performed in a real aircraft and for emergency circumstances in a simulator.

Like education iterations, training iterations must also occur in phases. However, to ensure the success, education phases must be completed first, before progressing to training. Specifically, the agent must be able to communicate an understanding of the know-how to perform a task before being trained. Without this understanding, the agent will not be able to meaningfully connect and interrelate lessons experienced during training to promote their situational awareness. To classify the level of training completed, assessing SAS against both the SRL and situational awareness level achieved in the training phase is required. Training iterates until sufficient readiness on both levels are delivered by the HAT. Such training is analogous to the pilots undergoing currency checks on their certified roles both in a real aircraft and simulator. The importance is to ensure the simulation environment is representative, realistic, and transparent (Tolk 2017).

In summary, SAS involved in M-HAT require governance that itself likely needs to be augmented with AI (watchdog) and similar levels of education and training oversight, such as through-life education of SAS itself, as well as high-performance human operators (i.e. pilot, surgeon, etc.). With regular team-building competency assessments focused on situation awareness, M-HAT teams will be better able to foster trust and other performance characteristics, such as task achievement on operations and low error rates.

4.6   Developing Case Study: Airborne Shepherding SAS

Strömbom et al. (2014) advanced research into the modeling of shepherding with the two-dimensional modeling of sheep behavior to herding agents. According to Singh et al. (2019), the modeling "offers a level of abstraction for Human-Swarm Interaction" or "a smart robot could act as a shepherd to replace biological shepherds with ground or air vehicles." The role of AI so far has been to "preserve the energy of the shepherd by modulating the shepherd's influence vector on the sheep." Other related research includes by Clayton and Abbass (2019) into machine-learning reward functions "to evolve a swarm controller for an agent shepherding a boids[3]-based swarm" and by Gee and Abbass (2019) on "a novel curriculum to teach an AI-empowered agent to shepherd." However, this research remains laboratory based, at least since the original study by Strömbom et al. (2014), needing applied research to put the SAS capability into practical use. Using the proposed SRL, such applied research aims to take the readiness level from "*2ABc*" to "*7ABCde*" (Table 4.4). Given geographical distances and terrain in Australia, practical research is ideally airborne, which reduces the technological readiness from the two-dimensional basis just given. An online search of videos disclosed an alarming number of farmers in Australia, New Zealand, and the United States exploring the use of remotely piloted drones for the herding of sheep, despite the lack of serious research and underpinning science. The potential inefficiency of using remotely piloted drones is of concern with these ad hoc trials, as is the ethical lens of drone pilots who are not familiar with sheep or the subject paddocks (fields) trying to herd. There also exists a potential of drone accidents, resulting in injury to farmers or livestock, due to farmers not familiar with complexities in drone piloting. Applying smart autonomy research from two-dimensional shepherding to an airborne capability through to an operational demonstration is an ideal case study for applying the SAS readiness taxonomy and proposed concepts for meaningful HAT. While this case study will take several years to apply, using the taxonomy now to illustrate the usefulness of the proposed method for developmental design will assist in the development and evaluation of an exemplar M-HAT system. Figure 4.1 distinguishes the four phases leading to the autonomous use of drones to herd a flock of sheep to a goal, such as an open gate on a paddock.

The research has so far identified four phases leading to the autonomous use of drones to herd a flock of sheep to a goal, such as an open gate on a paddock. Phase 0 commenced with the use of a commercial off-the-shelf drone to investigate the significant factors within the environment. Experiments were designed to have a pilot executing communicated by a farmer. Phase 1 sees the implementation of a research developed drone (Sky Shepherd). Both phase 0 and phase 1 have users operating the system, with considerations for animal ethics and the investigation of how to safely and efficiently herd flocks of sheep.

The evolution from phase 1 to phase 2 will see the introduction of an autonomous control system for operating the Sky Shepherd, coinciding with the first evolution in the socio-technology

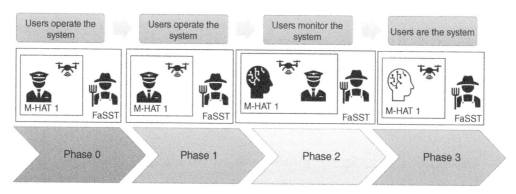

Figure 4.1 Technology evolution for successful M-HAT in Sky Shepherd research.

---

3 The term *boids* was initially used by Reynolds (1987) to model the behavior of biologically inspired swarms.

relationship, from *users operating* to *users monitoring*. The monitoring role for the pilot will be necessary to ensure ethical actions and decisions are achieved by the control station while herding flocks. The ethical actions will extend from animal welfare to upholding aviation regulations based upon human values. Further, the communication between the control station and pilot will be successful when a natural blending of human and machine is promoted. These meaningful, timely human interactions (K3) will shape the autonomous system into a SAS that is reliable and transparent (K1) and, by extension, provides accurate information for the monitoring user (K2), allowing accountability (K4). As the autonomous system learns, the monitoring may be completed by a farmer only, allowing for the evolution to phase 3, which is where the autonomous system is a transparent SAS. The users become the system, that is, a SAS and farmer teaming together with a Sky Shepherd or Farmer and Sky Shepherd Team (FaSST).

Simulations on how sheep behave in the presence of ground-based herder, including artificially intelligent autonomous collecting and driving of sheep, currently work in two dimensions. Future work focuses on taking the modeling to four dimensions, by first using a drone, which adds the vertical dimension, and second applying a speaker to the drone to add a fourth audio effect dimension. Another research line is exploring how to get the SAS sky shepherds to team as a team of working dogs would so that much larger flocks can be worked still using several teaming cost-effective drones. Finally, the last research thread is to make a straightforward mission planning system where a farmer can tell the drones from a smartphone where the property boundaries are, how many sheep to expect, where the goal is, and any paddock complexities like dams, tree lines, and the such. This approach seeks to overcome some of the limitations of AI documented by Zador (2019) by employing structured collaboration (i.e. M-HAT), as suggested by Goldberg (2019).

Phase 0 of our research begins more foundationally in establishing how sheep respond to drones, such as those pictured in Figure 4.2, carefully and systematically, to inform subsequent phases of the research by knowing what the safest and most effective maneuvers and noise for a drone are to use in sky shepherding. This phase starts with some heart-rate monitoring of sheep, pictured in Figure 4.3, to systematically model the impact of drones on the health and welfare of animals; thus, ensuring the design is ethical by anticipating potential harm and taking appropriate action to prevent it. Smart drones must learn to understand cues for sheep alertness, anticipate sheep response, and maneuver in the least distressing but still effective ways possible. Importantly, the ethical baseline and thus benchmark is current shepherding practice with sheepdogs, and this requires some testing to confirm from early research (Baldock and Sibly 1990) with our current instrumentation.

The initial phase ensures the credibility of the sheep response modeling for simulations, screens key influence parameters, but importantly grounds the research in what is *meaningful* for human control. To apply a SAS to such a problem involves a better understanding of not only farmer's current shepherding uses but how that can evolve with new capability. For example, much of the early focus of autonomous flight vehicles were on weapon delivery compared with a wide diversity of modern use (Floreano and Wood 2015). While that use remains controversial still Roff and Moyes (2016), the vast majority of autonomous flight vehicle time is spent on surveillance. Hence, making a shepherd airborne is likely to provide significant surveillance capability for the farmer, especially in remote stations where assessments of feed, water, fencing, and feral dog threats are essential precursors to an actual shepherding. Hence drones offer timely surveillance and action that should be a significant and ethically aligned advantage to farmers.

In research like this, it is tempting to think at the shallowest level of "autonomy" as replacing the human altogether and thus removing the pilot, farmer, and dogs. Such presumption leads to criticisms on nonmeaningful control. However, the next level of conjecture is that the HAT starts with

Figure 4.2 Sky Shepherd pilot sets up a DJI Mavic II Enterprise Duo ready for phase 0 testing. *Source:* The photos (Figures 4.2 and 4.3) presented in this chapter form part of the research approved by The University of New South Wales Animal Ethics Committee (ACEC 19/122B).

Figure 4.3 A flock of Dorper sheep (*Ovis aries*) responding to the presence of low speed, low altitude, Sky Shepherd, augmented with an aural cue. *Source:* The photos (Figures 4.2 and 4.3) presented in this chapter form part of the research approved by The University of New South Wales Animal Ethics Committee (ACEC 19/122B).

the laboratory or operational demonstration (TRL 6–7). This presumption leads to highly nonoptimal interfaces and the lack of trust described earlier. Working the HAT as soon as the concept stages, such as an infield response collection, offers a chance to focus HSE on whether design requirements meet farmer's needs and to support more rigorous simulations. Such simulations, in turn, mean farmers can be engaged on a user interface for mission planning with far less risk.

The proposed systematic approach to M-HAT readiness (Table 4.4), the CSG view, and HSE principles has helped focus the path of researchers to a team of stakeholders representing the governing influences for the envisaged complex system, including, farmers, animal welfare, pilots, drone designers, and commercial livestock agents. The CSG model has also highlighted the importance of information and communication. New ethical uses of drones need to be sought and also seen to be explored since to have practical effect requires significant collective farmer and consumer acceptance (Munoz et al. 2019). Ongoing public awareness programs will be vital to making the autonomy trusted by all concerned (*Proposed Australian Animal Welfare Standards and Guidelines – Sheep* 2013).

Furthermore, this application raises new ethical challenges, such as follows:

- What is the principal actor?
- Can phenomenology characterize the actor as executor of acts?
- Is unity discoverable in the tables of values, moral values, norms, and virtues such that any given action could be characterized consistently as either right or wrong?

Finally, the concept of a *watchdog AI* for this system appears as a simple presumption, to be an added complication to an already complex research task, and it may always face such concern. However, on deeper examination, the commercial and public acceptance hinges on trust. The likely most significant treatment for any abnormal event, whether it occurs in the laboratory development or each farmer's first use, is early timely feedback on whether the behavior is anticipated or emergent. How many farmers and videos of their use would it take to cause an issue with the SAS, whereas how many times could a resident *watchdog AI* cause a farmer to reread the mission planning instructions? Developing the *watchdog AI* from the inception of the SAS offers "learned" governance of meaningful in this context.

4.7 Conclusion

This chapter has presented the tenets of M-HAT as being transparent, human-autonomy communication, and interaction. Further, the chapter has proposed how to assess the capability of SAS in meeting these tenets. Combining current readiness assessments, for both technical systems and educational taxonomy, has provided human systems engineers, evaluators, and educators with a taxonomy to identify the overall readiness level of a SAS to team with the necessary human-autonomy two-way trust. To ensure new proposed competency levels are accurately assessed requires testing to be completed in matched systematic phases that include the simulated environment, hardware-in-the-loop, and operational environment using contemporary test and evaluation methods (Joiner et al. 2019). Throughout these phases, general and high-stress scenarios are assessed as necessary for the human-autonomy team to develop all the tenets of meaningful teaming fully and to identify both optimum performance and areas for improved performance or learning while also evolving the sociotechnical principle of *users are the system*. This rigor in stressing is entirely analogous to the cyber resilience trade-offs for through-life management documented by Joiner et al. (2018).

SAS involved in meaningful teaming with humans are likely to require governance through-life that itself is augmented with AI (*watchdog AI*). Further, the governance regime is likely to need similar levels of education and training oversight through-life as high-performance human operators (i.e. pilot, surgeon, etc.) and to include regular team-building competency assessments focused on situation awareness. Throughout the iterative lifelong learning, a team of human-autonomy experts, including the introduction of a smart autonomy educator, will be required to ensure iterative education and training. Together, the supervisor, *watchdog AI* agents, and educator are necessary to ensure the success of SAS to meaningfully team alongside the four critical human control tenets (Table 4.1).

Our research is verifying how the HAT methodology proposed can be applied to new Australian research on the herding of sheep using airborne drones. The proposed research will involve the teaming of farmers with the autonomous capability to assure MHC for the farmers and their public obligations. This application potentially solves extant ethical challenges in shepherding in Australia concerning the use of dogs and timely monitoring and action in vast sheep stations. Still, it also raises new ethical challenges around the continually evolving AI in the system.

The proposed systematic approach to the readiness of M-HAT can assist in meeting new autonomy design standards, rigorously assessing readiness levels for teaming, and promoting through-life situational awareness assessments against new proposed levels.

Acknowledgment

The photos (Figures 4.2 and 4.3) presented in this chapter form part of the research approved by the University of New South Wales Animal Ethics Committee (ACEC 19/122B).

References

Abbass, H., Petraki, E., Merrick, K. et al. (2016a). Trusted autonomy and cognitive cyber Symbiosis: open challenges. *Cognitive Computation* 8 (3): 385–408.

Abbass, H.A., Leu, G., and Merrick, K. (2016b). A review of theoretical and practical challenges of trusted autonomy in big data. *IEEE Access* 4: 2808–2830.

Abbass, H.A., Harvey, J., and Yaxley, K.J. (2018). Lifelong testing of smart autonomous systems by shepherding a swarm of watchdog artificial intelligence agents. arXiv preprint (airXiv:1812.08960).

Baldock, N. and Sibly, R. (1990). Effects of handling and transportation on the heart-rate and behavior of sheep. *Applied Animal Behaviour Science* 28 (1–2): 15–39.

Berens, M. (2007). Trust and betrayal in the workplace: building effective relationships in your organization. *Journal of Organizational Change Management* 20 (3): 463–465.

Brennan, G., Joiner, K., and Sitnikova, E. (2019a). Bespoke versus open architectures: impacts on cyber-resilience of mission-critical systems. Paper presented at the SETE 2019: Systems Engineering Test and Evaluation Conference 2019, Canberra.

Brennan, G., Joiner, K.F., and Sitnikova, E. (2019b). Architectural choices for cyber resilience. *Australian Journal of Multidisciplinary Engineering* 15: 68–74.

Brey, P. (2000). Method in computer ethics: towards a multi-level interdisciplinary approach. *Ethics and Information Technology* 2 (2): 125–129.

Chan, S. (2018). Prototype orchestration framework as a high exposure dimension cyber defense accelerant amidst ever-increasing cycles of adaption by attackers. Paper presented at the CYBER 2018: The Third International Conference on Cyber-Technologies and Cyber-Systems, Athens.

Clark, J.O. (2008). System of systems engineering and family of systems engineering from a standards perspective. In *2008 IEEE International Conference on System of Systems Engineering*, Singapore, pp. 1–6.

Clayton, N.R. and Abbass, H. (2019). Machine teaching in hierarchical genetic reinforcement learning: curriculum design of reward functions for Swarm Shepherding. Paper presented at the 2019 IEEE Congress on Evolutionary Computation (CEC), Wellington.

Costa, A.C. and Anderson, N. (2011). Measuring trust in teams: development and validation of a multifaceted measure of formative and reflective indicators of team trust. *European Journal of Work and Organizational Psychology* 20 (1): 119–154. https://doi.org/10.1080/13594320903272083.

Cummings, L.L. (1983). Performance-evaluation systems in context of individual trust and commitment. In: *Performance Measurement and Theory* (eds. F. Landy, S. Zedeck and J. Cleveland), 89–96. New Jersey: Erlbaum Hillsdale.

DeCostanza, A.H., Gallus, J.A., and Babin, L.B. (2015). Global teams in the military. In: *Leading Global Teams: Translating Multidisciplinary Science to Practice*, 295–322. New York, NY: Springer New York.

U.S. Department of Defense (2016a). *Defense Science Board (DS) Summer Study on Autonomy*.

U.S. Department of Defense (2016b). *Technology Readiness Assessment Guide: Best Practices for Evaluating the Readiness of Technology for Use in Acquisition Programs and Projects*. Washington: GAO.

Denisi, A.S. and Pritchad, R.D. (2006). Performance appraisal, performance management and improving individual performance: a motivational framework. *Management and Organization Review* 2 (2): 253–277.

Douglas, M.A. and Strutton, D. (2009). Going "purple": can military jointness principles provide a key to more successful integration at the marketing-manufacturing interface? *Business Horizons* 52 (3): 251–263.

Endsley, M.R. (1995). Toward a theory of situation awareness in dynamic systems. *Human Factors: The Journal of Human Factors and Ergonomics Society* 37 (1): 32–64.

Endsley, M.R. (1997). Level of automation: Integrating humans and automated systems. *Proceedings of the Human Factors and Ergonomics Society 41st Annual Meeting* 1: 200–204.

Endsley, M.R. (2013). Situation awareness. In: *The Oxford Handbook of Cognitive Engineering* (eds. J.D. Lee and A. Kirlik). New York: Oxford University Press.

Favero, N., Meier, K.J., and O'Toole, L.J. Jr. (2016). Goals, trust, participation, and feedback: linking internal management with performance outcomes. *Journal of Public Administration Research and Theory* 26 (2): 327–343.

Floreano, D. and Wood, R.J. (2015). Science, technology and the future of small autonomous drones. *Nature* 521: 460–466.

Gee, A. and Abbass, H. (2019). Transparent machine education of neural networks for Swarm shepherding using curriculum design. Paper presented at the 2019 International Joint Conference on Neural Networks (IJCNN), Budapest.

Goldberg, K. (2019). Robots and the return to collaborative intelligence. *Nature Machine Intelligence* 1: 2–4.

Gunkel, D.J. (2017). Mind the gap: responsible robotics and the problem of responsibility. *Ethics and Information Technology* https://doi.org/10.1007/s10676-017-9428-2.

IEEE (2015). Terms, definitions and abbreviated terms. In: *ISO/IEC/IEEE 15288 Systems and Software Engineering – System Life Cycle Processes*. Switzerland: IEEE.

IEEE (2017). *P7009 – Standard for Fail-Safe Design of Autonomous and Semi-Autonomous Systems*. https://standards.ieee.org/project/7009.html (accessed 8 July 2020).

IEEE Code of Ethics (2020). https://www.ieee.org/about/corporate/governance/p7-8.html (accessed 8 July 2020).

Jobin, A., Ienca, M., and Vayena, E. (2019). The global landscape of AI ethics guidelines. *Nature Machine Intelligence* 1: 389–399.

Joiner, K.F. and Tutty, M.G. (2018). A tale of two allied defence departments: new assurance initiatives for managing increasing system complexity, interconnectedness and vulnerability. *Australian Journal of Multi-Disciplinary Engineering* 14 (1): 4–25.

Joiner, K.F., Ghildyal, A., Devine, N. et al. (2018). Four testing types core to informed ICT governance for cyber-resilient systems. *International Journal of Advances in Security* 11: 313–327.

Joiner, K.F., Efatmaneshnik, M., and Tutty, M. (2019). Test and evaluation toolset. In: *Evolving Toolbox for Complex Project Management* (eds. A. Gorod, L. Hallo, V. Ireland and I. Gunawan), 339–370. Boca Raton, FL: Auerbach Publications.

Katina, P.F. (2016). Systems theory as a foundation for discovery of pathologies for complex system problem formulation. In: *Applications of Systems Thinking and Soft Operations Research in Managing Complexity*, 227–267. Cham: Springer.

Katina, P.F., Tolk, A., Keating, C.B., and Joiner, K.F. (2020). Modelling and simulation in complex system governance. *International Journal of System of Systems Engineering* 10 (3): 262–292.

Keating, C.B. and Bradley, J.M. (2015). Complex system governance reference model. *International Journal of System of Systems Engineering* 6 (1–2): 33–52. https://doi.org/10.1504/ijsse.2015.068811.

Keating, C.B. and Katina, P.F. (2020). Enterprise governance toolset. In: *Evolving Toolbox for Complex Project Management* (eds. A. Gorod, L. Hallo, V. Ireland and I. Gunawan), 153–181. New York: Auerbach Publications.

Kelly, E. (2011). *Material Ethics of Value: Max Scheler and Nicolai Hartmann*. Dordrecht: Springer.

Klemas, T.J. and Chan, S. (2018). Harnessing machine learning, data analytics, and computer-aided testing for cyber security applications. Paper presented at the CYBER 2018: The Third International Conference on Cyber-Technologies and Cyber-Systems, Athens.

Krathwohl, D.R. (2002). A revision of Bloom's taxonomy: an overview. *Theory into Practice* 41 (4): 212–218.

Laukyte, M. (2017). Artificial agents among us: should we recognize them as agents proper? *Ethics and Information Technology* 19 (1): 1–17.

Lyons, J.B., Ho, N.T., Hoffman, L.C. et al. (2018). Trust in sensing technologies and human wingmen: analogies for human-machine teaming. *Augmented Cognition: Intelligent Technologies* 10915: 148–157.

Matessa, M., Vu, K.-P., Strybel, T.Z. et al. (2018). Using distributed simulation to investigate human-autonomy teaming. In: *Human Interface and the Management of Information. Information in Applications and Services*, vol. 10905, 541–550. Cham: Springer.

Mayer, R.C., Davis, J.H., and Schoorman, F.D. (1995). An integrative model of organizational trust. *The Academy of Management Review* 20 (3): 709–734.

Muller, H.A. (2017). The rise of intelligent cyber-physical systems. *Computer* 50 (12): 7–9.

Munoz, C.A., Campbell, A.J.D., Hemsworth, P.H., and Doyle, R.E. (2019). Evaluating the welfare of extensively managed sheep. *PLoS One* 14 (6): e0218603. https://doi.org/10.1371/journal.pone.0218603.

Petraki, E. and Abbass, H.A. (2014). On trust and influence: a computational red teaming game theoretic prospective. Paper presented at the 2014 Seventh IEEE Symposium on Computational Intelligence for Security and Defence Applications (CISDA), Hanoi.

Animal Health Australia (AHA) (2014). *Australian Animal Welfare Standards and Guidelines – Sheep*. Version: 1.0, January 2016 Endorsed. www.animalwelfarestandards.net.au (accessed 23 February 2020).

Quigley, J. (2013). The leadership cycle: how to enable collective leadership: James Quigley, CPA, details strategy for creating successful teams. *Journal of Accountancy* 215 (6): 28–30.

Reina, D.S. and Reina, M.L. (1999). The need for trust. In: *Trust & Betrayal in the Workplace: Building Effective Relationships in Your Organization*, 1e, 3–12. San Francisco: Berrett-Koehler Publishers Inc.

Reynolds, C. (1987). Flocks, herds and schools: a distributed behavioral model. *ACM* 21 (4): 25–34.

Robbins, C. and Steffen, M.R. (2018). The future of autonomous ground and surface systems testing. *The ITEA Journal of Test and Evaluation* 39: 82–85.

Roff, H. and Moyes, R. (2016). Meaningful human control, artificial intelligence and autonomous weapons. In: *Briefing Paper Prepared for the Informal Meeting of Experts on Lethal Autonomous Weapons Systems*, 1–6. Geneva: UN Convention on Certain Conventional Weapons.

Ryan, M. (2018). *Human-Machine Teaming for Future Ground Forces*. Washington: Center for Strategic and Budgetary Assessments.

Scholtz, J. (2003). Theory and evaluation of human robot interactions. Paper presented at the Proceedings of the 36th Annual Hawaii International Conference on System, Sciences, 2003, Big Island.

Sharkey, A. (2017). Can we program or train robots to be good? *Ethics and Information Technology* 2017: 1–13.

Silver, D., Huang, A., Maddison, C.J. et al. (2016). Mastering the game of Go with deep neural networks and tree search. *Nature* 529: 484–489.

Silver, D., Schrittwieser, J., Simonyan, K. et al. (2017). Mastering the game of Go without human knowledge. *Nature* 550 (7676): 354–359.

Singh, H., Campbell, B., Elsayed, S., et al. (2019). Modulation of force vectors for effective shepherding of a swarm: a bi-objective approach. Paper presented at the 2019 IEEE Congress on Evolutionary Computation (CEC), Wellington.

Stearns, J.P. (2018). Motivating teams through the impossible. *Strategic Finance* 100 (2): 23–24. 26.

Strömbom, D., Mann, R.P., Wilson, A.M. et al. (2014). Solving the shepherding problem: heuristics for herding autonomous, interacting agents. *Journal of the Royal Society* 11: 20140719. https://doi.org/10.1098/rsif.2014.0719.

Strybel, T.Z., Keeler, J., Mattoon, N. et al. (2018). Measuring the effectiveness of human-autonomy teaming. Paper presented at the Advances in Neuroergonomics and Cognitive Engineering, Cham.

The Australian Code for the Responsible Conduct of Research (2018). www.nhmrc.gov.au/sites/default/files/documents/attachments/grant%20documents/The-australian-code-for-the-responsible-conduct-of-research-2018.pdf (accessed 28 August 2020).

Tolk, A. (2017). Code of ethics. In: *The Profession of Modeling and Simulation: Discipline, Ethics, Education, Vocation, Societies and Economics*, 35–52. Hoboken, NJ: Wiley.

Tolk, A., Madhavan, P., Jain, L.C., and Tweedale, J.W. (2009). Agents and decision support systems. In: *Agent-Directed Simulation and Systems Engineering* (eds. L. Yilmaz and T. Ören), 399–431. Hoboken, NJ: Wiley.

Tolk, A., Adams, K.M., and Keating, C.B. (2011). Towards intelligence-based systems engineering and system of systems engineering. In: *Intelligence-Based Systems Engineering*, 1–22. Springer.

U.S. Government Accountability Office (2018). *Weapon Systems Cybersecurity: DOD Just Beginning to Grapple with Scale of Vulnerabilities*. Washington, DC: U.S. Government Accountability Office.

Wickens, C.D., Lee, J., Liu, Y., and Gordon-Becker, S. (2014). Human-computer interaction. In: *An Introduction to Human Factors Engineering*, 2e, 363–397. New York: Pearson Prentice Hall.

Zador, A.M. (2019). A critique of pure learning and what artificial neural networks can learn from animal brains. *Nature Communications* 10 (1) https://doi.org/10.1038/s41467-019-11786-6.

Zanero, S. (2017). Cyber-physical systems. *Computer* 50 (4): 14–16.

Section 2

Domain Deep Dives

5

Modeling the Evolution of Organizational Systems for the Digital Transformation of Heavy Rail

Grace A. L. Kennedy, William R. Scott, Farid Shirvani, and A. Peter Campbell

SMART Infrastructure Facility, University of Wollongong, Wollongong, NSW, Australia

5.1 Introduction

The engineering problems of the twenty-first century require innovative approaches that deal with the increasing complexity of contemporary problems and their system solution spaces; novel methods are also required if the organizations that manage and govern these systems are to remain aligned (Keating et al. 2014). Meeting the customer's needs while maintaining value becomes progressively more challenging as the focus shifts from product-based development to service and capability-based solutions that are holistic and integrated across the whole life cycle and supply chain (Kemp and Daw 2014; Purchase et al. 2011). This book chapter describes a case study application of how model-based systems engineering (MBSE) can be used to manage the introduction of new technologies, with a focus on the "soft enterprise" or organizational issues. Within the human systems engineering (HSE) framework described in this book, the case study is positioned within the transportation (Australian heavy rail systems) domain. The modeling approach presented can be applied to many sociotechnical system (STS) types depending on the scope of the system of interest. The user of the new technologies may be considered as an operator of a technical system (e.g. a rail protection officer operating a tablet) or as an entity that is contained by a system (e.g. a train driver utilizing new onboard train control systems). Finally, the STS can be viewed from an abstract organizational system level where the humans involved form a system in itself. As a primary aim of the research was to address the impact of new technology introduction on human factors (HF) concerns, it follows that many of the assessments presented in this chapter speculate on the earlier stages of the future system life cycle (e.g. concept and design) and comparison with the equivalent functionality of the existent system (usually in operational stage). However, the models can consider the various technological systems at any stage in the system life cycle and enable insight into the management of concurrent systems with staggering life cycles as the organization evolves over time.

An *enterprise* can be described as "one or more organizations sharing a definite mission, goals, and objectives to offer an output such as a product or service" (ISO 2019). INCOSE defines an *engineered system* as "a system designed or adapted to interact with an anticipated operational environment to achieve one or more purposes while complying with applicable constraints" (Silitto et al. 2019). This definition has recently been updated to reflect that although engineered systems

A Framework of Human Systems Engineering: Applications and Case Studies, First Edition.
Edited by Holly A. H. Handley and Andreas Tolk.
© 2021 The Institute of Electrical and Electronics Engineers, Inc. Published 2021 by John Wiley & Sons, Inc.

are traditionally technological in nature, abstractions exist that are nontechnical and can be engineered for a purpose. Given these two definitions, it can be inferred that enterprises can be systems and that they can be engineered. Enterprise systems are not a new concept (Rouse 2005; Valerdi et al. 2008). However, within the field of systems engineering (SE), there is some disagreement around their epistemology (in particular, the overlaps between systems of systems [SoS] and STS [Pennock and Rouse 2016]). Pennock and Rouse (2016) define an enterprise system as "a collection of interacting organizations that construct and/or operate one or more technological systems to achieve a goal." The capabilities of the enterprise system depend on both the technological systems and the organizations involved. Improvements in capability require some form of transformation, and it is essential to consider the enterprise system if businesses are to attain the value planned successfully. Optimizing design improvements, reducing costs, streamlining processes, and managing the effectiveness of supply chains are commonplace in business. These initiatives, however, are often performed in isolation, while generating improvement in one area may be less optimal in others (Rouse 2005). Enterprise architecture frameworks (AF) enable enterprise systems to be modeled from various viewpoints in an integrated manner while managing the complexity of the information (ISO 2019). A vast plethora of enterprise AF exist and vary in scope of interpretation (Schekkerman 2004); these differences are due to the focus and application of the framework (from genericity and domain to national differences). Within the last decade, there have been efforts in the field of human systems integration (HSI) to include a complete view of the organizational aspects and how they integrate into the enterprise architecture. Bruseberg (2008) first developed the human view as part of the Ministry of Defense Architecture Framework (MODAF) in the United Kingdom, while around a similar epoch, Handley and Knapp (2014) were developing the human view in NAF and later Department of Defense Architecture Framework (DoDAF). Most recently, the Unified Architecture Framework (UAF) by the Object Management Group (OMG) included a new personnel view to explicitly address the integration of these concepts (Hause and Wilson 2017).

This case study describes an MBSE tool developed through funding from the Australasian Centre for Rail Innovation (ACRI) to enable business stakeholders to consider the evolution of organizational systems as new technologies are introduced named the Organizational Capability Maturity Model (OCMM). This chapter discusses how the framework has been applied specifically for organizational system evolution (within the context of the framework developed) and considers the organizational aspects and how the organizational system(s) must both adapt to and inform the design and introduction of technologies to provide new capabilities for the enterprise system. The chapter is structured as follows: Section 5.2 introduces the characteristics of organizational systems and how such systems evolve. Section 5.3 provides a brief introduction to the benefits of MBSE and how MBSE principles have been utilized in this case study. Section 5.4 summarizes the authors' approach. Section 5.5 discusses the implementation and how portals have been created to balance utility with usability. Section 5.6 provides an example case study application. Finally, Section 5.7 summarizes the chapter and draws conclusions.

5.2 Organizational System Evolution

5.2.1 Characteristics of Organizational Systems

HSI is concerned with ensuring that human capabilities and limitations are considered within the SE activities of a project throughout the system life cycle in order to optimize the performance of the system (while also supporting the well-being of the human) (INCOSE 2015). Within the field of HF (also known as ergonomics), the organizational aspects represent one of three interlinked domains (the other two domains being cognitive and physical). The organizational system concerns the interacting set of people working toward a common purpose and how the interfaces to technological systems and other organizations are managed. The organizational system considers the various aspects of the humans involved:

- The organizational system is an abstract concept, and it exists within one or more enterprise systems (ISO 2019; Pennock and Rouse 2016).
- Organizational systems are recursive or scalable, depending on the scope and boundaries considered. Organizational systems may include suborganizations within them or may consider other organizations separately in scope (ISO 2015).
- Organizational systems support the top-down enterprise strategy (the strategy should be propagated, decomposed, and instantiated throughout the organizational system) (Galbraith and Lawler 1993).
- Organizational systems can be designed but also exhibit variability, learning, and self-organizing properties.
- Organizational systems support the enterprise at the different stages of the system life cycle (ISO 2015). The organizational system may include suborganizations or interface with other organizations that perform system development, installation, operations, maintenance, and disposal.
- Organizational systems consider the impact of the individual, team, and structure on the behavior of the organization (Robbins et al. 2017).
- Organizational systems can be structured in different ways. These structures are not confined to the traditional hierarchical organization chart but may represent formal and informal networks, as well as static and dynamically changing relationships. Organizations display all these types of structures. The formal static structure is required to understand control structures and maintain governance. However, the informal dynamic structures enable the organization a degree of compensatory fit (Gulati and Puranam 2009) and hence flexibility.
- Organizational systems consider the "soft" organizational characteristics such as role interactions, team working, cultural values, knowledge distribution, competencies, training, and decision making (Hubbard et al. 2010).

HSE is concerned with various STS types. Although it would seem contrary in HSI to consider the organizational and technical as separate concepts, an STS consists of an organization interacting with a technical systems (Pennock and Rouse 2016). The organizational system is scalable and can be viewed with different boundaries; this can be anywhere from the users being an embedded part of the STS, a user of the STS, or at levels of abstraction where the users themselves are the STS.

5.2.2 The Organization in Flux

Organizational systems are not closed systems. They are not static institutions that exist in an independent microcosm from society. Organizational systems are subject to macro-environmental factors – in this context, the systems must be flexible to simultaneously seek out strategic opportunities and exhibit resilience to changes in the political, economic, societal, technological, legal, and environmental landscapes. The dynamic nature of the environment requires organizations to display agility to change, and thus the organization is in a state of flux as it adapts and learns (Galbraith and Lawler 1993). Conversely, organizations are also in flux from an internal perspective as organizations seek continuous improvement strategies, from new management paradigms, efficiency, and lean principles to troubleshooting recognized problematic areas.

The workforce in an organization is also transient. New workers commence employment, some retire, and some will leave employment – it is estimated that during their working life, Australians will make 17 changes in employers across five different careers (FYA 2017); this, in turn, yields challenges in the retention of skills and organizational knowledge. Workers also tend to move around within organizations as part of career development plans. Although organizations may have a semi-stable, formal organizational structure, the informal networks, teaming, and shared experiences are composed of a history of day-to-day momentary situations.

Enterprise SoS, or extended enterprises, are created when organizations are required to work closely with other organizations but maintain separate goals, governance, and structures. Supply chains and collaborations add a layer of complexity to the organizational system as interfaces must be managed, and each organization may need to compromise its individual goals to meet the overall goal of the enterprise SoS. The enterprise SoS is dynamic in nature as component organizations change to meet the capability needs from phase to phase, and the boundaries may become blurry.

5.2.3 Introducing New Technologies

In this section the epistemology of enterprise systems (Pennock and Rouse 2016) is revisited to consider the hypothetical introduction of new technologies. Figure 5.1 expands on some of the concepts from Pennock and Rouse's notional view of enterprise systems by considering their evolution over time. The top plane represents an abstract enterprise system at an arbitrary point in time (t_1) that acts as the initial point of interest. The middle plane is the same enterprise system at a further point in time (t_2), and the bottom plane (t_3) chronologically advanced from there.

At any point in time, an enterprise system will consist of a number of organizations that are dealing with a number of systems that are at different phases of their system life cycle (e.g. from concept, development, production, operations, support/maintenance, and disposal). Table 5.1

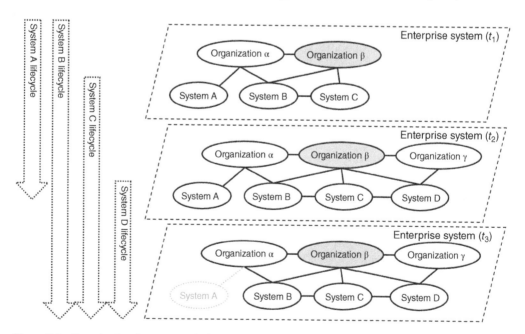

Figure 5.1 Organizational system evolution.

Table 5.1 Organizational evolution example.

Time	System life cycle stage			
	System A	**System B**	**System C**	**System D**
t_1	Operate	Install	Develop	—
t_2	Disposal	Operate	Install	Develop
t_3	{Legacy}	Upgrade	Operate	Install

summarizes the systems that are being considered at each transition point. Note that this table does not include all system life cycle stages, but a simplified subset to introduce the concept that organizations will have different needs at different stages.

At t_1 the enterprise system consists of two organizations that are developing or operating three systems. At this stage, System A is in the operational stage, System B has been developed and is being installed, and System C is being developed. Systems B and C are coupled together as an SoS. In our example, organization β is the primary organizational system of interest, while organization α represents a maintenance subcontractor. Organization α is maintaining System A and is involved in the planning for future maintenance of System B. Organization β has been developing and is installing System B, and simultaneously developing System C.

At t_2 a new Organization γ and a new System D are introduced to the existing enterprise system. On this plane level, the previous systems (A–D) are now further along their system life cycle. System A is reaching the end of its system life cycle, and Organization α is decommissioning it. System B is now in operation, and Organization α is maintaining it, while Organization β is still concerned with managing the system as they have planned to upgrade the functionality in stages. System C is now being installed, and as identified earlier, Systems C and B are tightly coupled. System C will operate on parts of System B. This added layer of complexity must be managed by Organization β. To make matters more complicated, a new Organization γ has been introduced as a vendor for a future System D, which is at the conceptual stage of systems development. It will be integrated with the upgraded System B and operational System C. Organization β decides not to outsource the maintenance for System C and must start resourcing workforces that can operate and maintain the future system.

Finally, at t_3, System A is now part of Organization α's legacy and organizational history. Although System A no longer exists, the components of the organization still exist with knowledge and experience of working within that STS. System B is upgraded by Organization β, and Organization α will continue to maintain System B into the future after the upgrade. System C has now been deployed and is in operational stage. Organization β can use the knowledge gained through installing and operating System B to integrate System C. System D is being installed by Organization γ and will be part of an SoS (B, C, and D) that will be operated by Organization β. Table 5.1 summarizes the enterprise system stages at all three points in time.

This hypothetical enterprise system seems complicated but is not an uncommon situation in large-scale complex enterprises where there may be SoS, diversity of systems, collaborations, and convoluted supply chains. All the organizational systems (α, β, and γ) must include different competencies, training, tasks, safety procedures, and resourcing for the current stages of systems they are concerned with while at the same time planning for the future stages. In addition to evolving the organizational systems to meet the system life cycle changes, they must also adapt and evolve with the changing interfaces between the other organizations to ensure seamless transitions between plancs. The planes depicted in Figure 5.1 have been simplified and are unlikely to be conveniently synchronous as system life cycle stages do not necessarily all transition simultaneously. Different systems stages have different durations, and slippage invariably occurs. Although organizational change appears continuous, it is not feasible, nor time and labor efficient to model every detailed shift and change. Businesses will need to carefully select a set of discrete points (planes) that will be analyzed. As a first step, consideration should be made of the extent of the envisaged change and possible impact on the enterprise. Purchase et al. (2011) suggest that there are three candidate criteria for distinguishing such transformations:

1) Response to radical changes in the environment
2) Fundamental alteration of context
3) Step change in performance

In order to understand the nature of human work in the future of an engineered system, it is necessary to consider how the human systems will be integrated throughout all the SE activities. To do this, human systems engineers need to build models or representations of the various aspects of STS for analysis and simulation. From an organizational perspective, all these HSE activities and models need to then be integrated across the portfolio of engineering projects. Modeling of the evolution of organizational systems pose specific challenges such as:

- Dealing with enterprise system complexity (both from the environment and internally).
- Integrating the organizational system and the other parts of the enterprise system.
- Ensuring consistency between parts of the integrated organizational system.
- Forward-looking planning to ensure that the organizational system is positioned effectively to meet future goals.
- Feedforward of changes to assess the extent of adaptability for different solutions.
- Maintaining traceability of organizational decisions made.
- Storage of organizational history, knowledge, and experience for the future.
- Level of effort required to model and maintain the evolving models.
- Information overload as the size of the models increases over time, meaning that the information will require careful management so that the content of the views can be constrained.

SE aims to manage changes in systems to ensure that they are fit for purpose; therefore these practices may be applied (with tailoring) to manage an organizational system.

5.3 Model-Based Systems Engineering

MBSE is defined as "the formalized application of modeling to support system requirements, design, analysis, verification, and validation, beginning in the conceptual design phase and continuing throughout development and later life cycle phases" (INCOSE 2007). One of the major trends in the field of SE this century is the movement toward being a model-based discipline, as engineering practices evolve toward the digital transformation of enterprise systems. So, MBSE must also transform to meet these new challenges to become a key component in managing the increasing complexity of the systems (Madni and Sievers 2018; Peterson 2019). MBSE enables businesses to model the enterprise system and create digital versions of the SE artifacts that are integrated and provide a "single source of truth" (thus improving traceability and consistency).

MBSE also creates a digital copy of the project history and decisions made. Madni and Purohit (2019) discuss the promise of MBSE:

- Capture and harmonize information across multiple disciplines.
- Create an integrated digital model of the system.
- Auto-propagation of changes through dependencies in the system.
- Consistent and current information.
- Replication of models in the hierarchy and opportunities for reuse/tailoring.
- Consistent, complete, and traceable.

Ultimately, MBSE (when done well) improves the efficiency of the organization through helping to manage complexity and reducing cost and risk. To manage the complexity of the models, an AF is required to organize the categorization and storage of the information as well as the mechanisms for retrieving the correct amount of information from across the model to ensure that it is fit for purpose.

MBSE tools utilize object-orientated databases to contain information that is commonly presented in one of several formats, including diagrams, tables, matrices, and Gantt charts. The database is used to control the information with changes made in one part of the model automatically conveyed to other parts of the model using common model elements, thereby reducing effort and ensuring consistency. Visual modeling is a practical and pragmatic method for participatory modeling and leads to effectively capture and review of information about the system from various subject matter experts (SMEs). MBSE enables modelers not only to consider the holistic system but also to consider different levels of abstraction and decomposition, from overlapping alternative viewpoints, to understand the nature of the interactions and to break down the complexity of the system.

The tool has been developed using a combination of the Systems Modeling Language (SysML) and a tailored form of the UAF. SysML was established and formally specified by OMG in 2006 and was recently released as version v1.6 (OMG 2019). The purpose of SysML is given as:

> A standard modeling language for systems engineering to analyze, specify, design, and verify complex systems. It is intended to enhance systems quality, improve the ability to exchange systems engineering information amongst tools, and help bridge the semantic gap between systems, software, and other engineering disciplines. (OMG 2019)

While SysML provides a language for "how" to express information, UAF provides a baseline of "what" to model. UAF (OMG 2017) began as the Unified Profile for DoDAF and MODAF (UPDM). This framework enables the capture of a wide variety of aspects about a system, including the organization (with associated vision and required capabilities), risk, operational concept, and project information. Utilizing this standard ensures that the underlying model is based on international best practice but is tailored specifically to meet the identified stakeholders' concerns and information needs. Although the UAF profile formed the basis of the initial framework in the work described in this chapter, expansion and tailoring were required to create additional conceptual elements that would enable the users to meet their HF modeling needs.

5.4 Modeling Approach for the Development of OCMM

The OCMM was designed to manage the evolution of the organizational system in response to technology changes using MBSE technologies. The approach – to understand a technology change and how the organization will change in response – consists of a five-step process, as shown in Figure 5.2. The approach is presented as sequential to depict the flow of how the changes are

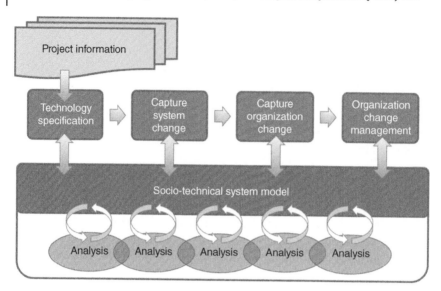

Figure 5.2 OCMM development approach.

managed. However, in practice, these processes are parallel efforts feeding each other through the evolving understanding contained within the STS model (the STS being composed of one or more technical systems interacting with one or more organizations [Pennock and Rouse 2016]). Enterprises at a given point in time will have a project portfolio. The "starting point" in the authors' approach is to consider the projects of interest (these could be of varying maturity, for an existing or future state), and pertinent information was gathered about the projects (whether from a technical or management perspective). Each of these processes is described in the following subsections. Concept demonstrators were created and reviewed at regular intervals with the rail operator project stakeholders to gather information, elucidate feedback, and validate the models and functionality of the tool being developed.

5.4.1 Technology Specification

The first step in the process captures the specific information about the new technology and the timeline about its development and deployment, which is commonly found within the volume of information developed in the originating project. This information is necessary to gain an understanding of the technology being introduced and the view of how the technology will be operated and maintained. Projects often focus on the information directly associated with the technology and leave the receiving organization responsible for understanding how the technology will be incorporated into its broader systems and processes. These processes have been separated from an understanding of how the wider system that envelopes the new technology will change. The data associated with understanding the specific technology is often "owned" by the companies (or alliance depending on the acquisition strategy) that is responsible for the technology's development. In contrast, the recipient organization controls the understanding of the wider system. For example, a project will develop new technology, but the recipient organization will be the operator/maintainer that is responsible for the operation and maintenance of the system. Based on this understanding of the technology, the organization needs to plan how the organization will change in response to the technology needs, such as incorporating the technology processes within the organization processes (discussed below).

5.4.2   Capture System Change

This part of the model development process captures how the wider system will change with the introduction of new technology and how the associated processes need to be modified. For an extensive, complex system such as heavy rail, understanding the introduction of new technology can be difficult as there can be a variety of configurations where the latest technology touches upon the existing infrastructure and the shift toward the adoption and integration of technologically mature COTS solutions. As heavy rail systems have evolved in Australia for over a century (the first railway in NSW was installed in 1849), there are often variations in the interfaces that need to be considered to maintain interoperability. This is particularly acute with the introduction of digital control systems into what have been customarily analog-based systems.

The broader functions also need to be considered as to how the specific mechanisms provided by the technology will fit strategically into the capabilities of the system. Some local processes may have to be tailored for particular circumstances; for example, there are locations (such as tunnels and bridges) where workers cannot be used as lookouts to detect approaching rail traffic and, therefore, other (often higher) forms of track worker protection are required.

5.4.3   Capture Organizational Changes

The introduction of new technology will stimulate changes in the organization that need to be captured, such as changes in roles, responsibilities, and competencies. At this step, formal information around the organizational structure can be gathered and captured, alongside any envisaged future configurations. Various models of the organization are developed from the information about key processes, organizational units, or roles of interest. This provides a vehicle for demonstrating how the humans and tasks involved will transition as the system changes (as captured in Section 5.4.2).

5.4.4   Manage Organization Change

This step examines the necessary change in the organization to develop plans to realize the changes. These plans need to ensure that the transformation is completed on time that is aligned with the technology transformation. These activities may be incorporated into broader training management processes associated with other objectives, such as competency maintenance.

5.4.5   Analyze Emergent System

Once the STS model is of an appropriate level of maturity (this could be aligned to the planned phase gates of the project but should be of sufficient detail to provide meaningful results that will support the decision makers), analyses can be performed to provide insight into the various stakeholders' HF concerns (e.g. competencies, training, situational awareness, workload, organizational culture, safety assurance, and cybersecurity). This step in the approach is iterative and can be fed back into the STS model to improve the design and be available in the central repository for other types of analyses. The analyses are stored "live" alongside the models and become a formal part of the project documentation (improving the traceability of any decisions made).

Many analyses need to be undertaken to validate the emerging STS after all changes have been made. Each of these analyses is intended to address a specific concern but can have an overlap in their content or effect on one another. For example, a risk assessment (Shirvani et al. 2019) may include the examination of the effects of people's competency in various circumstances or generate new competency requirements for risk mitigation. Similarly, activities may be appended to processes for risk mitigation that may affect other analyses such as workload and situation awareness (SA) that consider a wider set of tasks being undertaken. The OCMM tool contains a set of example analyses to demonstrate how the integrated STS model can be assessed. However, the framework is intended to be flexible to accept other HF assessment techniques and methods (on a needs basis as determined by the HSI specialists who would use the tool).

The five steps outlined in this approach have a threefold benefit for HSE. Firstly, they enable the HSI SMEs (i.e. the HSI practitioners or HSI domain specialists who are experts in their relative domains) to access the current system development at any stage as it matures and create HF models and analyses that can be fed back into the development of the system; in particular, this brings the HSI earlier into the system life cycle from as early as concept through to disposal. Secondly, the HSI integrator (the role that integrates all the HSI activities for the projects) will be able to get an overview of the current HSI activities and ensure consistency between shared artifacts between HF models and analyses for the system of interest. The HSI integrator (as part of the SE team) will also be able to act as a conduit between the HSI activities and the SE activities to ensure project planning and management is fed both ways. Where MBSE evolved from a need for seamless SE modeling for development of the system alongside the technical modeling, so too should the HSI activities. The third benefit is to the HSI manager or policy makers in a business. The OCMM approach enables the full project portfolio to be assessed at various states in time (i.e. the planes in Figure 5.1); the HSI manager can ensure that governance is being adequately enacted for each project at its appropriate phase and that the overall business plans and resourcing requirements will be met.

5.5 Implementation

The OCMM was developed as an AF that employed ISO 42010 principles (ISO 2011) and based on UAF (OMG 2017) was implemented to manage the models, organize information effectively, and address the complexity emerging from the interconnected, evolving models (Shirvani et al. 2019). This framework was developed to be used by various stakeholders (usually managers within the business) who would be responsible for managing the specific areas of concern resultant from the technological changes. Using a standardized AF also enables the models created within OCMM to be aligned with other MBSE artifacts being created within the business and vice versa. The intention is that as businesses adopt MBSE activities as part of the digital thread, the OCMM would fit within the greater suite of models created (rather than a stand-alone tool).

The AF uses the concept of viewpoints and views to provide stakeholders with their particular capability needs while shielding them from the full complexity of the whole system. At the same time, the AF structure ensures that interdependencies and linkages are maintained across the entire system being modeled. The ISO 42010 standard was adopted to create a method for developing the proposed AF in this study. The method steps are as follows:

1) Identifying the stakeholders who are concerned about the system of interest.
2) Identifying the concerns of each stakeholder.
3) Identifying the viewpoints (types of information) required for addressing the concerns.
4) Developing the metamodel representing the data structure of the viewpoints:
   - Borrowing from existing metamodels and frameworks (e.g. UAF/UPDM).
   - Creating new elements and relationships that will be fit for purpose to meet the stakeholder's needs.
5) Defining the model kind(s) for the viewpoint.

The OCMM tool has been developed in MagicDraw™, an MBSE software environment (Dassault Systems 2020). MagicDraw was selected due to the authors' familiarity with the software; in addition to built-in profiles for UAF and SysML, MagicDraw also provides multiple timeline and configuration functionality, which enables simple management of the evolutionary nature of the models and change.

5.5.1 User Portals

The OCMM was developed with the set of key stakeholders identified in the AF, and their envisaged use cases for the tool were identified and explored with the SMEs within the rail operator organizations. Thirteen types of users of the tool were identified. To present the viewpoints of each kind of user, *user portals* were developed as an interface to present the pertinent information to address their concerns but also to manage the input of information from that user and to allocate it in the correct areas of the database underlying the architecture. Sets of libraries were created to manage the information that populates the models. As the concerns are often overlapping, users will require access to information from a subset of the set of libraries. Table 5.2 shows a mapping of the user portals and the libraries of information that each user requires access to. Access levels can be set for each user portal, enabling libraries (and certain determined models) to be shared with other user portals. A generic set of organizational models was first created to demonstrate the functionality of the tool. These generic models were then extended and tailored to create organizational-specific models using information specific to two Australian heavy rail operators.

5.5.2 OCMM Metamodel

The first step in analyzing a domain utilizing a model-based approach, regardless of the purpose of analysis, is to capture the domain information in a standard format that will be interpreted in a common manner by the domain stakeholders. Using an abstract model of the domain (aka a metamodel) allows for consistent generation of domain models that provide a common understanding of the domain concepts. The OCMM metamodel was developed where possible using elements from within the UAF profiles (e.g. organization, function, post, competence, resource) that would provide the building blocks of the models. Abstract organizational models were created for each of the stakeholder concerns to provide additional elements (not available within UAF) and the relationships between the concepts. Each of the models within the OCMM contains an overlapping subset of the elements from the metamodel. Figure 5.3 shows an excerpt of the OCMM metamodel, showing a subset of the pertinent aspects across the HF concerns spectrum, and their interactions. Each small box shows an element (or class), which provides a type of building block from which models can be created, and the connectors between them are the relationships (or associations). The small text along the connectors describes the relationship in the direction of the arrow. An additional layer of information is overlaid on top of the metamodel to show how groupings of subsets of elements might appear in one type of model but are also shared between types of models. Five example metamodel groupings are shown in Table 5.3.

While some models have clear boundaries and interactions (such as the relationships between competency and training), other types of models are more interconnected (the process modeling and safety assurance are overlapping because the OCMM safety assurance is focused on task-based analyses). To show the complete OCMM metamodel would be too complicated to convey in one diagram easily. It should be noted that the metamodel is not exhaustive of all possible components and relationships within a generic organizational system, the metamodel is explicitly tailored to

Table 5.2 Mapping the user portals to information libraries.

User portals to information libraries	Competencies library	Training library	Safety hazards library	Risks library	Cyber library	Culture library	Project details	Regulations and Standards library	Project timeline	Generic OCMM model	Organization-specific model
Project manager				✓			✓		✓	✓	✓
Change manager				✓			✓	✓	✓	✓	✓
Culture manager	✓	✓				✓	✓	✓	✓	✓	✓
Training provider					✓		✓		✓	✓	✓
Training manager	✓	✓							✓	✓	✓
Workforce manager	✓						✓		✓	✓	✓
Competency manager	✓			✓	✓				✓	✓	✓
Internal auditors/assessors	✓	✓	✓			✓		✓			✓
Cybersecurity manager	✓		✓	✓	✓		✓	✓	✓	✓	✓
Regulations manager								✓	✓		✓
Safety assurance	✓		✓	✓	✓			✓	✓	✓	✓
Various roles affected	✓	✓	✓	✓				✓	✓		✓
Tool manager	✓	✓	✓	✓	✓	✓	✓	✓	✓	✓	✓

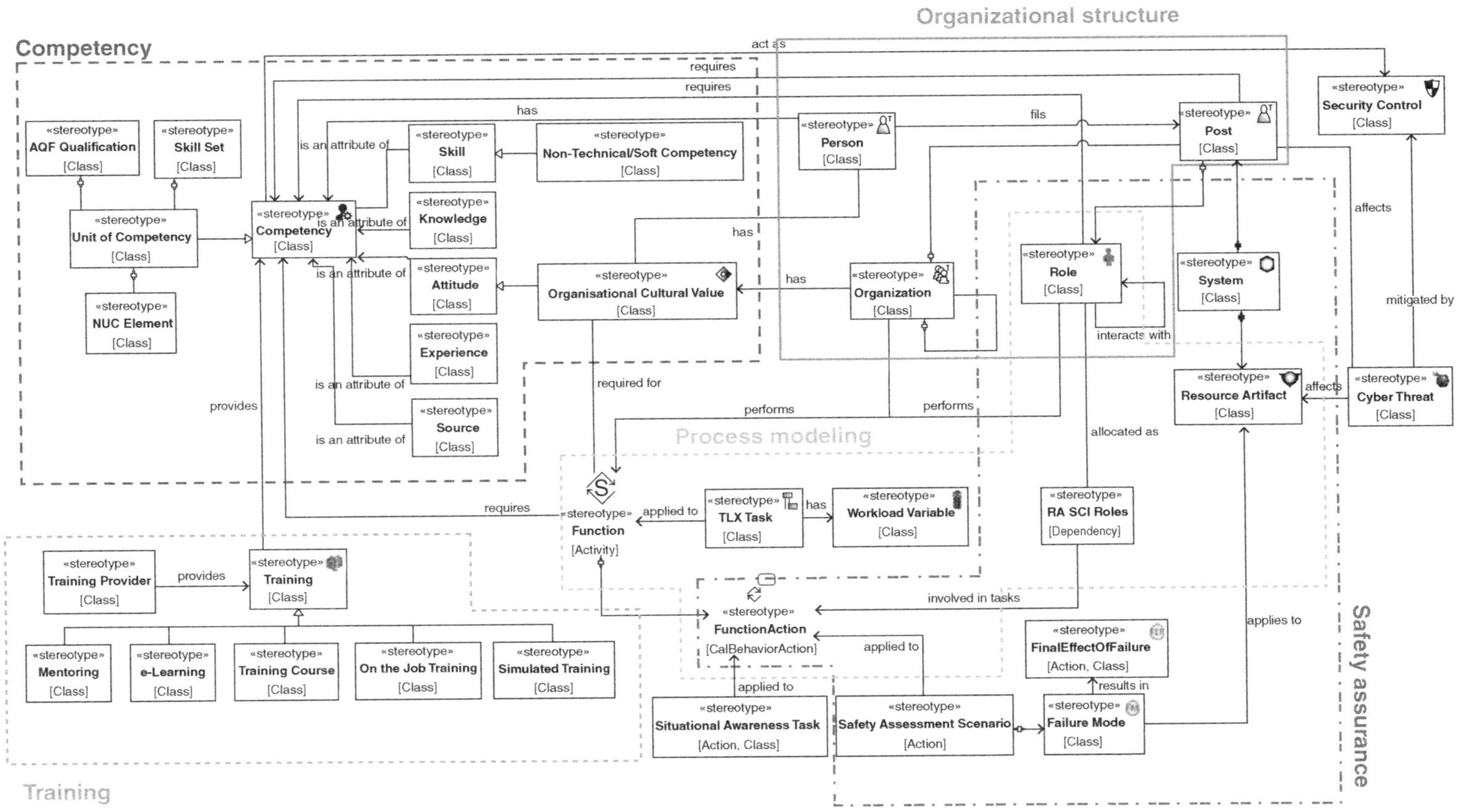

Figure 5.3 OCMM metamodel (excerpt).

Table 5.3 Model groupings in metamodel.

Model grouping	Position in Figure 5.3 and key	
Competency	Upper left	— — — — ·
Training	Lower left	‑‑ ‑‑ ‑‑ ‑‑ ‑‑ ‑‑
Organizational structure	Upper right	──────────
Process modeling	Central	····················
Safety assurance	Lower right	──── · ──── ·

address the stakeholders' concerns and was built from the bottom up, as an integrated aggregation of various smaller-scale HF models.

5.6 Case Study: Digital Transformation in the Rail Industry

Demonstrator versions of the OCMM were created for two heavy rail organizations as part of the ACRI project PF12. The Australian rail industry is currently focused on introducing a range of new digital technology systems over the next 5–10 years (Australian Railway Association and Deakin University 2018). These demonstrator versions scoped the technologies, processes, and concerns (held within user portals) based on stakeholders' input and the amount of information available from which to build the model. Table 5.4 shows the final scope of the demonstrators. However, for brevity, the focus in this case study section will be placed on the mechanisms created to support the various HFs concerns. Note that the user portals scoped within this section are a subset of interest to the stakeholders of the case study from those listed in Table 5.2.

Note that the models and analyses shown are only for demonstration and have been sanitized not to show any proprietary information. They should not be reused as assessments without first being tailored and reassessed by SMEs of the system on which it is to be applied. The example to be presented here focuses on the introduction of the European Train Control System (ETCS) technologies. The following subsections follow the same steps as outlined in the approach section.

Table 5.4 Demonstrator scope.

Technologies	Processes	User portals

• European Train Control System (ETCS)
• Drones
• Tablets
• Customer Information Systems (CIMS)
• Advanced Train Management Systems (ATMS)
• Enterprise asset management (EAM)
• Data storage cloud
• Customer interactions
	• Track work procedures
	• Train operation
	• Train control
	• Drone operation
	  – Overhead wire Inspection
	• Customer information management
		• Track worker
		• Competency manager
		• Training manager
		• Safety assurance
		• Workforce manager
		  – Role modeling
		  – Resourcing
		  – Workload
		  – Situational awareness
		• Cybersecurity manager
		• Culture manager

5.6.1 Technology Specification

As with all major transformation projects, ETCS will have an associated timeline of when systems will be installed and subsequently brought into service. Figure 5.4 is an example of a generic set of timelines for two systems to be introduced. The project managers can create the stages that are appropriate to the projects and update the durations and milestones for the various locations. These timelines are crucial to enable understanding of when transitions are being made. The timelines within this step will be used throughout the STS model, with assessments linked directly to a timeline in a row. The timelines are critical to also understanding the effects of multiple technology introductions. Figure 5.1 described how organizational systems should evolve alongside the introduction of different systems with different life cycles. At any two defined points in time, comparisons can be made between the current STS state and the future STS state. Sub-timelines can also be used to represent staggered introductions or alternative configurations (see Section 5.6.2 for more details on system configurations).

The technology introduction can also include other information such as the design of the system and the use cases of how the systems are intended to be used. These are typical process products generated by the SE process to ensure that the system is fit for purpose. At this step, existing information sources about the new technology and key stakeholder managers/representatives were identified.

5.6.2 Capture System Change

While the project focuses on the technology itself, the absorption of the new technology into the preexisting system is likely to be far more complex. The introduction of ETCS into a heavy rail system is a pertinent example as there are many combinations of systems that ETCS will interact

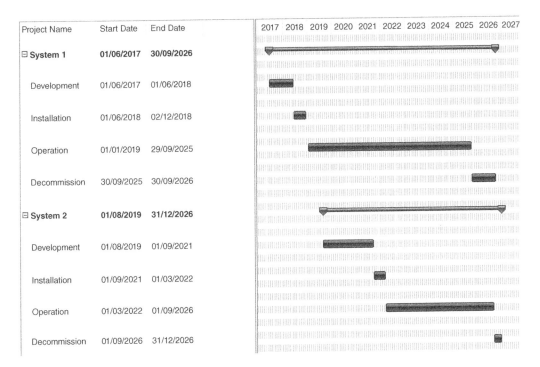

Figure 5.4 Example system introductions.

with across the breadth of a rail network. A model of the rail network system (e.g. by lines, depots, or rail segments) is required to understand how the technology will be rolled out. The various combinations need to be captured to enable validation across the network.

Figure 5.5 shows an example of the system resources (modeled within the resources taxonomy part of the UAF), which depicts an instance of ETCS level 2 in one configuration of systems. Each configuration presents the system elements within a specific context and their relationships, which are subsequently linked to their physical locations within the rail network. This approach enables multiple locations to be examined collectively due to the common combination of systems, thereby reducing effort yet allowing the separation to of problematic network locations.

The physical system configurations are then augmented by the modified processes required to operate and maintain the system. The processes can be modeled at various levels of decomposition, from high-level use cases through to functions, operational procedures, activities, sequence diagrams, and task actions. The processes are stored hierarchically to different levels of decomposition, depending on how much detail is required. The OCMM supports decomposition of the high-level functions and use cases to processes and then down to activity, action, and task level, and this provides traceability to the models, which are also easily traversable by the user. Figure 5.6 shows the example high-level use cases for the ETCS L2 onboard operational modes as specified in AS7711 (RISSB 2018), while Figure 5.7 shows an activity diagram for a decomposed part of the process.

The processes that have been modeled augment the use cases (provided by a project) by placing them in various contexts. For example, software on a tablet being used by a protection officer to interact with the network controller to indicate when track work begins or ends needs to be aligned with the wider track safety processes that include crew safety briefings, the layout of warning signs, and deployment of lookouts to monitor for rail traffic. These processes feed into the overall STS model as they include how the technical system elements interact with various roles as they undertake specific tasks.

5.6.3 Capture Organization Changes

The understanding of how the new system will behave drives change in the organization (and vice versa). For example, automation often replaces roles that were previously occupied by people, thereby reducing the need to employ a human in that role. The organization may, however, wish to retain the skills in a set of people to be able to undertake the role in specific circumstances, or the role may shift to be more supervisory, which will require consideration of system design, handover of control, new procedures, and user interfaces. It is necessary to understand how the organizational HF is then changing in response to the technology and how the HF can shape how the technical system is designed and used.

The OCMM captures the elements of the organizational system in three ways:

1) The organizational structure identifies the posts people fill, the roles that they may be called upon to undertake, and the hierarchy of control and communication between people. Organization charts are commonly used to model the formal structure; however, the informal structures provide a complementary view of the organizational behavior within different contexts.
2) Role allocation links the roles to the activities that are to be undertaken as a part of that role. The use of activity diagrams (such as Figure 5.7) enables the traceable allocation of activities to roles.
3) Specification of the competencies required to undertake a role or task. This is powerful as allocation of competency to a role or activity can then be automatically allocated to all people that undertake that role or activity.

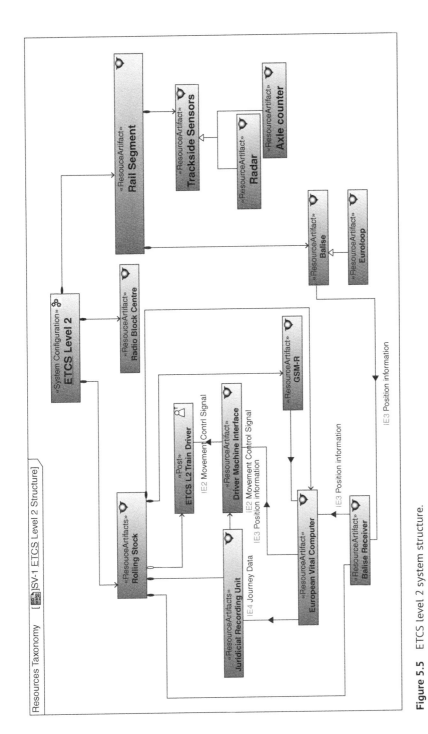

Figure 5.5 ETCS level 2 system structure.

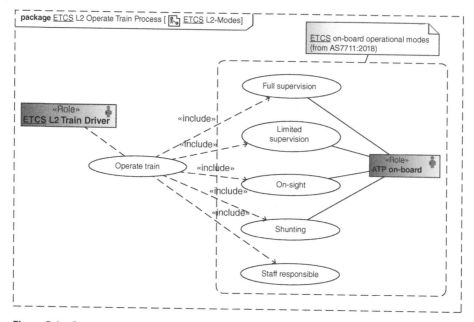

Figure 5.6 Operate train (ETCS L2 onboard operation modes).

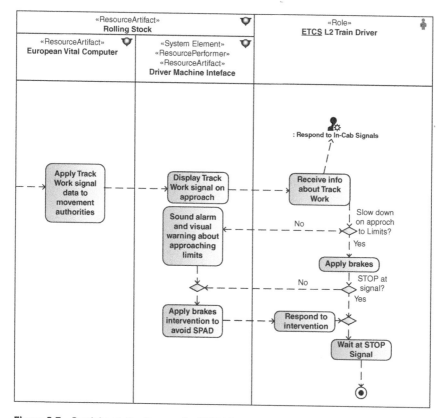

Figure 5.7 Partial activity diagram for ETCS L2 stop at signal.

Libraries of information and models of the following organizational elements are contained within the OCMM:

- Formal organizational structure plus informal structures.
- Sets of elements for posts (jobs) and role requirements under different technologies.
- Models of competency requirements for tasks or roles.
- Roles allocation models linked to tasks and processes.
- Models of training mechanisms linked to providers, competencies, and roles.
- Models of organizational culture at individual, group, and organizational level.
- Models for the workload at both task level and system level.
- Model for situational awareness linked to roles performing tasks.

Figure 5.8 shows an output from the model using one of the methods for allocating roles to the tasks. In this example, the roles pre- and post-ETCS introduction are shown and allocated via a dependency that specifies the nature in which the role is involved in that task. The RASCI (Smith et al. 2005) roles are defined as follows:

Responsible (R) – The role that is responsible for performing the task (i.e. the "doers").
Accountable (A) – The role that ultimately is accountable for or controlling a task being completed.
Supporting (S) – Those roles that are supporting the responsible role.
Consultant (C) – The roles that provide advice and information (usually a two-way communication) within a task.
Inform (I) – The role that is a recipient of the outcome or information at the end of the task.

The assessor allocates roles to the processes that will transform as new technology is introduced. Roles may be allocated directly (as shown in Figure 5.8) or allocated via visual models of the processes. Once the roles have been allocated, role structure diagrams are automatically generated (Figure 5.9), which provide a visual summary/overview of which roles are involved and the nature

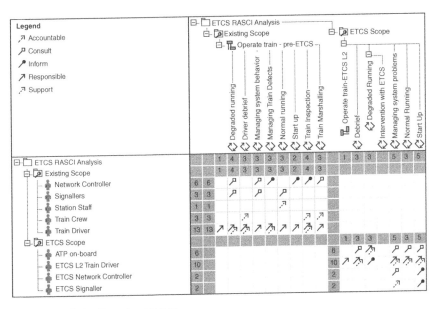

Figure 5.8 Role allocation (RASCI).

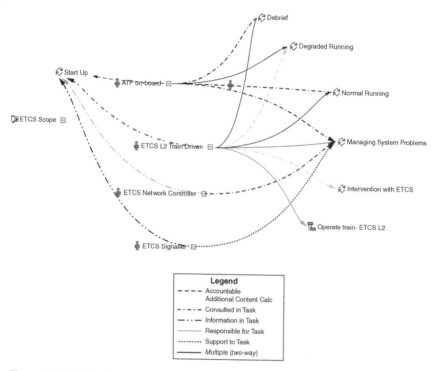

Figure 5.9 RASCI role structure for operate train process under ETCS.

of their involvement within the set of processes of interest. Different visual patterns can be indicative of a variety of issues. For example, if the responsibility is shared among some different roles, then attention may be required to coordinate these processes. These diagrams are alternate representations that show the role allocations to the various tasks, which illustrates the flexibility in OCMM's ability to represent information depending on the assessor's needs and preferences.

The organizational models for this case study demonstrator have been developed for the existing (or pre-change) state and the post-implementation state (i.e. pre-change and post-change); however, it is acknowledged that such change will not necessarily occur in a "big bang" fashion. Business managers and modelers should use the timelines in Figure 5.4 to specify the granularity of the transition by specifying specific phases as individual timelines (e.g. staggered transitional stages). These models should also evolve as the system development matures, and further decisions are made on how to manage the change (in this case study, the system being introduced is still at a conceptual stage).

5.6.4 Organization Change Management

Given the changes being made to the organization, activities need to be planned to ensure that changes are made promptly. This is supported by several gap analyses that need to be aligned with each other to ensure consistency. First is the identification of training required by personnel. OCMM supports this by first identifying the changes in competency requirements for personnel through the automatic generation of a report of the changes in competency needs driven by the new technologies. The result is a model of how the organization needs to be shaped in response to the changes.

With the improved and traceable understanding of which roles require additional competencies, the OCMM enables the specification of valid mechanisms to attain the identified competencies. While the focus is mostly on training, other required mechanisms can be captured, such as mentoring, certification, or experience. A new report can then also be reissued to determine the training required by individuals holding (or planned for) specific roles, so they are suitably prepared for the introduction of new technology in a timely manner. This information can also feed the longer-term plans necessary to ensure the retention of the competencies over long technology transition periods.

OCMM also contains functionality to ensure the training is provided in a window aligned with the deployment of the system elements. Training timelines can be linked to the project timelines (Figure 5.4) so that the timing of the activities can be assessed. OCMM automatically compares the completion dates to determine if training is not completed before the systems are brought into service or, conversely, the training is completed more than a nominated period (such as 90 days) before the competencies are needed, which may cause unused skills to atrophy. When these occur, an error is raised for the user to investigate. Further specifics and examples of this step in the approach can be found in the authors' previous publication (Shirvani et al. 2019).

5.6.5 Analyze Emergent System

While some assessments have been created such as safety assurance (Shirvani et al. 2019) and cybersecurity, the focus in this section will be on two HF concerns, namely, SA and workload analysis (WA). This case study demonstrates how commonly used HF assessment methods and techniques can be integrated into an MBSE environment (the framework is, however, flexible in accepting other assessments and techniques). It is essential that SMEs undertake this step of the approach with appropriate working knowledge of the HF techniques, the organization, and the system of interest; liaison with SME teams responsible for other interacting parts of the model or overlapping assessments will likely be required.

5.6.5.1 Situation Awareness

SA is described as "the perception of elements in the environment within a volume of time and space, the comprehension of their meaning and the projection of their status in the near future" (Endsley and Garland 2009). This definition allays itself to a critical aspect of STS that enable the users to be adequately informed of their environment and the context within which they currently exist. SA allows users to make appropriate actions, and poor SA is often a factor in incorrect or unsafe operation by users.

SA evaluation techniques were first developed in the context of air crew assessment. When considering suitable techniques for assessing SA for OCMM, it was important to account for the nature of the information that could be modeled, as well as its suitability for the rail domain. Literature (Bearman et al. 2013; Salmon et al. 2009) provides an excellent summary of SA theories and measures for the Australian rail industry, whereas Gulati and Puranam (2009) discuss the pros and cons of the application of SA techniques in rail signaling in particular. Some candidate techniques were considered including the Rail Ergonomics Situation Awareness (RESA) tool (Wilson et al. 2001) and Endsley's (1995) Situation Awareness Global Assessment Technique (SAGAT) as well as techniques such as the Situation Awareness Rating Technique (SART) (Taylor 1990). It was decided to keep the analysis within OCMM as simple as possible and to consider which technique best fits the application to a conceptual design (rather than getting into the cognitive assessment of individuals). Hence, the OCMM uses the initial stages of the SAGAT.

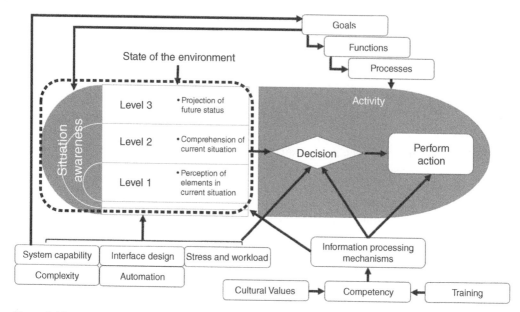

Figure 5.10 OCMM model of situation awareness. *Source:* Based on Endsley (1995).

Endsley's model of SA is well accepted by peers and has suitable genericity to be considered for any industry and application. OCMM enables the user to consider the conceptual design and outputs of any task analyses to assess and record information about any tasks of interest being undertaken pre- and post-technology introduction for the SA factors at the three levels of SA provided in the model (see the area bounded by the dashed rounded rectangle in Figure 5.10). This knowledge may then be used to derive the survey questions for SAGAT (the actual survey and performing of SA assessment on individuals are outside of the scope of OCMM).

Endsley's model of SA contains contextual information around the SA levels. An assessor needs to be able to use the context of the decision being made and how they affect the decisions and performance of an operator. Figure 5.10 shows a conceptual model of SA for use within OCMM; it is based on Endsley's model (Endsley 1995) but shows further linkages to other parts of the organizational models. Within OCMM, the assessor can access the evolving system capability and design models, interfaces, information interactions, goals, workload, cultural values, competencies, and training from within the same tool. OCMM provides the assessor with a means to consider the SA levels for various tasks pre- and post-technology introduction. The assessor is provided with simple activity diagrams to guide their use of the tool, with annotations to describe each step. Figure 5.11 shows an example activity diagram for a SA assessment of the operate train process during pre-ETCS and post-ETCS introduction.

Not all processes are associated with SA, and it may not be practicable to assess every possible process to the nth degree. Therefore, the assessor begins by reviewing the existing processes (provided from other user's processes such as defining the system's concept of operations) and identify which processes need be included for task analysis (or existing task analyses may be imported into the tool). The assessment may be applied at any level of decomposition of the function, process, or action/task level, and it is left to the assessor to determine the level that best gives the appropriate detail required.

The next step involves assessing the situational awareness factors for tasks for the various states of the system (such as current, during the introduction of the technology, and once the technology

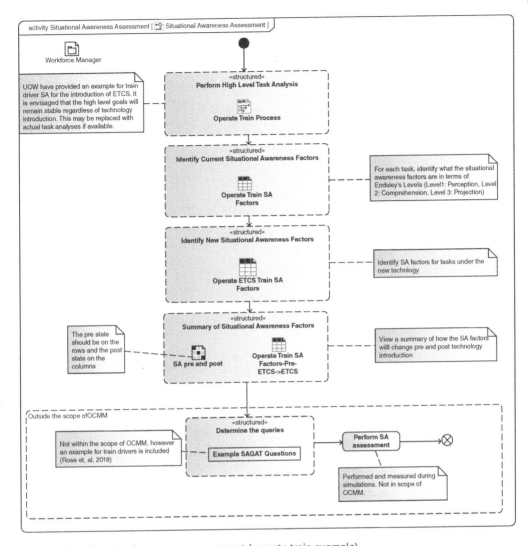

Figure 5.11 Situational awareness assessment (operate train example).

insertion is completed). The SA levels for each task used are described in Endsley (1995). Table 5.5 shows an example of the SA factors assessment for the pre-ETCS operate train process. The figure illustrates the separation of aspects required to perceive, comprehend, and respond to their understanding of the situation.

Once the assessor has completed the assessment of the SA factors in the various states, the OCMM automatically generates a summary table to view the SA factors side by side for direct comparison. Table 5.6 shows the summary of SA factors for the operate train tasks and how they will change when the technology is introduced. In this example, the maintain correct speed task (shown in the bold dashed rectangle) has more aspects under consideration after ETCS has been introduced to improve the situational awareness of the train driver. In particular, the automatic train protection (ATP) is supporting the driver in not exceeding the speed limits and providing digital displays of not just the train speed but the location-specific factors. The assessor can build up their situational awareness model from these perceptions into the comprehension of these inputs and finally into a decided projection of the actions to be undertaken to anticipate these changes.

Table 5.5 Task-based SA assessment (pre-ETCS example).

#	Name	◇ perception	◇ comprehension	◇ projection
1	⊟ ⊞ Operate train-pre-ETCS			
2	⊟ ◇ Normal running			
3	⊡ Communicate with Train Control	Awareness of incoming communication	Ascertain what the communication is	
			Ascertain what actions to take	
4	⊡ Maintain correct speed	Current speed	Amount of difference in speed and limit	Adjust speed
		Current speed limit	Assessment of whether under or over speed	Anticipate speed limit changes coming up
		Speed limit changes		
5	⊡ Monitor for external threats	Obstruction on line	Ascertain whether obstruction is a threat	Apply brakes at correct level to stop in time
		Distance to obstruction	Ascertain what obstruction is	Contact Train Control check what obstruction may be
			Ascertain what actions are required	
6	⊡ Operate for correct track profile	Current location	Ascertain what actions are required	Anticipate imminent track profile changes
		Track profile details	under different track profile scenarios	Apply brakes to adjust speed before track profile change
		Operating procedure		
7	⊡ Respond to alarms	Audible alarm	Ascertain which alarm	Respond to the alarm
		Visual alarm	Ascertain what the alarm means	
8	⊡ Respond to environment conditions	Perception of buff and draft forces on train	Ascertain whether conditions will impact	Reduce speed to match conditions
		Outside track temperature	how train is operated	Check environment conditions with Train Control
		Weather conditions		
		Track condition		
9	⊡ Respond to signals	Authority limits		
		Signal state		

Table 5.6 Summary of SA factors (operate train example).

#	Name	◇ perception	New Perception	◇ comprehension	New Comprehension	◇ projection	New projection
1	Operate train-pre-ETCS						
2	Normal running						
3	Communicate with Train Control	Awareness of incoming communication	Awareness of incoming audible communication Awareness of incoming visual communication	Ascertain what the communications is	Ascertain the source of the communica Ascertain what the communication is Ascertain what actions to take		Form appropriate plan of action Confirm communication received
4	Maintain correct speed	Current speed Current speed limit Speed limit changes	Current speed ATP warning on speed DMI Current speed limit DMI Speed limit changes DMI warning	Amount of difference in speed and limit Assessment of whether under or over	Amount of difference in speed and limit Assessment of whether under or over Understand the ATP speed warning Understand the DMI information	Adjust speed Anticipate speed limit changes coming u	Respond to ATP speed warning Adjust speed Anticipate speed changes coming up
5	Monitor for external threats	Obstruction on line Distance to obstruction	Obstruction on the line Distance to the obstruction Track work movement authority displayed	Ascertain whether obstruction is a thre Ascertain what obstruction is Ascertain what actions are required	Ascertain whether obstruction is a thre Ascertain what the obstruction is Ascertain what actions are required	Apply brakes at correct level to stop in Contact Train Control to check what obj	Apply brakes at correct level to stop in time Contact Train Control to check what obstruction
6	Operate for correct track profile	Current location Track profile details Operating procedure	Current location DMI track profile details Operating procedures	Ascertain what actions are required under different track profile scenarios	Ascertain whether ETCS level change Ascertain what actions are required under different track profile scenarios	Anticipate imminent track profile change f Apply brakes to adjust speed before track	Anticipate imminent changes in track profile Apply brakes to adjust speed before track profile
7	Respond to alarms	Audible alarm Visual alarm	Audible alarm Visual alarm	Ascertain which alarm Ascertain what the alarm means	Ascertain which alarm Ascertain what alarm means	Respond to the alarm	Respond to alarm Use other knowledge to build big picture of the tr
8	Respond to environmental conditions	Perception of buff and draft f outside track temperature Weather conditions Track condition	Perception of buff and draft outside track temperature Condition monitoring sensor information Track condition Weather conditions Condition beginning Condition ending	Ascertain whether conditions will impact how train is operated	Ascertain whether conditions will impact hoe train is operated	Reduce speed to match conditions Check environmental conditions with Train	Reduce speed to match conditions Check environmental conditions with Train Control Estimate duration of conditions
9	Respond to signals	Authority limits Signal state	DMI movement authorities Movement authority limits Current location		Ascertain what the DMI is conveying Understand what changes to train open		Confirm signal is received Alter train operation plans

The final stages of the SA assessment enable the assessor to use the SA factors to derive the questions in a SAGAT, which in turn can be used to design a practical SA assessment (the questions are typically used alongside simulator trials to gauge whether the operator has the correct perception, comprehension, and projection as expected).

5.6.5.2 Workload Analysis

With the introduction of new technologies, understanding is needed on how staff workloads will change as a result. This evaluation needs to provide understanding not only of the resulting workload but the workload during the transition phase between the introduction of the new technology until the retirement of the existing technologies.

The workload can be defined as "the effort demanded from people by the tasks they have to do. It can be the effort demanded at a single point in time, or over a whole shift" (RSSB 2008). It can be challenging to decide on suitable workload assessment methods because there are differing interpretations of what constitutes workload among practitioners. Wilson et al. (2005) outline a number of workload-related areas recognized within the rail industry. Alternative, interrelated views of workload considerations can be:

- Task complexity (number and interactions between tasks).
- Performance and capacity based (balancing of the load in terms of the amount of work under time constraints such as time pressures or % hours worked or job design) (Smith et al. 2005).
- Synonymous with the manifestation of physiological stress and fatigue measurement (Hancock and Desmond 2000).
- Most commonly, perceptions of the mental or cognitive demands and ISO (2017) (Young et al. 2015).

Wilson et al. (2005) further suggest that there is an essential distinction in understanding workload that is imposed by a system and the perception of workload on individuals. There are arguments that one cannot consider workload purely at system level as the individuals within the system will have variability and, alternately, that workload cannot be assessed validly through survey of individuals, because it is based on the perception of workload, not actual workload, or can be mistaken for level of performance (i.e. an operator may perceive that the workload was high if they did not manage to perform the task, and vice versa) (Hart and Staveland 1988). It would seem beneficial to provide then a selection of workload tools that can be used to yield insight from these various viewpoints. OCMM contains two additional methods of assessing workload for comparison in addition to the RASCI analysis previously described and to illustrate how MBSE can support alternative assessment approaches. Each method uses a different approach and understanding of workload, which may be used independently or in conjunction to give different viewpoints. These methods are:

- In2Rail Workload Predesign – Considers the effects on the workload of a system encountering a step change (i.e. a holistic assessment including people, process, and technology dimensions) (Evans 2017).
- NASA TLX – Considers the workload at an individual process or task level (Hart 2006).

Each of these approaches is described below.

5.6.5.2.1 *In2Rail Workload Predesign* This assessment method evaluates the impact on workload due to the introduction of a system change. A holistic systems approach may be taken when assessing the workload for rail traffic management systems (Evans 2017). OCMM utilizes the prediction

part of Evans' method to forecast the likely workload changes from a people, process, and technology perspective by considering the various states of the system as the technology is introduced and guides the assessor to consider multiple variables that may affect workload.

This workload assessment method considers the changes to workload at a system level as a new technology is introduced. The method assumes that the design of the future state of the system is mature enough to be assessed. However, this could be applied at various stages of the system development life cycle to assess early design concepts and inform design choices as the design matures further. The assessor (and SMEs) should consider each of the variables, in turn assessing what the effect on workload (positive, negative, or static) will be for each transition between states, how safety or performance is impacted, and recommendations for further assessment or design considerations. Table 5.7 shows an example assessment of how aspects will change for both the better and worse for the case of introducing ATMs.

5.6.5.2.2 **NASA TLX Workload Survey** The NASA TLX workload survey is a quantitative task-based WA that considers the cognitive demands of the operator performing the task. The NASA TLX workload survey (Hart 2006) is one of the most well-known workload assessment tools. The survey consists of six dimensions of workload (mental demand, physical demand, temporal demand, overall performance, effort, and frustration level). After a task has been performed, the operator rates each of the dimensions based on their perceptions of workload. Pairwise comparison is used to identify the relative weightings/importance of each of the dimensions. The workload index is calculated as a percentage of the weighted sum of the measured dimensions. OCMM uses these scales to enable the assessor to forecast an estimated workload for an operator using an envisaged design.

Although this method is usually applied as a survey during or shortly after an individual performs a set of tasks, the workload demand scales are useful to apply to a "generic" role holder undertaking an envisaged task. As the design matures and simulators and finally trials are run, the NASA TLX survey may be applied repeatedly to assess how the workload is changing (in particular, there is value in considering whether design changes are mitigating risks as intended). Table 5.8 shows an example of a NASA TLX assessment of workload for different processes under different technologies. Once the assessor completes the demand scales and weightings, the workload index is calculated automatically for the assessor using the NASA TLX formula. In this example, it can be seen that the introduction of ETCS L1 represents a higher workload due to the overlaying of information upon existing infrastructure, which subsequently decreases once ETCS L2 is introduced. Hely et al. (2015) suggest that ATP does indeed increase the workload for train drivers due to overlaid information. It should be noted that these workload assessments consider that the operator is experienced in using the technology and undertaking the tasks (the assessor can confirm this by viewing the information held within the OCMM competency portals). The assessor may add additional NASA TLX stereotyped tasks to other existing processes of interest that are anticipated to be affected by new technology introduction.

5.7 OCMM Reception

The OCMM was demonstrated and subsequently delivered to several Australian heavy rail operators during which many salient points were raised. First was a general acceptance and acknowledgment that the use of such a system would resolve some of their existing issues, particularly concerning information management and sharing. However, there were concerns about the level

Table 5.7 Workload by variables (ATMS example).

#	Name	Effect On Workload	Summary Of Workload Change	Effect On Safety Or Performance	Legacy Systems To Use	Recommendations
1	⊟ 🗀 Variables that affect the objective					
2	⊟ 🗀 Control room environment					
3	🮰 Temperature					
4	🮰 Lighting	lighting and workstation design ergonomically supportive	−1	Improved performance		Iterate assessments in design process to ensure that lighting design remains appropropriate for any changes ATMS brings in
5	🮰 Noise	Consider levels of noise large open plan	+1	Decreased performance and safety		Undertake workload assessments on tasks, receive feedback from users for improvement
6	⊟ 🗀 Infrastructure					
7	🮰 Track					
8	🮰 Assets	ATMS will provide tracking of trains to aid the train controller in coordinating	−1	Improved performance		
9	⊟ 🗀 Operations					
10	🮰 Tasks	It is assumed that the number and frequency of tasks will be decreased as automation takes over	−1	Increased performance		Will need to consider effects of complacency and boredom
11	🮰 Areas of control	Areas of control will be scatable		Need more information		
12	🮰 Degraded modes	Will need to consider various degraded mode scenarios	+1	Possible safety issue		Consider failure modes, compromised operations and recovery of the system.
13	⊟ 🗀 Traffic					
14	🮰 Traffic volume	Increased traffic volume will increase workload, however automating some coordination functionalties will reduce train controller workload				May need to perform workload assessments for the pre and post ATMS tasks in order to understand the workload change
15	🮰 Timetable	ATMS will provide optimisation and dynamic timetabling	−1			
16	⊟ 🗀 Variables that affect system's a					
17	⊟ 🗀 Process					
18	🮰 Rules and standards					
19	🮰 Process and procedures	New procedures will be created, workload will be high until operator is experienced	+1			
20	🮰 Team working	Better situational awareness between train controller interfaces	−1			Provide team based training
21	🮰 Communication	Improved communications, more efficient	−1	Improved performance		Provide training to use new comms
22	⊟ 🗀 People					
23	🮰 Roles	Need more information on what the new roles will be (e.g. there was creation of roles)				
24	🮰 Culture	Is there a culture or overloading or underloading? How do they support these scenarios?				Need further investigation
25	🮰 Shift patterns	Need more information				
26	🮰 Training	Training to use the new systems	−1			Provide training and perform workload assessments on tasks

of effort required to introduce and maintain such a model. The OCMM is intended to be scalable so that the model can not only be used to address the current set of concerns but later be expanded to the wider user group. The OCMM was also designed to support existing processes, and therefore, only minimal additional effort above the existing processes was anticipated.

The proliferation of different MBSE and digital thread software tool sets yield software and data interoperability issues for the engineering industry (Aerospace and Defense PLM Action Group 2019; Shani et al. 2016). Within the Australian heavy rail industry, there was inconsistent

Table 5.8 Workload assessment by process (excerpt).

#	Name	Responsible for task	Workload Index	Mental Demand Scale	Mental Demand Weighting	Physical Demand Scale	Physical Demand Weighting
1	☐ Workload by NASA TLX						
2	☐ ☐ ETCS						
3	⊞ ☐ Operate train - pre-ETCS	Train Driver	63.33	80	5	40	1
12	⊞ ☐ Operate train - ETCS L1	ETCS L1 Train Driver	72.00	90	5	40	1
18	⊞ ☐ Operate train - ETCS L2	ETCS L2 Train Driver	60.00	70	5	40	1
25	☐ ☐ ATMS						
26	☐ Existing Train Control Process	Network Controller	75.33	90	4	10	1
27	☐ ATMS Train Control Process	ATMS Network Controller	73.00	75	4	10	1

adoption and maturity of application of MBSE software during the research project's duration; while some rail operators were in the process of selecting a software tool that could be mandated throughout the supply chain, others were adopting different MBSE tools in an ad hoc manner from suborganizational unit to unit. Others were investigating the use of third-party software to translate and interface from vendor to vendor. As a part of the adoption process, the OCMM is expected to be interfaced with existing tools rather than replacing them. Therefore, users can use their preferred tool sets that exchange data with the OCMM for better data consistency and transparency.

The user portals interface was well received for its ease of use and constraining of complexity by restricting the information provided to that which the user needs to address their concerns. The user guide was embedded as annotations directly within the user activities and models so that novice users could quickly navigate around the models with confidence.

5.8 Summary and Conclusions

This chapter has discussed the characteristics of organizational systems and why it is important to consider how they will evolve within the context of the enterprise system. The authors' approach to modeling the organizational system evolution has been introduced, and the implementation of the OCMM tool described using AF principles. A case study populated with tailored models for an Australian rail operator was provided to demonstrate for a set of examples to inform the reader further.

The OCMM should be tailored and maintained by businesses and aligned to existing HSI activities to offer practitioners context to the current and forecast future states of the organizational system and to provide a mechanism to record the information and decisions as the system evolves. The OCMM has been validated through stakeholder feedback and by creating proof-of-concept demonstrators for two rail organizations. Further development will be required to expand and align the HF activities with those in the existing businesses. Industry engagement with the model has been positive and recognizes the potential for the improved integration of information while reducing the amount of effort.

The use of MBSE technologies can be time and labor intensive (transferring from legacy systems, training users, cost of software, integration of data, and risk) to initiate within a company, and decisions will need to be made as to the extent required to leverage the most value. Madni and Purohit (2019) propose a methodology for assessing the economic value of MBSE that enables organizations to objectively decide whether to invest in MBSE by quantifying the gains based on various measures. As digital thread practices become ubiquitous across the fields of engineering, MBSE will be an essential activity to manage the ever-increasing complexity of the enterprise system. In turn, it will become commonplace for engineering companies to embed the practices into their day-to-day business, and model-based approaches such as OCMM will become easier to embed and integrate the HSE into MBSE activities.

References

Aerospace & Defense PLM Action Group (2019). *Model-Based Systems Engineering Data Interoperability*. CIMdata.

Australian Railway Association & Deakin University (2018). *Smart Rail Route Map*. ARA.

Bearman, C., Naweed, A., Dorrian, J. et al. (2013). *Evaluation of Rail Technology – A Practical Human Factors Guide*, 1e. Surrey, England: Ashgate.

Bruseberg, A. (2008). Human views for MODAF as a bridge between human factors integration and systems engineering. *Journal of Cognitive Engineering and Decision Making* 2 (3): 220–248.

Dassault Systems (2020). MagicDraw. https://www.nomagic.com/products/magicdraw (accessed 6 February 2020).

Endsley, M. (1995). Toward a theory of situation awareness in dynamic systems. *Human Factors* 37: 32–64.

Endsley, M. and Garland, D. (2009). *Situation Awareness Analysis and Measurement* (e-Book Ed.). Mahwah: Lawrence Erlbaum Associates.

Evans, J. (2017). A systems approach to predicting and measuring workload in rail traffic management systems. In: *H-Workload 2017: The First International Symposium on Human Mental Workload*. Dublin, Ireland: Dublin Institute of Technology.

FYA (2017). *The New Work Mindset*. The Foundation for Young Australians.

Galbraith, J.R. and Lawler, E.E. (1993). *Organizing for the Future – The New Logic for Managing Complex Organisations*. San Francisco: Jossey-Bass.

Gulati, R. and Puranam, P. (2009). Renewal through reorganization: the value of inconsistencies between formal and informal organization. *Organization Science* 20 (2): 422–440.

Hancock, P. and Desmond, P. (2000). *Stress, Workload, and Fatigue*. Boca Raton: Taylor & Francis.

Handley, H.A. and Knapp, B.G. (2014). Where are the people? The human viewpoint approach for architecting and acquisition. *Defence ARJ* 21 (4): 852–874.

Hart, S. (2006). NASA-task load index (NASA TLX) 20 years later. *Proceedings of the Human Factors and Ergonomics Society 50th Annual Meeting*, San Francisco (16–20 October 2006).

Hart, S. and Staveland, L. (1988). Development of NASA-TLX: results of empirical and theoretical research. In: *Human Mental Workload* (ed. H.A. Meshkati), 77–106. New York: Elsevier.

Hause, M. and Wilson, M. (2017). Integrated human factors view in the unified architecture framework. *INCOSE International Symposium*, Adelaide (15–20 July 2017).

Hely, M., Shardlow, T., Butt, B. et al. (2015). Effects of automatic train protection on human factors and driving behaviour. *Proceedings of the 19th Triennial Congress of the IEA*, Melbourne (9–14 August 2015).

Hubbard, E., Siemieniuch, C., Sinclair, M., and Hodgson, A. (2010). Working towards a holistic organisational systems model. *5th International Conference on Systems of Systems Engineering*, Loughborough (22–24 June 2010).

INCOSE (2007). *Systems Engineering Vision 2020*. INCOSE.

INCOSE (2015). *Systems Engineering Handbook – A Guide for System Lifecycle Processes and Activities*. Hoboken, NJ: Wiley.

ISO (2011). *ISO/IEC/IEEE 42010:2011 Systems and Software Engineering – Architecture Description*. Geneva: ISO.

ISO (2015). *ISO/IEC/IEEE 15288:2015 Systems and Software Engineering – System Life Cycle Processes*. Geneva: ISO.

ISO (2017). *ISO 10075-1:2017 Ergonomics Principles Related to Mental Workload*. Geneva: ISO.

ISO (2019). *ISO 15704:2019 Enterprise Modeling and Architecture – Requirements for Enterprise-Referencing Architectures and Methodologies (ISO 15704:2019)*. Geneva: ISO.

Keating, C., Polinpapilinho, F., and Bradley, J. (2014). Complex system governance: concept, challenges, and emerging research. *International Journal of System of Systems Engineering* 5 (3): 263–288.

Kemp, D. and Daw, A. (2014). *Capability Systems Engineering Guide*. INCOSE UK.

Madni, A. and Purohit, S. (2019). Economic analysis of model-based systems engineering. *Systems* 7 (1): 12.

Madni, A. and Sievers, M. (2018). Model-based systems engineering: motivation, current status and research opportunities. *Systems Engineering* 21 (3): 172–190.

OMG (2017). *Unified Architecture Framework – Version 1.0 Appendix C*. OMG.

OMG (2019). *OMG Systems Modeling Language v1.6*. OMG.

Pennock, M.J. and Rouse, W.B. (2016). The epistemology of enterprises. *Systems Engineering* 19: 24–43.

Peterson, T. (2019). Systems engineering: transforming digital transformation. *INCOSE International Symposium* 29: 434–447.

Purchase, V., Parry, G., Valerdi, R. et al. (2011). Enterprise transformation: why are we interested, what is it, and what are the challenges? *Journal of Enterprise Transformation* 1 (1): 14–33.

RISSB (2018). *AS 7711 Signalling Principles*. Canberra: RISSB.

Robbins, S.P., Judge, T.A., Millett, B., and Boyle, M. (2017). *Organisational Behaviour*. Melbourne: Pearson Australia.

Rouse, W.B. (2005). Enterprises as systems: essential challenges and approaches to transformation. *Systems Engineering* 8: 138–150.

RSSB (2008). *Understanding Human Factors – A Guide for the Railway Industry*. Rail Safety Standards Board.

Salmon, P., Stanton, N., Walker, G. et al. (2009). Measuring situation awareness in complex systems: comparison of measures study. *International Journal of Industrial Ergonomics* 39: 490–500.

Schekkerman, J. (2004). *How to Survive in the Jungle of Enterprise Architecture Frameworks*. Victoria, Canada: Trafford.

Shani, U., Jacobs, S., Wengrowicz, N., and Dori, D. (2016). Engaging ontologies to break MBSE tools boundaries through semantic mediation. *Conference on Systems Engineering Research*, Huntsville, AL.

Shirvani, F., Scott, W., Kennedy, G., and Campbell, A. (2019). Enhancement of FMEA risk assessment with SysML. *Australian Journal of Multi-Disciplinary Engineering* 15: 52–61.

Shirvani, F., Scott, W., Kennedy, G., Rezaibagha, F., & Campbell, A. (2019). Developing a modelling framework for aligning the human aspects to the physical system in large complex systems. *IEEE SysCon*, Orlando.

Silitto, H., Martin, J., McKinney, D. et al. (2019). *Systems Engineering and System Definitions*. INCOSE.

Smith, M., Erwin, J., and Diaferio, S. (2005). Role & responsibility charting (RACI). Project Management Forum (PMForum).

Taylor, R. (1990). Situational awareness rating technique (SART). In: *Situational Awareness in Aerospace Operations*. Neuilly sur Seine: NATO-AGARD.

Valerdi, R., Nightingale, D., and Blackburn, C. (2008). Enterprises as systems: context, boundaries, and practical implications. *Information Knowledge Systems Management* 7 (4): 377–399.

Wilson, J., Cordiner, L., Nichols, S. et al. (2001). On the right track: systematic implementation of ergonomics in railway network control. *Cognition, Technology and Work* 3 (4): 238–253.

Wilson, J., Pickup, L., Norris, B. et al. (2005). Understanding of mental workload. In: *Rail Human Factors Supporting the Integrated Railway* (eds. J.R. Wilson, B. Norris, T. Clarke and A. Mills), 309–318. Aldershot: Ashgate.

Young, M., Brookhuis, K., Wickens, C., and Hancock, P. (2015). State of science: mental workload in ergonomics. *Ergonomics* 58: 1–17.

6

Human Systems Integration in the Space Exploration Systems Engineering Life Cycle

George Salazar[1] and Maria Natalia Russi-Vigoya[2]

[1] *Johnson Space Center, NASA, Houston, TX, USA*
[2] *KBR, Houston, TX, USA*

6.1 Introduction

Space travel risks are present in every phase of the mission, from the launch to the reentry and recovery operations. Elimination or mitigation of crew risks and hazards will help ensure mission success. An effective Human systems integration (HSI) program as part of the National Aeronautics and Space Administration (NASA) Systems Engineering (SE) process is critical from the initial stages of the design to the final design. As the programs evolve, it is crucial to learn to enhance how the crew and the ground support interact with systems and products. HSI has been successfully adopted by several federal agencies (e.g., US Department of Defense (DoD), Nuclear Regulatory Commission (NRC), and the Federal Railroad Administration) as a methodology for reducing system Life Cycle Costs (LCCs) and increased in safety in operating systems. NASA has been working on HSI requirements since the early 2000s and has cared about focusing on the human (despite the lack of formal mandate) safety and health. An HSI practitioner's guide (NASA, 2015b) enables the incorporation of agency HSI policies and processes into the development and deployment of NASA systems. The practitioner's guide has information about the importance of following HSI practices and aids to implement HSI, how it embeds into the SE processes, and ways for planning and implementing of HSI in NASA practices. Salazar and Russi-Vigoya (2019) explained that future missions need to have iterative system maturity innovation and adaptation; HSI needs to be embedded into the system design at all levels, and support from the organization culture is important to ensuring successful missions, reducing LCCs while maintaining the safety of the human operating complex systems. NASA has a history of considering astronaut health and performance in spacecraft and mission design, particularly in mission planning and system design.

NASA is actively planning for future long-duration missions. Communication latencies, crew size, and vehicle complexity will pose new challenges for NASA human missions. Researchers (Holden et al., 2020) have identified and validated a list of tasks that the crew will have to perform in future long-duration missions that will require autonomous interactions with smart spacecraft. The ground supports many tasks right now. However, future spacecraft will need more onboard autonomy to enhance the crew capabilities (Vera et al., 2019). NASA has investigated the preliminary onboard capability requirements for future missions (Wu and Vera, 2019) and also has

A Framework of Human Systems Engineering: Applications and Case Studies, First Edition.
Edited by Holly A. H. Handley and Andreas Tolk.
© 2021 The Institute of Electrical and Electronics Engineers, Inc. Published 2021 by John Wiley & Sons, Inc.

identified potential human performance standards and guidelines for autonomous missions that are available in the literature that could apply to space research (Holden et al., 2019).

Future long-duration missions will also have the challenges that current missions have, but on a larger scale. These challenges include isolation and confinement, changes in gravitational fields, radiation, and operation paradigms (Salazar and Russi-Vigoya, 2019). This chapter discusses future mission challenges. It identifies how the HSI technical and management of the domains integrated into the SE process can help mitigate mission risks and hazards and optimize total system performance while reducing cost and schedule.

6.2 Spacecraft History

Since the Mercury program, NASA human spaceflight missions have grown in complexity. With time, NASA has learned to master mission design/planning, training, and spacecraft system development processes to take the crew to space. As technologies evolve to enhance crew interactions with the space vehicle, the crew interactions become more determinant of mission success. To be able to learn about our challenges of future missions, it is important to understand what has been done in the past that could enhance crew capabilities in future long-duration missions. The following sections describe a historical evolution of NASA spacecraft, including Mercury/Gemini/Apollo, Space Shuttle, International Space Station (ISS), and Orion Spacecraft.

6.2.1 Mercury/Gemini/Apollo

The early spacecraft displays and controls (D&C) subsystems from Mercury to Skylab pretty much consisted of the D&C of the same type switches (toggle, rotary, and pushbutton), displays (moving/fixed pointer), numeric, and status lights. As more subsystems and features were added, more switches and displays were added to support the complexity of the vehicles. Mercury had approximately 120 D&C components (Swenson et al., 1966). It had no onboard computer but rather relied on the ground for uplink commanding.

Gemini had over 200 D&C interfaces (McDonnell Company, 1965) as well as having the first onboard computer as well as added D&C capability for rendezvous and docking maneuvers. In addition, rotational and translational hand controllers were added to control the spacecraft to provide a linkage from the astronaut to the stabilization and control to override the automatic control mode. Figures 6.1 and 6.2 show the progression of D&C complexity going from a one-person cockpit to a two-person cockpit, respectively.

Figures 6.3 and 6.4 show an increase in D&C layout complexity for the Apollo program (North American Rockwell Corp. 1968). Figure 6.3 shows the layout of the Apollo command module (CM) that had nearly 700 D&C interfaces. The overall layout is composed of the forward section of the pressurized cabin that contained the crescent-shaped main display panel composed of three panels. Each of the three panels was associated with the responsibilities of each of the three crewmembers. The mission commander's panel (left side) dealt with the vehicle velocity, attitude, and altitude indicators as well as the primary flight controls. The pilot handled the center panel. The panel included the guidance and navigation computer controls and the service propulsion system. The Lunar Module (LM) pilot that served as the systems engineer handled the right-side panel. He had access to the fuel cell gauges and controls, the electrical and battery controls, and the communications controls. In total, the D&C are composed of 24 instruments, 566 switches, 40 event indicators, and 71 lights.

Figure 6.1 Mercury spacecraft displays and controls. *Source:* Credit NASA.

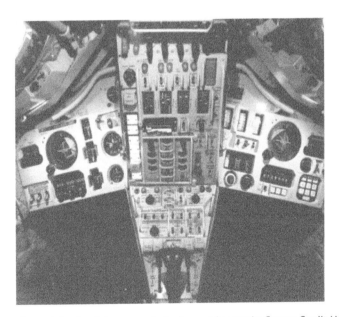

Figure 6.2 Gemini spacecraft displays and controls. *Source:* Credit NASA.

Figure 6.4 shows the layout for the lunar lander that only accommodated two astronauts. The LM display and control subsystem consists of a variety of equipment. All controls were of two basic types: (i) switches and (ii) variable controls. Displays were of four basic types: (i) Moving pointer, (ii) Fixed pointer, (iii) Numeric, and (iv) Status indicators. In all, there were a total of 388 switches,

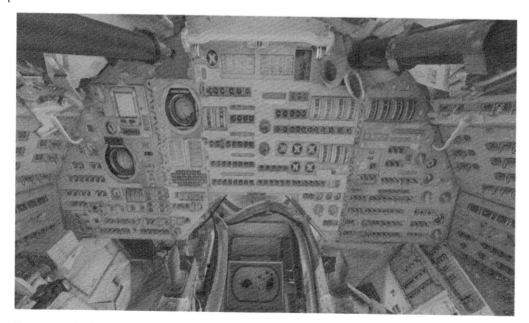

Figure 6.3 Apollo command module displays and controls. *Source:* Credit NASA.

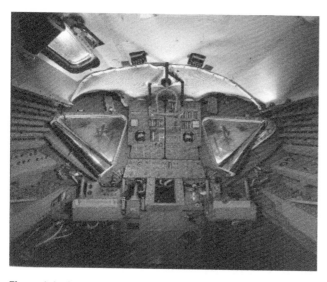

Figure 6.4 Lunar lander displays and controls. *Source:* Credit NASA.

indicators, instruments, and lights. Though not discussed in this paper, Skylab D&C comprised much of the same type of D&C interfaces as Apollo.

6.2.2 Space Shuttle

According to NASA (2013b), the Space Shuttle was a complex spacecraft that launched like a rocket, orbited like a satellite, and returned to Earth like a plane. It comprised well over 2500 displays and controls (D&C) switches, displays, hand controllers, and status lights. These were spread throughout the pressurized cabin area but mostly in the flight deck. The flight deck contained the D&C used

Figure 6.5 Space Shuttle flight deck. *Source:* Credit NASA.

to pilot, monitor, and control the orbiter, the integrated shuttle vehicle, and the mission payloads. Up to four crewmembers could be seated in the flight deck. The midsection contained seating for up to three crewmembers. It served as the living area where the astronauts ate and slept. The mid-deck as contained the airlock and avionics equipment compartments. Figure 6.5 shows the flight deck.

6.2.3 International Space Station

Unlike the previous spacecraft that contained dedicated D&C panels, the ISS was designed with a multipurpose workstation as well as distributed control of the vehicle via laptops called PCS (portable computer system). They run on a Linux operating system that can operate as bus controllers or remote terminals. On the Russian side, the laptops called "Russian laptops" run/command the Russian segment. Both the PCS and Russian laptop used graphical interfaces for their respective ISS module to enables the user to select the ISS module they wish to interact with, the system, and then the specific piece of hardware. Similarly, both the Japanese modules provided by the Japanese space agency and the European modules are controlled with laptops. In addition, there are also laptops associated with the less critical Ethernet operational local area network – used for low-criticality applications such as viewing procedures, performing supply inventory, recording notes, sending an e-mail, and video conferencing as well as social media.

Figure 6.6 shows the ISS Cupola workstation as a representation of how ISS is controlled by the crew using laptops. Its seven windows in the Cupola are used to conduct experiments, dockings, and observations of Earth. The space station remote manipulator system is controlled from there. In addition, the large viewing area provides an opportunity to spend time viewing earth.

The ISS still contains switches and pushbutton to control some of the subsystems. However, for the most part, the laptops have dramatically reduced the number of dedicated D&C devices.

6.2.4 Orion Spacecraft

After the Space Shuttle retirement, the Orion became the next American spacecraft to take astronauts to station, moon, Mars, and beyond. The spacecraft is the continued work done from the canceled Constellation Program. The first crewed mission will be Artemis Mission 2 (AM-2)

Figure 6.6 ISS Cupola workstation. *Source:* Credit NASA.

Figure 6.7 Mockup of the Orion D&C glass cockpit. *Source:* Credit NASA.

designed for four-person crew support for up to a 21-day independent mission. As shown in Figure 6.7, the AM-2 will contain a crew console that includes the "glass-cockpit" station for the crew to monitor and operate the vehicle.

The glass cockpit replaces on the order of 2000 manual switches to operate the vehicle. The crew console still contains several pushbutton and toggle switches necessary for critical operations. However, they are primarily for outside contingency scenarios. The displays provide the status of the spacecraft systems with the capability to display vehicle status and to command vehicle systems via menu-selected display formats. Each of the screens can display two formats. Therefore, there will be six display formats shown across the console. Between the upper and lower display formats, a title bar with six simultaneous display formats is displayed to the crew.

Given the discussion of increased vehicle complexity, for the Artemis Gateway (the orbiting moon station) and Mars vehicles, it is not hard to see that the increased mission complexity will result in vehicle control design new to NASA. The NASA Human Research Program (HRP) that is investigating gaps in technology to support these future space missions has identified one of the 33 gaps being "inadequate human–computer interaction." This gap was based on the HRP HCI risk states that "Given that human-computer interaction and information architecture designs must support crew tasks, and given the greater dependence on HCI in the context of long-duration spaceflight operations, there is a risk that critical information systems will not support crew tasks effectively, resulting in flight and ground crew errors and inefficiencies, failed mission and program objectives, and an increase in crew injuries" (NASA, 2015a).

6.3 Human Systems Integration in the NASA Systems Engineering Process

The formal mandate of HSI as part of NASA's human spaceflight program/project development processes did not occur until 2013. However, since the Mercury program, NASA has always considered the human in the safety, health, and performance of human spaceflight missions. In the early 2000s, NASA developed HSI requirements at the technical level (NASA, 2010). To optimize aerospace systems and reduce human error, Human-In-The-Loop (HITL) training and testing are an integral part of the spacecraft system development effort. These activities form part of the human-centered design (HCD) process to be discussed later in this chapter. The NASA practitioner's guide (NASA, 2015b) provides a formal framework to implement HSI for NASA projects as part of the SE process. Figure 6.8 shows the six domains defined by NASA, along with a short description of each domain. HSI is not just the technical process to ensure human performance issues related to hardware (H/W) and software (S/W) are addressed as part of the SE process but also the management process to ensure human-related concerns are addressed at all aspects of the program/project life cycle phases related to human system requirements, risk management, planning, etc.

As shown in Figure 6.9, the heart of the HSI effort is HFE. The HFE activity ensures all domains are addressed in the SE process and *integrated* to factor the human element. Also, note that all domains are not necessarily organizational/engineering entities but rather technical disciplines that have Subject Matter Experts (SME) that participate in HSI activities. The integration of the HSI domains occurs at unified requirements level to ensure all domains are considered in development of requirements and at the trade off level where requirements are analyzed to ensure no conflicting domain/system requirements. Each of the domains affects the crew, ground operators, and trainers, as well as shipping and receiving personnel. It is important that as the system design begins early in the life cycle, the SMEs ensure the design does not put constraints on both the system and the users.

6.3.1 NASA Systems Engineering Process and HSI

The NASA SE process is composed of three sets of conventional technical processes as defined in NPR 7123.1B (NASA, 2013a), NASA Systems Engineering (SE) Processes and Requirements: system design, product realization, and technical management. For Human Systems Integration to be effective, it must be included early on of the project in each of those sets of processes. Figure 6.10 shows the NASA Systems Engineering (SE) engine composed of 17 conventional technical processes to engineer system products. These conventional technical processes, along with Systems Engineering (SE) tools and methods and a knowledgeable/skilled workforce, make up the NASA Systems Engineering (SE) capability. At a high level, the NASA Systems Engineering (SE) engine involves analysis, design, and development activities along with technical management of those

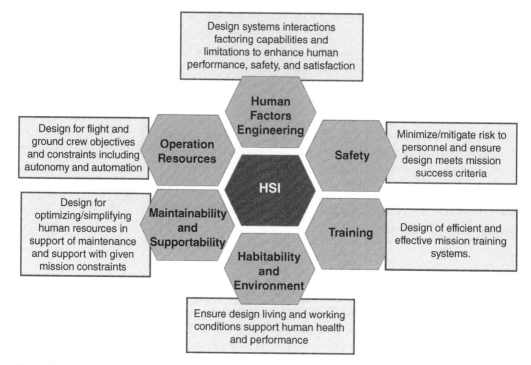

Figure 6.8 NASA HSI domains. *Source:* Adapted from NASA (2015b).

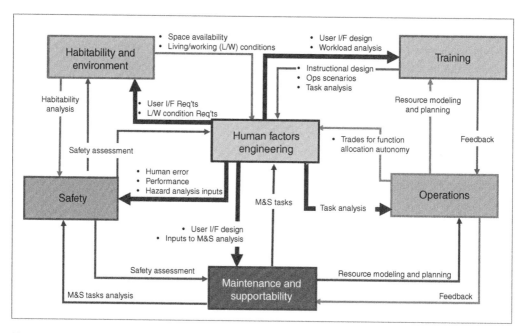

Figure 6.9 Notional HSI domain interaction. *Source:* Adapted from NASA (2015b).

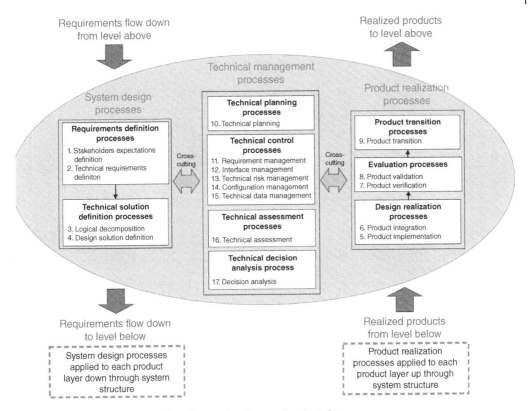

Figure 6.10 NASA systems engineering engine. *Source:* Credit NASA

activities. The requirements definition process defines the *analysis* aspect of the program/project. Technical solutions encompasses the *design and development* activities of the program/project. Moreover, the design realization and product evaluation delineate the *test and evaluation* aspect of the program/project.

HSI and the NASA SE process also involve a considerable amout of interaction during system development activities. Figure 6.11 shows a high-level notional interaction between HSI and the NASA SE process for the development of a system/subsystem. It is important early on that HSI and SE personnel agree on the processes, activities, and milestones to develop the system/subsystem. Similarly, an important element of both HSI and SE working together is the system's technical requirements definition. A sound technical requirement development effort that factors the capabilities and limitation of the human element ensures that an effective design solution and system verification and validation are produced.

HSI is a vital part of SE and must be conducted concurrently with the SE process. This simultaneous activity means that communication effectiveness between HSI and SE practitioners is important as design and human factors personnel may have communication difficulties – designers typically are driven by requirements, specifications, and design documents, and the HFE personnel design viewpoint is from a human user perception. Early planning in the HSI/SE effort ensures different perspectives of system development are aligned – hardware, software, and the human.

Figure 6.12 shows a notional holistic view of NASA HSI and SE process in the NASA human spaceflight SE life cycle. Early on, both the SE and HSI technical management engage in defining

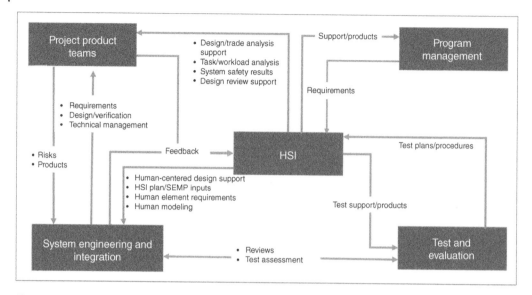

Figure 6.11 Notional systems engineering and HSI interaction. *Source:* Adapted from NASA (2015b).

the Roles and Responsibilities (R/Rs) of SE and HSI. For large projects, an HSI Plan (HSIP) or a Systems Engineering Management Plan containing the HSIP would include defining the R/Rs as well as where HSI processes/activities in the SE process occurs. By engaging HSI early on, risks associated with the design that affects the crew can be identified, controlled, eliminated, or mitigated. The HSI key process areas related to analysis, design, and test/evaluation are performed as part of the SE effort to ensure objectives in crew requirements, risks/performance associated with human–computer interface optimization, maintenance, operations, training, and safety are met. Note that HSI is not only involved with system design, product realization, and technical management but also heavily involved in the major milestone life cycle reviews (SRR, SDR, PDR, CDR, etc.). Though not shown, the iteration continues even after deployment, where feedback on the operational use of the system provides additional opportunities to improve the system – particularly if it is software related.

A critical HSI process to the SE effort is HCD. This is a methodology used to ensure that a design accommodates human capabilities and limitations. Figure 6.13 shows the high-level process use in the development of human interactive systems that iterate on the design until the usability of the system meets the crew acceptance and capabilities operating the system while maximizing system performance and reducing risks and costs. HCD is an iterative activity that uses data gathered from the users of the system and evaluations to update the design. Key aspects of the human-centered effort are given below (ISO 9241-210 2019) and adapted for space applications:

- The design is based on an explicit understanding of the crew, the task, and the environments.
- The crew throughout the design and development process.
- The design and requirements are defined by the evaluation of the prototype system.
- The process is iterative.

Mission success is optimized when attention is paid to human interfaces that provide operational clarity and consistency and reduce the potential for human error, performance failure, or injury.

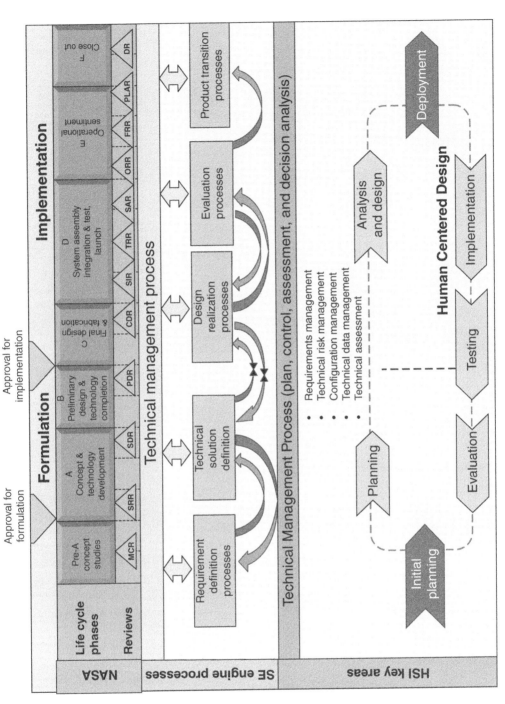

Figure 6.12 Notional holistic view of NASA HSI and the SE life cycle. *Source:* Adapted from NASA (2015b).

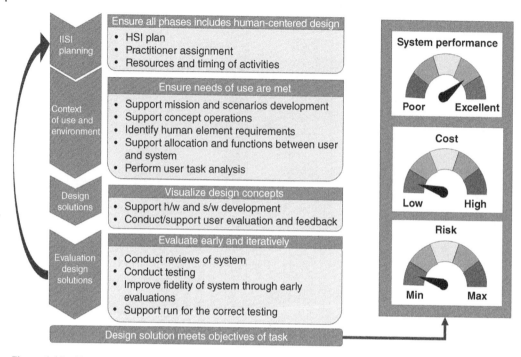

Figure 6.13 Human-centered design. *Source:* Adapted from NASA (2014) and Salazar and Russi-Vigoya (2019).

6.4 Mission Challenges

6.4.1 Innovation and Future Vehicle Designs Challenge

The space industry is growing, and as it grows and bold space vehicle concepts pursued, there will be more opportunities for future innovative vehicle designs. However, as vehicle complexity increases, the demand to add automation to help the crew control the vehicle will also increase. These vehicles will involve many different system/subsystem developers and the means of controlling them. Potentially, the variability of system/subsystem control could affect human performance. Crewmembers can be highly trained in one system, but dealing with controls/usage of different system interfaces and interactions may create performance issues. Autonomous system control must always provide the crewmember situational awareness of the system/subsystem state in the event autonomy override is necessary.

As noted, the vehicles and systems of the future will be much more complicated than NASA has ever built. Therefore, it is important to work closely with human factors practitioners on task analysis, human performance measures (workload, usability, and situation awareness), Human Factors Engineering (HFE) design (anthropometry and biomechanics, crew functions, habitat architecture), HITL evaluation, human error analysis, human system interface, system design, and HFE analysis. The simplicity of the design plays a very important role, along with usability, safety, and ways in which the system enhances the users (crew and ground) performance.

NASA has extensive experience with Low Earth Orbit (LEO) space psychology and the effects of microgravity and living/working in small spaces. However, interplanetary travel is a new area for NASA. Therefore, each of the HSI domains will also need to work with medical and design engineers to arrive at ways of helping to keep the crew psychologically healthy. Monitoring the crew and making sure the crew is physically and mentally healthy will be a challenge. Researchers and practitioners need to work together to identify stress-relieving systems. An example of such systems is XR technologies (i.e. virtual reality, augmented reality, hybrid realities, and physical realities) and biometrics relations to enhance human health and performance.

Deep space missions may have different effects on the human body that could aggravate the crew's health. For instance, radiation will be higher than on Earth. Therefore, the crew needs to be prepared to protect themselves to avoid cognitive and physical problems. The crew will also be working in hostile closed environments (i.e., controlled temperature, oxygen, pressure, lighting, noise, and space). Preparing for long-duration missions, the vehicles should account for environmental support to enhance the crew life and mission; the isolation and confinement of the crew could have behavioral issues, limited communication, circadian desynchronization, and changes on sleep schedules. Technologies must be able to support long-duration isolation and confinement to maintain crew mental and physical health. The gravitational field changes can create changes in bone densities and changes in the internal body fluid pressures. Long-duration missions have multiple gravitational targets. Future designs need to account for gravitation factors for job designs.

These interplanetary missions will have long communication delays, resulting in lesser communications with Earth only getting worse as the missions extend deeper into space. Innovations in higher communication data rates and experience and decision making aids to respond to anomalies or to manage resources will be very important. The complexity of these new missions demand that all of the HSI domains are considered in the design to ensure the vehicle and systems factors the crew health, safety, and mission performance.

6.4.2 Operations Challenges

Future deep space missions will need to have powerful considerations of the resources required for operations planning and execution. The crew will be isolated from all resources they can find in Earth and will be in a confined environment (spacecraft and suit) in space. Coming back from Mars is expected to take about three years. The crew will not be able to come back quickly to Earth to get what they need. There is a need to plan for the long term and plan for unexpected situations. If resources are not wisely used or if there are issues that limit the number of resources (e.g. anomalies that result in limited resources), the crew may need to be more frugal while using the resources. Operations process design for both ground and flight crew, human–machine resource allocation, mission operations, resource modeling, and complexity analysis. Vera et al. (2019) pointed out the importance of intelligent systems that facilitate crew decision making in future missions. These intelligent systems may also need to look at the onboard capabilities and provide the crew with simpler interfaces than the ones available for the ground operations. Also, these interfaces should help navigate the crew in an efficient and effective matter for troubleshooting and decision making.

The crew may need to be more empowered for flight operations and procedure development. The crew may have to make decisions on equipment without Mission Control (MC) support. It is possible that while learning about the systems, the crew will learn new ways to improve their work. Therefore, these smart systems should be flexible to give the crew the capability to improve the system and to recover from changes in the system that did not work. Also, these systems should be able to provide a data-driven means to inform decision making and performance risk mitigation on anomalies during an exploration mission.

6.4.3 Maintainability and Supportability Challenges

As NASA's missions extend beyond LEO, having access to resources for maintainability and support will place a burden on the crew as well as the mission. The distance of the missions and the time it will take to get additional resources to perform maintenance and support for operations will call for good logistics. Maintenance procedures need to be intuitive and simple to strategically utilize all resources (i.e., human resources, equipment, spares, system repurposing, and consumables). In-flight maintenance and housekeeping technologies are going to require greater utility and usability while minimizing system complexity. Ground maintenance and assembly should not be much different.

Traditionally, developers have only considered space missions as a weightless environment. However, future missions will have destinations where gravitational environments will change. Biomechanic theories may need to be evaluated for those environments and investigate the optimum interaction possibilities given the environment limitation changes. As we learn from new environments and challenges, these systems must be sustainable and able to adapt and enhance capabilities and support crew limitations.

Maintenance and supportability verification and validation are rarely done before certification of a vehicle/system. New ways of performing maintainability and supportability ground training using flight-like mock-ups to ensure that the vehicle/system was designed properly for the human to repair and maintain.

6.4.4 Habitability and Environment Challenges

Currently, NASA has identified multiple physical and virtual/augmented/mix reality developments that enhance human life in space. However, it is imperative to consider human capabilities and limitations with respect to the interface design of the space habitats, vehicles, and other systems and how these interactions and interfaces may need to change on long-duration journeys beyond LEO. Environmental factors such as changes in temperature, availability, changes in microgravity, acceleration, vibration, availability of air and quality of the air, sounds, and noise, radiation, isolation, and confinement characterize these missions. Therefore, the development of habitat is a very important factor that helps to determine mission success. It is expected that these factors are not going to get better as we go further and beyond into deep space missions. When these factors are not considered for future exploration, there is a risk of vehicle/habitat design that is incompatible with human capabilities and limitations. Future space missions shall have a habitat that is adaptable and modular. The habitats have to enable crew not only to survive but also to live and expand developments. Humans perform better when their environment allows them to perform better. Future missions may need to identify crew architecture and countermeasures that enhance their existence.

6.4.5 Safety Challenges

Safety is paramount for mission success. The usability of the system contributes to enhancing system safety. Other times, the usability of the system takes second place to enhance safety. For instance, looking at the spacesuit, you will note that user mobility is very restricted. In this case, the usability of the spacesuit is limited by the need to keep the crew safe. The spacesuit helps the astronaut to be protected from extreme changes of temperature, allows the astronaut to breathe as it provides oxygen, provides water to the astronaut, and protects the crew from getting hurt

(NASA, 2019). The suit design has improved, but it is clear that the design most important factor is safety. Future missions will have to think about deeper/more in-depth safety analysis. For instance, not only usability and utility may need to be improved but also system reliability for crew safety. For instance, if the suit rips open for any reason, it will put the crew at risk. It is important to identify mechanisms that identify system anomalies and rapidly recover from them. Future long-duration missions may have limited communication with the ground. So technical support may not be time available. Systems need to have reliability and quality. Perhaps, developing materials and systems that can regenerate will be important to support the crew. Factors of survivability for long-duration missions may change. There may be more research needed to enhance the capabilities of survivability and a need to evaluate if the systems are assessed with human analysis and hazard analysis in mind.

6.4.6   Training Challenges

Training systems will play a very important role. Researchers have shown that many capabilities on the ground will be to be transferred to onboard activities, especially to respond to anomalies. Vera et al. (2019) and Holden et al. (2020) stated the importance of advanced training systems, including the need to know the core vehicle/habitat systems and accurately be able to predict issues, training to anticipate issues, and training for maintenance and support. The researchers also noted the importance of refresher and Just-In-Time (JIT) training for long-duration missions. Many times crews have come back from their mission and stated that they were not trained on activities that they did not experience often. After three years on an isolated mission, they should be able to retrain or train to resolve anomalies. Vera et al. (2019) identified that MC currently has many capabilities and training that are reinforced to enhance their decisions. Their research showed that training on troubleshooting and problem solving might need to be on board to facilitate operations when there is no communication with MC. In addition, there is maybe the need for high-fidelity systems that facilitate learning stress-inducing simulation training in analogs.

Additionally, crew time and staffing/qualifications analysis may not be able to be done by MC on time. Future missions need to aid the crew to optimize their schedule and support new capabilities.

6.5   Conclusions

This chapter summarizes the space history of the evolution of spacecraft system design. It briefly details the HSI process and its domains. All vehicles that NASA has developed for human space flight (i.e. Mercury, Gemini, Apollo, Shuttle, and ISS) have presented different challenges and learning opportunities. As we prepare for future long-duration missions, new challenges will have to be addressed to satisfy the mission requirements. Going back to the moon and interplanetary missions will bring new challenges that could affect human capabilities and limitations. Safety and crew health is paramount in all space vehicles. The crew and the ground will be supported for operations with an autonomous mission and new ways to address the mission challenges. It is expected that future spacecraft will have smarter systems to support autonomous missions. Training methodologies and strategies will need to satisfy new mission requirements. All HSI domains will need to address future challenges supporting the crew. This chapter describes the importance of HSI, given the different mission challenges that future long-duration missions will face. Each of the domains defines some of the challenges.

References

Holden, K., Russi-Vigoya, M.N., Adelstein, B., and Munson, B. (2019). *Human Capabilities Assessment for Autonomous Missions (HCAAM). Phase I: Human Performance Standards and Guidelines* (February). Internal NASA Report Prepared for the Human Factors and Behavioral Performance Element of the Human Research Program.

Holden, K., Munson, B., Russi-Vigoya, M.N. et al. (2020). *Human Capabilities Assessment for Autonomous Missions (HCAAM). Phase II: Development and Validation of an Autonomous Operations Task List* (January). Internal NASA Report Prepared for the Human Factors and Behavioral Performance Element of the Human Research Program.

International Organization for Standardization (ISO) 9241-210 (2019). *Ergonomics of Human-system Interaction, Part 210 Human-centered Design for Interactive Systems (ISO 9241-210)*.

McDonnell Company (1965). *NASA Project Gemini Supplement Familiarization Manual* (Control no. 119162). McDonnell Company https://www.ibiblio.org/apollo/Documents/GeminiManualVol1Sec2.pdf (accessed 28 August 2020).

National Aeronautics and Space Administration (NASA) (2010). *Human Systems Integration Requirements Revision C (HSIR Revision E)* (November). NASA-CXP70024. Houston, TX: Lyndon B. Johnson Space Center https://history.nasa.gov/SP-407/part3.htm (accessed 28 August 2020).

National Aeronautics and Space Administration (NASA) (2013a). *NASA Procedural Requirements (NPR) 7123.1B (2013). NASA Systems Engineering Processes and Requirements* (April). https://nodis3.gsfc.nasa.gov/displayDir.cfm?t=NPR&c=7123&s=1B (accessed 28 August 2020).

National Aeronautics and Space Administration (NASA) (2013b). *What Was the Space Shuttle?* (November). https://www.nasa.gov/audience/forstudents/k-4/stories/nasa-knows/what-is-the-space-shuttle-k4.html

National Aeronautics and Space Administration (NASA) (2014). *Human Integration Design Processes (HIDP)* (NASA/TP-2014-218556) (September).

National Aeronautics and Space Administration (NASA) (2015a). *Human Research Program Integrated Research Plan* (Plan: HRP-47065) (March).

National Aeronautics and Space Administration (NASA) (2015b). *Human Systems Integration (HSI) Practitioner's Guide* (Report No.: NASA/SP–2015-3709) (November). https://ntrs.nasa.gov/archive/nasa/casi.ntrs.nasa.gov/20150022283.pdf (accessed 28 August 2020).

National Aeronautics and Space Administration (NASA) (2019). *What Is a Spacesuit?* (May). https://www.nasa.gov/audience/forstudents/k-4/stories/nasa-knows/what-is-a-spacesuit-k4.html (accessed 28 August 2020).

North American Rockwell Corp (1968). *Apollo Spacecraft* (September). North American Rockwell Corp.

Salazar, G. and Russi-Vigoya, M.N. (2019). Human systems integration relationships between deep space and deepwater exploration challenges. *Off-Shore Technology Conference 2020*, Houston, TX.

Swenson, L.M., Grimwood, J.M., and Alexendar, C.C. (1966). *The New Ocean-The History of the Mercury Program*, The NASA Historical Series. Washington, DC: National Aeronautics and Space Administration (NASA).

Vera, A., Holden, K., Dempsey, D. et al. (2019). *Contextual Inquiries and Interviews to Support Crew Autonomous Operations in Future Deep Space Missions: Preliminary Requirements and Proposed Future Research* (February). Internal NASA Report Prepared for the Human Factors and Behavioral Performance Element of the Human Research Program.

Wu, S.C. and Vera, A.H. (2019). *Supporting Crew Autonomy in Deep Space Exploration: Preliminary Onboard Capability Requirements and Proposed Research Questions* (Report No.: NASA/TM-2019-22034). https://ntrs.nasa.gov/archive/nasa/casi.ntrs.nasa.gov/20190032086.pdf (accessed 28 August 2020).

7

Aerospace Human Systems Integration

Evolution over the Last 40 Years

Guy André Boy

CentraleSupélec, Paris Saclay University, Gif-sur-Yvette, and ESTIA Institute of Technology, Bidart, France

7.1 Introduction

For the last forty years, aerospace systems evolved tremendously, mainly due to constant increasing automation, improvement of design, and development methods and tools, and most importantly under the constant search for more safety, efficiency, and usability. Air traffic management (ATM) is a matter of system of systems (SoS), which requires a more solid systemic approach where technology, organizations, and people are properly considered in an integrated framework. The search for such an approach is the main objective of this chapter. What are the purposeful attributes of ATM systems? What are purposeful nominal and off-nominal contexts of ATM systems? What are ATM disruptive factors? What are ATM intrinsic and extrinsic factors? What is the best ATM topological model that can support analysis, design, and evaluation of the growing real ATM system? What are the new ATM metrics in terms of operational performance, decision making, trust, and collaboration, for example? What is the role of human-in-the-loop simulation (HITLS) in SoS design?

This chapter first presents the evolution of aviation and the difficult issue of separability in the overall aeronautical community. Section 7.2 then discusses how software took the lead on hardware during the last fifty years or so, introducing a drastic shift from automation to autonomy in the aerospace domain. In addition to human models and related human factors approaches, human roles drastically changed in the real world. Human models shifted from single-agent to multi-agent representations and related approaches. This evolution contributed to the emergence of new disciplines (Section 7.3). Digitalization of industrial processes during the whole life cycle of products brought forward tangibility issues (Section 7.4). The consideration of unexpected situations urges us to augment, and in some cases replace, rigid automation produced during the twentieth century by flexible autonomy (Section 7.5). All these observations lead to the development of an appropriate human-centered systemic framework in the form of a conceptual framework, which will be useful for human systems integration (HSI) (Section 7.6). The conclusion will emphasize future endeavors of HSI as a generic process in our increasingly digitized society.

A Framework of Human Systems Engineering: Applications and Case Studies, First Edition.
Edited by Holly A. H. Handley and Andreas Tolk.
© 2021 The Institute of Electrical and Electronics Engineers, Inc. Published 2021 by John Wiley & Sons, Inc.

7.2 Evolution of Aviation: A Human Systems Integration Perspective

At beginning of aviation, air traffic controllers (ATCOs) had to guess both current and future aircraft positions in order to reduce uncertainty (Figure 7.1).

During the second phase, based on the use of radar technology, ATCOs know aircraft positions but still have to guess future aircraft positions. We remain in this era despite saturation that in bustling airports demands shifting to the next phase, which is trajectory management, also called trajectory-based operations (TBO). In TBO, ATCOs know both current and future aircraft positions, which is intended to reduce uncertainty considerably. While this looks great, a new problem emerges from the implementation of the TBO solution, which is necessary planning (e.g. 4D trajectories). Planning involves rigidity. Whenever rule-bound procedures are applicable, everything runs smoothly, but when unexpected situations occur, air traffic managers require flexibility. Rigidity and flexibility are contradictory concepts! This is the reason we now need to think in terms of flexible trajectory planning.

Another observation worth noting is that even if pilots and ATCOs are interacting at operations time, aircraft and ATC manufacturers and suppliers are very different institutions that rarely talk to each other. By contrast, human-centered design (HCD) requires participatory design of complex systems (Boy 2013). The space shuttle design and development is a good example of such an HCD approach where onboard and ground systems were designed and developed in concert. TBO requires such a multi-agent design approach. Methods were developed to this end (Boy 1998, 2011), and the TOP model should be used toward the best articulation of technology, organizations, and people in the HCD of the overall aeronautical system (Figure 7.2).

ATM complexity and automation are two major concepts in contemporary aviation. Aeronautical automation effectively started in commercial aviation in the 1930s (e.g. the Boeing 247 commercial aircraft flew with an autopilot in 1933). Aeronautics has then a long experience in automation. However, a big jump happened during the 1980s when glass cockpits started to be developed. This kind of digital automation drastically changed the way pilots interacted with their aircraft (Wiener 1989). Technologically speaking, we moved from analog mechanical instruments to digital displays (e.g. primary flight display and navigation display) and controls (e.g. sidesticks and auto-throttle). If the number of instruments in the cockpit drastically decreased, the quantity of information exponentially increased. A typical question is: how could such information be organized within the limited space provided by cockpit screens? Substantial research efforts have been carried out to answer this question (Boy 1995; Doyon-Poulin et al. 2012; Letondal et al. 2018). More generally, complexity exponentially increased (i.e. the number of processed parameters and

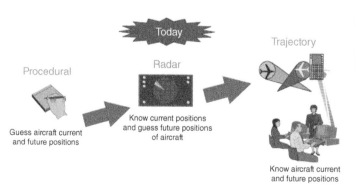

Figure 7.1 ATC-to-ATM evolution from procedural control to trajectory management.

related systems increased), shifting from mechanical to digital complexity.

These digital systems were developed to take care of handling qualities, navigation, and many other mechanical things. Systems invaded aircraft to the point that pilots ended up with an system of systems (SoS), not externally yet, but internally to the cockpit. Such a digitalization induced incremental production of layers and layers of electronics and software, added on top of each other, and resulting in the deeper separation of pilots from the real physical world. For example, two decades ago, we started to talk about "interactive" cockpits, not because pilots

Figure 7.2 The TOP model. *Source:* Boy (2020). © 2020, Taylor & Francis.

interacted mechanically with engines, flaps, and slats as they used to do since the beginning of aviation, but because they are now interacting with a pointing device on cockpit screens. We are talking about human–computer interaction (HCI) in the cockpit (Boy 1993).

In addition to cockpit automation (the technological side), the number of technical aircrew members in transoceanic flights was reduced over the last sixty years or so: five until the 1950s when the radio navigator was removed (the radio navigator was dedicated to voice communication equipment), four until the 1970s when the navigator was removed (when inertial navigation systems were introduced), three until the 1980s when the flight engineer was removed (new monitoring equipment for engines and aircraft systems were introduced), and two to date. The next change will shortly happen if the single-pilot operations (SPO) goal is reached. Reducing crews involves organizational changes that need to be seriously considered, again in terms of safety, efficiency, and comfort. At this point, it is crucial to understand whether such changes are evolutionary or revolutionary. Circa the end of the 1980s, we became conscious of this sociotechnical distinction when the Airbus 320 was certified. The A320 was highly automated compared with other commercial aircrafts of its category. Many experts thought that it was easier to fly. Some others claimed that such automation was dangerous. We did not realize at that time that even if we thought that automation was developed incrementally, almost linearly (we thought!), the nature of pilot's job radically changed from control to management (i.e. from control of aircraft trajectory to management of systems, which were controlling the trajectory). Once this job shift was understood, everything went right. This is the reason why systems engineering (SE) requires taking into account HSI since the beginning of the design and development process.

Historically, the design of a system was done in silos, and, in many cases, systems were only connected just before operations, which is not a problem when subsystems in the system are separable (Figure 7.3). Clumsy integration, often done too late in the development process, is likely to cause surprises and, sometimes, a few catastrophes. This is the reason why adjustments are always required, either operationally via adapted procedures and/or interfaces, and in the worst case more drastic redesign of the system itself. The separability concept has been used for a long time by physiologists to denote the possibility of separating an organ from the human body to work on it separately and put it back. Some organs (i.e. systems) are separable, that is, the overall body (i.e. an SoS) does not die from this momentary separation. Some other organs, such as the brain, cannot be separated because the human being could die from this separation. Therefore, those organs have to be investigated and treated while connected to the rest of the body.

In addition, twentieth-century engineering involved technicians who were, and still are, working in isolation and focused on a specific field or discipline. They were, and still are, barely aware of the integration of the overall complex SoS, i.e. "the whole picture" that they build. HSI has to

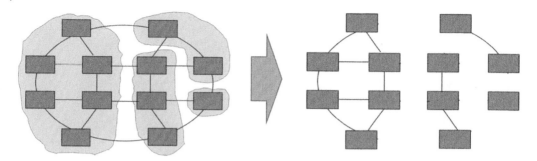

Figure 7.3 Example of separability property of a system of systems (SoS). Four sub-SoSs of the SoS are separable and therefore can be analyzed, designed, and evaluated in isolation. *Source:* Boy (2020). © 2020, Taylor & Francis.

consider interconnected SoSs, where systems include people and machines. This kind of requirement does not fit the current urgency of fast market economy. . . or fast anything for that matter! It takes time to get complex systems working well and maturing from three points of view: technology, people's experience, and society. Anticipation, which involves creativity, and HITLS using appropriate scenarios enable to explore and test possible futures.

The lack of consideration for the separability issue in ATM is a good example, where, for a long time, most air and ground technologies have been designed and developed in isolation. Recent programs, such as Single European Sky ATM Research (SESAR), try to associate air and ground stakeholders. The TOP model is a good framework for design and development teams to understand and rationalize interdependencies of technology, organizations, and people. This objective requires that various activities be observed and analyzed using modeling and simulation, in order to discover emerging properties and functions of the airspace technologies under development, and not await the discovery of these emerging properties and functions at operations time.

7.3 Evolution with Respect to Models, Human Roles, and Disciplines

7.3.1 From Single-Agent Interaction to Multi-agent Integration

For a long time in aeronautics, we were essentially centered on single-agent interaction with cockpit instruments and controls in the cockpit and with traffic displays and radio on the ground. Focusing on cockpits, for several decades of the twentieth century, electrical engineering, computer science, and information technology incrementally penetrated commercial aircraft. Many kinds of embedded systems (ESs) were developed in four steps that Captain Etienne Tarnowski called "the four loops of automation" (Figure 7.4).

Everything started with automation around the center of gravity using yoke or sidestick and thrust levers. The first loop consisted in a single agent, the autopilot, regulating parameters, such as speed and heading, one parameter at a time. Time constant of the feedback is around 500 milliseconds. Pilots had to adapt to this control loop by changing from the control of flight parameters to supervising the behavior of flight control with respect to a set point.

The guidance loop was developed circa the early 1980s. This second feedback loop involves several parameters. Its time constant is around 15 seconds. Note that this feedback loop has been typically implemented on top of the flight control loop. High-level modes of automation appeared

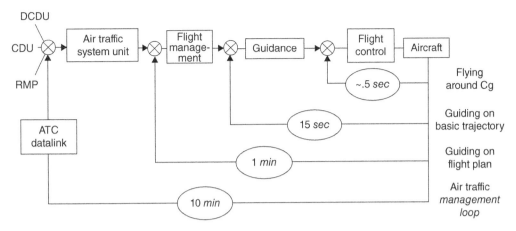

Figure 7.4 The four loops of automation of the airspace. *Source:* Tarnowski (2006).

and were managed on the flight control unit panel. At the same time, integrated and digital autopilot and auto-throttle were installed.

The third loop concerns navigation automation with a time constant of about one minute. Guidance and flight management was integrated circa mid-1980s. This was the first real revolution in the development of avionics systems. We were shifting from control of flight parameters to management of avionics systems, which grew exponentially as software became dominant. For that matter, pilots now have to deal with a variety of avionics systems that qualify as software-based agents, which generate problems as they become more interconnected.

Technology continued to improve and aircraft cockpits became more computerized. As already described elsewhere (Boy 1998, 2011), commercial aircrafts were equipped with autopilots for a long time (since the 1930s), but what drastically changed is the amount of software in avionics systems. The notion of systems quickly became persistent, and pilots' work radically changed from handling flight qualities (manual control) to aircraft systems management. A pilot's role shifted from control to management, exactly like when someone becomes a manager in an organization and has a team of agents to manage. In this case, pilots had to learn how to manage very advanced systems and coordinate their activities. It was not obvious when suddenly a pilot had to become a manager (of systems). This new emerging cognitive function (i.e. systems management) had to be learned and stabilized.

The fourth loop concerns ATM with a time constant of about 10 minutes. We now deal with airspace "automation." This is about air-ground integration. This is a new revolution, where new considerations, such as authority sharing, air traffic complexity management (Hilburn 2004), and organizational automation, have become essential (Boy and Grote 2009). Note that this ATM loop is multidimensional and multi-agent.

7.3.2 Systems Management and Authority Sharing

There was a big controversy during the late 1980s when the first highly automated glass cockpits of commercial aircraft were delivered and used (i.e. integration of the first three loops described above). This controversy started with social issues in the beginning of the 1980s because the commercial aircraft industry went from three-crewmen cockpits to two-crewmen cockpits, and downsizing the technical crew was not universally accepted. The role that was previously performed by a flight engineer was shared among the captain, first officer, and avionics systems. Function allocation was at stake and we had to find out how to certify these new cockpits. Therefore, we developed human factors methods that enabled the evaluation of aircrew workload and performance, comparing various types of configurations. We needed to demonstrate that a forward-facing cockpit with two crewmembers was as safe as the previous types of cockpits.

At the same time, the number of aircrafts grew exponentially and induced new issues related to traffic density and airspace capacity. New systems came on board, such as the traffic alert and collision avoidance system (TCAS), which provided a new kind of information to the pilot. Not only the TCAS provided traffic alerts, but also advice to climb or descend to avoid a converging aircraft. Not only were the collision avoidance orders from ATCOs replaced by the TCAS, but also these orders were, and remain, on board and given by a machine. Authority shifted from the ground to the aircraft and from humans to machines. We can see here that technology has resulted in the emergence of a different practice related to a different authority allocation.

The PAUSA[1] project investigated practical issues in authority sharing in the airspace (Boy and Grote 2009; Boy et al.,2008). In this project, we extensively analyzed the 2002 Überlingen mid-air collision due to a wrong TCAS usage. TCAS introduced a gradual shift from central control to decentralized self-organization (Weyer 2006). The PAUSA approach to authority was grounded in a multi-agent approach that enables the expression of function distribution, the notion of a common frame of reference, task delegation, and information flows among agents. The metaphor of the shift from the army to an orchestra (Boy 2009), which represents the ongoing evolution of the airspace multi-agent system, emerged from multiple experience-based investigations within PAUSA.

7.3.3 Human-Centered Disciplines Involved

At this point, it is important to clarify the relationship between the task/activity distinction and distinctions between sociotechnical disciplines. More specifically, we need to define the following concepts: task, activity, human factors, ergonomics, HCI, HCD, and HSI. A task is what is prescribed to human operators or users. An activity is what is effectively performed by human operators or users.

For the last sixty years, sociotechnical evolution can be decomposed into three eras in which three communities[2] emerged (Figure 7.5):

- HFE was developed after World War II to correct engineering production and generated the concepts of human–machine interfaces or user interfaces (UI) and operational procedures; activity-based evaluation could not be holistically performed before products were finished or almost finished, which enormously handicapped possibilities of redesign. Sometimes, activity analyses were carried out prior to designing a new product, based on existing technology and practice; however, this HFE approach forced continuity, reduced risk taking, and most of the time prevented disruptive innovation.

---

1 PAUSA is a French acronym for "authority distribution in the aeronautical system." This national project was sponsored by the French Aviation Administration (DPAC) and the aeronautical industry. Nine organizations participated in this project. This work was carried out when the author was the director of the PAUSA project at the European Institute of Cognitive Sciences and Engineering (EURISCO).

2 The author is qualified to talk about these three communities. He is still the chair of the Aerospace Technical Committee of the International Ergonomics Association (IEA), which encapsulates most HFE societies around the world. From 1995 to 1999, he was the executive vice-chair of the Association for Computing Machinery (ACM) Special Interest Group on Computer–Human Interaction (SIGCHI) and senior member of the ACM. He is currently co-chair of the Human Systems Integration Working Group of the International Council on Systems Engineering (INCOSE).

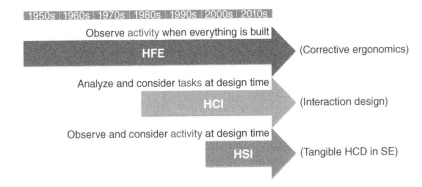

Figure 7.5 Human-centered design evolution.

- HCI was developed during the 1980s to better understand and master human interaction with computers; it contributed to the shift from corrective ergonomics to interaction design mainly based on task analysis. Activity-based analysis was introduced within the HCI community by people who understood phenomenology (Winograd and Flores 1986) and activity theory (Kaptelinin and Nardi 2006).
- HSI emerged from the need to officially consider human possibilities and necessities as variables in SE; incrementally combined, SE and HCD lead to HSI to take care of systems during their whole life cycle (Boy and Narkevicius 2013). HCD involves looking beyond human factors evaluations and task analyses. It involves activity analysis at design time using virtual prototyping and HITLS (e.g. we can model and simulate an entire aircraft, fly it as a computing game, and observe pilot's activity). HCD also involves creativity, systems thinking, risk taking, prototype development using agile approaches, complexity analyses, organizational design, and management, as well as HSI knowledge and skills.

7.3.4 From Automation Issues to Tangibility Issues

During the 1980s and 1990s, automation drawbacks emerged from several HFE studies, such as "ironies of automation" (Bainbridge 1983), "clumsy automation" (Wiener 1989), and "automation surprises" (Sarter et al. 1997). These studies considered neither technology maturity nor maturity of practice. Automation can be modeled as cognitive function transfer from people to systems (Boy 1998). If automation considerably reduced people's burdens, it also caused problems such as complacency, which is an emerging cognitive function (i.e. not predictable at design time, but at operations time).

Today, we can develop an entire aircraft on computers from inception of design to finished product. Therefore, we are able to test its operability from the very beginning and along its life cycle using HITLS and an agile approach. Consequently, we are able to observe, and therefore analyze, human activity using systems in a simulated environment at design time. The operability of complex systems can then be tested during design. This is the reason why HCD has become a discipline in its own right. HCD enables us to better understand HSI during the design process and then have an impact on requirements before complex systems are concretely developed.

HITLS enables activity analysis at design time, albeit in virtual environments. Even if these environments are very close to the real world, their tangibility must be questioned and most importantly validated.

What is tangibility? It has two meanings. First, physical tangibility is the property of an object that is physically graspable (i.e. you can touch it, hold it, sense it, and so on). Second, figurative tangibility is the property of a concept that is cognitively graspable (i.e. you can understand it, appropriate it, feel it, and so on). If I try to win an argument with you, you may argue back, "what you are telling me is not tangible!" This means that you do not believe me; you cannot grasp the concept I am trying to provide. We also may say that you do not have the right mental model to understand it or that I do not have enough empathy to deliver the message appropriately.

Tangibility is about situational awareness both physically and cognitively. For example, Tan developed the first versions of the Onboard Context-Sensitive Information System (OCSIS) for airline pilots on a tablet PC (Tan 2015). Physical tangibility considerations led to a better understanding of whether OCSIS should be handheld or fixed in the cockpit. Other considerations led to the choice of figurative displays of weather visualization going from vertical cylinders to more realistic cloud representations (figurative tangibility). A set of pilots gave their opinions on various kinds of OCSIS tablet configurations. It is interesting to note that the pilots always naturally used the term "tangible" to express their opinions.

Therefore, tangibility metrics should be developed to improve the assessment of complex systems operability. This is where subject matter experts and experienced people enter into play. We absolutely need such people in HCD to help assess HSI tangibility. For example, very realistic commercial aircraft cockpits, professional pilots, and realistic scenarios are mandatory to assess tangibility incrementally. OCSIS was tested from the early stages of the design process using HITLS by recording what pilots were doing while using it and analyzing their activity. Such formative evaluations lead to system modifications and improvements. HCD is iterative and agile[3] in the SE sense.

While the twenty-first-century shift from software to hardware is not necessarily straightforward, it is the next dilemma we must address, especially now that we can 3D print virtual systems and transform them into physical systems. We will denote resulting systems, tangible interactive systems (TISs) (Boy 2016). The TIS concept is very close to the cyber–physical systems (CPSs) concept, which are usually defined as a set of collaborative elements controlling physical entities (Lee 2008). CPSs are often qualified as ESs, but there is a distinction between ESs as purely computational elements and CPSs as computational and physical elements intimately linked (Wolf 2014). We can say that CPSs are extended avionics systems. Both TISs and CPSs are strongly based on the multi-agent concept, unlike twentieth-century automation that was based on the single-agent concept. TISs cannot be considered without an organizational approach. More generally, the coevolution of people's activities and technology necessarily lead to a tangible organizational evolution.

7.4 From Rigid Automation to Flexible Autonomy

When we know the domain well, both technically and operationally, it is possible to define appropriate procedures that can either be implemented as computer programs (i.e. automation of machines) or followed by human operators (i.e. automation of people). Figure 7.6 shows how, in a

---

3 The Manifesto for Agile Software Development (www.agilemanifesto.org) has been written to improve the development of software. It values individuals and interactions over processes and tools, working software over comprehensive documentation, customer collaboration over contract negotiation, and responding to change (flexibility) over following a plan (rigidity).

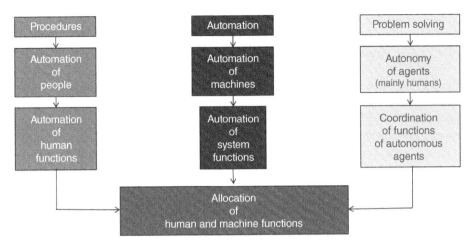

Figure 7.6 Procedures, automation, and problem solving leading to the allocation of human and machine functions. *Source:* Boy (2020). © 2020, Taylor & Francis.

very well-known and/or expected validity space, procedures and automation lead to automation of, respectively, human and machine functions. Everything goes fine within the validity space (i.e. in nominal situations). However, outside the validity space, the rigidity of both procedures and automation rapidly leads to instability.

In off-nominal situations (i.e. unexpected, unknown, abnormal, or emergency situations), people need to solve problems (Pinet 2015). Problem solving is a matter of knowledge and know-how. The more they have such knowledge and know-how, the more people are autonomous. They also need to have appropriate technological and/or organizational support. Altogether, autonomy is a matter of appropriate technological support enabling flexibility, coordinated organizational support, and people's knowledge and know-how. Off-nominal situations management involves functions of autonomous human and machine agents that need to be coordinated.

In all cases, functions from automated or autonomous agents need to be correctly allocated to the right agents whether they are people or machines. Such allocation cannot be done entirely statically, but dynamically with the evolution of context.

A fully autonomous agent, whether human or machine, does not exist, or may exist in very limited contexts. Indeed, no agent can always be aware of the overall situation, nor is always capable of making the right decision at the right time, nor act appropriately in the right context. The reason is that there are too many parameters entering into play. Flexible autonomy of an agent will have to be investigated using the following claims:

- An agent is defined as a society of agents (Minsky 1985).
- Each agent owns appropriate knowledge and processing capabilities, in terms of function(s) and structure(s), to perform a given task.
- The more agents become autonomous in a society of agents, the more coordination rules are needed to keep the stability and sustainability of the overall system.

In other words, agents become increasingly autonomous once they have acquired appropriate knowledge and improved capacities of coordinating with others. The concept of "appropriate knowledge" can be defined as knowledge suitable to have in a given context. The concept of "coordination" can be defined as both prescribed and effective coordination in a given context.

Prescribed coordination is a matter of coordinated tasks (i.e. like symphony scores produced by the composer to coordinate the various musical instruments). Effective coordination is a matter of coordinated activity (i.e. a conductor coordinating the various musicians of the orchestra).

Aerospace experience shows the need for deeper support in terms of systemic framework. Over the years, we accumulated ESs, now extended to CPSs, Internet of Things, and other systems. Even if each of these systems can be useful and usable, when they are put together and operated by people, they can become extremely difficult to manage. In critical situations, operational complexity, situation awareness, and workload are directly impacted. Consequently, there is a need for improving HSI. This need leads to a more fundamental requirement, that is, defining what a system is really about.

7.5 How Software Took the Lead on Hardware

During most parts of the twentieth century, hardware and mechanical machines were primary engineering concerns. We built washing machines, cars, aircraft, and industrial plants using mechanical engineering methods and tools. At the end of the twentieth century, we started to introduce electronics and software into these machines, to the point that we manage to shift practices from mechanical manipulation to HCI. Computers invaded our lives as mediators between people and mechanical systems. The computer-based UI became a primary issue and solution everywhere in HCI. In aeronautics, we ended up with the concept of interactive cockpits, not because pilots interacted with mechanical things, but because they now interact using pointing devices on computer displays. HFE combined with HCI and cognitive science, as a new discipline often called cognitive engineering, supported machine developments by studying, designing, and evaluating UIs, added after a mechanical machine was developed. Software contributed to adding automated functions to machines, with these functions offering new capabilities to people, who had to learn how to interact with automation safely, efficiently, and comfortably. In addition, automation came in the form of incrementally added layers of software that increased situation awareness issues (Endsley 1996).

It is interesting to notice that, since the beginning of the twenty-first century, most projects start on computers, with a PowerPoint slide deck, a computing model and visualization, a simulation, and in some cases a computer-based HITLS. For example, the Falcon 7X, developed by Dassault, was entirely built as a giant interconnected piece of software that led to a sophisticated computer game flown by test pilots. For the time, we were able to observe end user activity at design time, even before any piece of hardware was developed. Once such software models and simulation are tested, they enable design teams to develop appropriate requirements for making hardware parts and, in many cases now, 3D printed software models. We have moved from software to hardware, inverting the twentieth century's approach that moved from hardware to software. As a result, automation is less of an issue as it was before because HITLS enables testing functional safety, efficiency, and usability since the beginning of the design process. The problem is then to better understand the tangibility issue induced by the shift from software models and simulations to the concrete structural world. Tangibility is not only a matter of physics, but it is also a matter of intersubjectivity (i.e. mutual understanding) between end users and designers. In other words, end users should be able to understand what machines are doing at appropriate levels of granularity. This is the reason why complexity analysis has become tremendously important in our increasingly interconnected world.

System knowledge, design flexibility, and resource commitments are three parameters that should be followed carefully during the whole life cycle of a system. HSI aims to increase the following sufficiently early (Boy 2020):

- System knowledge that is knowing about systems at design, development, operations, and closeout times and how the overall system, including people and machines, works and behaves.
- Design flexibility that is keeping enough flexibility for systems changes later in development and usages.
- Resource commitments that is keeping enough "money" for choosing adapted resource management during the whole life cycle of the overall system.

When a technology-centered approach is used (typically what we have done up to now), system knowledge increases slowly in the beginning, growing faster toward the end of the cycle. Design flexibility drops very rapidly, leaving very few alternatives for changes, because resource commitments were too drastic too early during design and development processes.

Instead of developing technology first, software models are used for the development and use of HITLS, which enables activity observation and analysis, and therefore the discovery of emergent properties and functions, which in turn can be considered incrementally in design. This is a typical HCD process. Consequently, system knowledge increases more rapidly during such an agile design and development process. At the same time, design flexibility drops much more slowly, with an inverted concavity, enabling possible changes later in the life cycle. Software-based modeling and HITLS enable testing various kinds of configurations and scenarios, enabling softer resource commitments in the beginning and leaving more comfortable space for appropriate changes.

At this point, we see the need for modeling and simulation for HCD and consequently HSI. If twentieth-century automation, that is, putting software into hardware or transferring human functional knowledge into a machine, brought human functional issues, twenty-first-century tangibilization, that is, transforming software models into real-world tangible (concrete) human machine systems, is currently raising architectural issues. Before we provide a topological approach to HSI in Section 7.4, let us present and discuss the shift from rigid automation to flexible autonomy.

7.6 Toward a Human-Centered Systemic Framework

7.6.1 System of Systems, Physical and Cognitive Structures and Functions

Historically, engineers used to think about a system as an isolated system, or a quasi-isolated system, which has an input and produces an output (Figure 7.7). Comparatively an agent, in the artificial intelligence (AI) sense, has sensors and actuators, and a system, in the SE sense, has sensors to acquire an input and actuators to produce an output.

Today, systems are highly interconnected, and we talk about a system as an SoS, which means that a system can be represented as an organization of other systems. More generally, a system belongs to a bigger system and is interconnected with other systems (Figure 7.8). The same holds for agents in AI, which can be organized within a society of agents (Minsky 1985).

Following up on the separability concept for an SoS, it is now crucial to have a clear definition of what the system means. A system is a representation of:

1) A human or more generally a natural entity (e.g. a bird, a plant).

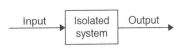

Figure 7.7 An isolated system.

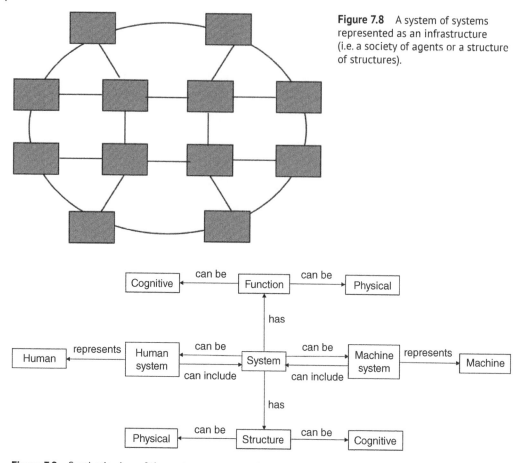

Figure 7.8 A system of systems represented as an infrastructure (i.e. a society of agents or a structure of structures).

Figure 7.9 Synthetic view of the system representation.

2) An organization or a social abstraction (e.g. a team, a community, a law, a legally defined country, a method).
3) A machine or a technological entity (e.g. a car, a motorway, a washing machine, a chair).

Consequently, this system definition breaks the traditional meaning of system, conceived as a machine only, but instead encapsulating humans, organizations, and machines.[4] Figure 7.9 presents a simple ontological definition of the "system" conceptual representation.

A system can be either cognitive (or conceptual), physical, or both (Boy 2017). It also has at least one structure and one function. Today, machines have software-supported cognitive functions (e.g. the cruise control function on a car enables the car to maintain a set speed). In practice, a system has several structures, and several functions articulated within structures of structures and functions of functions. It is interesting to recall the analog definition of an agent in AI provided by Russell and Norvig (2010), which is an architecture (i.e. structure) and a program (i.e. function).

---

4 This systemic view takes Herbert Simon's view of the Sciences of the Artificial (Simon, 1996), in the sense that he rejected treating human sciences using the exclusive model of the natural sciences (i.e. submission to natural laws) and to break between science and humanities by looking for a common ground that links them. The science of the artificial seeks new constructs that would explain things, which were not previously understood. These artificial constructs could be a language, an ontology, a conceptual model, or any kind of representation that makes sense.

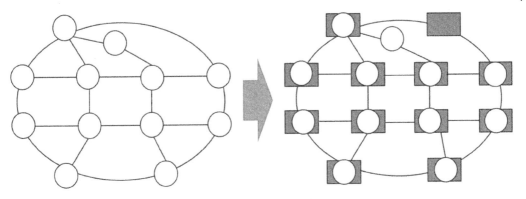

Figure 7.10   A function of functions mapped onto a structure of structures.

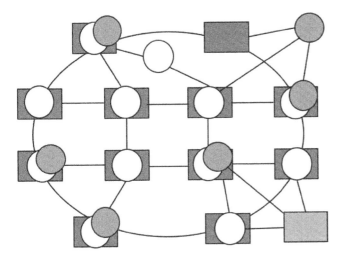

Figure 7.11   Emerging functions (yellow) and structures (pink) within an active system of systems.

Each system is interconnected to other systems either structurally (in terms of a systemic infrastructure) or functionally (in terms of functions appropriately allocated to systems). Summarizing, a system, as an SoS, is represented by an infrastructure where a network of functions could be dynamically allocated (i.e. a function of functions mapped onto the structure of structures; Figure 7.10).

7.6.2   Emergent Behaviors and Properties

At this point, a distinction should be made between deliberately established functions allocated onto an infrastructure and functions that necessarily emerge from system activity. Indeed, systems within a bigger system (i.e. an SoS) interact with each other to generate an activity. von Bertalanffy (1968)[5] said "a system is a set of elements in interaction." Emerging functions are discovered from such activity (Figure 7.11). The integration of such emergent functions into the SoS may lead to the generation of additional structures, which we also call emerging structures.

---

5 https://www.sebokwiki.org/wiki/What_is_a_System%3F

7.6.3 System of Systems Properties

The system's purpose is logically defined by its task space (i.e. all tasks the system can perform successfully). Each task is performed by the system using a specific function that produces an activity that can be fully or partially observable (Figure 7.12).

A system's function is teleologically defined by three entities:

1) Its role within the related system.
2) Its context of validity that frames the boundaries of the system's performance.
3) Its set of resources required to perform its role within its context of validity. Resources are systems themselves that have their own cognitive and/or physical functions.

Therefore, according to these definitions, a system can be represented by the recursive schema presented in Figure 7.13.

Let us consider a postman represented as a system (or an agent in the AI sense) with the function of delivering letters. The postman as a system is part of an SoS, which is the postal services. The role of this system is "delivering letters." The context of validity is, for example, seven hours a day five days a week (i.e. a time-wise context in France, for example) and a given neighborhood (i.e. space-wise context). Resources can be physical (e.g. a bicycle and a big bag) and cognitive (e.g. a pattern-matching algorithm that enables the postman to match the name of the street, the number on the door, and the name of the recipient). The corresponding pattern-matching algorithm is a cognitive function. Let us consider now that there is a strike and most postmen are no longer available for delivering letters. Remaining postmen should have longer hours of work in more significant neighborhood until this expansion is so extreme that the postmen need helpers to achieve the delivery task successfully.

In this case, a tenure postman should have cognitive resources such as "training," "supervising," and "assessing" temporary personnel. We see that the cognitive function of "delivering letters" owned by a postman (i.e. an agent or a system) has to be decomposed into several other functions allocated to temporary postmen. We start to see an organization developed as an answer to a strike. More generally, a function of functions can be distributed among a structure of structures.

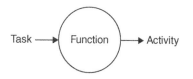

Figure 7.12 A function logically transforms a task into an activity.

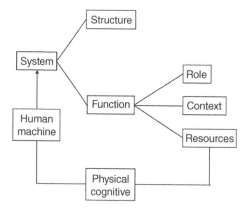

Figure 7.13 HSI recursive definition of a system.

7.7 Conclusion and Perspectives

This chapter benefited from years of aerospace experience that contributed to the genesis of a HSI conceptual framework, useful for the HCD of complex systems, where safety, efficiency, and comfort are most required. The HSI experience, concepts, and methods that have been developed in aerospace can be extended to other industrial and public sectors, such as mobility and medicine. More specifically, the concept of the system described in this chapter can be used for the rationalization of HSI in a large variety of domains. In addition, contemporary technology and organizations are becoming digital, and modeling and simulation will naturally develop over the next few years, providing tremendously useful capabilities to HSI endeavors in engineering design and other processes of the life cycle of systems (i.e. SE).

A question worth asking is: should we continue to talk about human-centered or life-centered technology and organizations? In this chapter, we introduced the first contribution to an HSI ontology (see Figure 7.10). However, it would be great to expand this approach to life in general, including natural entities and environmental issues. Indeed, aeronautics is increasing in sensitivity to climate change, for example, by trying to find solutions to this major planetary issue. From a general standpoint, we need to address appropriate life-critical constraints and goals for technology design and development (e.g. safety, efficiency, and comfort). More specifically, sustainability should be part of these constraints and goals, including social, economic, and environmental factors.

A shift from the old army pyramidal model to the Orchestra model is currently emerging (see the Orchestra model in Boy 2013). For example, technology and emergent practices have led people to change ways of communicating with each other. The army model induced mostly descendent vertical communication. Transversal communication (e.g. using telephone, e-mail, and the Web) contributed to the emergence of the Orchestra model (Boy 2009; Boy and Grote 2009). This functional evolution is now changing organizations themselves (i.e. structures). For example, smartphones and the Internet have contributed to change in both industrial and everyday life organizations.

The Orchestra model provides a usable framework for HSI. It requires a definition of a common frame of reference (music theory), as well as jobs such as the ones of human-centered designers and systems architects (composers) who provide coordinated requirements (scores), highly competent sociotechnical managers (conductors) and performers (musicians), and well-identified end users and engaged stakeholders (audience). Having this organizational model in mind, it is now crucial to use it in HCD. More specifically, in the framework of this chapter, it can be very useful for ATM research and design.

It is time to further develop methods and tools for the integration of people and organizations in the life-critical design and development of new technology. To do this, HSI foundations need to be further developed to support HCD. This chapter presented conceptual solutions to this endeavor. More is to come, and HSI research needs to be promoted and supported.

References

Bainbridge, L. (1983). Ironies of automation. *Automatica* 19 (6): 775–779.

von Bertalanffy, L. (1968). *General System Theory: Foundations, Development, Applications*. New York: George Braziller revised edition 1976.

Boy, G.A. (1993). Human-computer interaction in the cockpit. Invited lecture at Inter-CHI'93, Amsterdam, NL.

Boy, G.A. (1995). "Human-like" system certification and evaluation. In: *Expertise and Technology: Issues in Cognition and Human-Computer Cooperation (Expertise Series)* (eds. J.M. Hoc, E. Hollnagel and P.C. Cacciabue). Hillsdale, NJ: Lawrence Erlbaum Associates, Inc. Publishers.

Boy, G.A. (1998). *Cognitive Function Analysis*. USA: Praeger/Ablex.

Boy, G.A. (2009). The orchestra: a conceptual model for function allocation and scenario-based engineering in multi-agent safety-critical systems. *Proceedings of the European Conference on Cognitive Ergonomics*, Otaniemi, Helsinki area, Finland (30 September–2 October).

Boy, G.A. (2011). Introduction. In: *Handbook of Human-Machine Interaction: A Human-Centered Design Approach*. UK: Ashgate.

Boy, G.A. (2013). *Orchestrating Human-Centered Design*. UK: Springer.

Boy, G.A. (2016). *Tangible Interactive Systems*. UK: Springer.

Boy, G.A. (2017). Human-centered design of complex systems: an experience-based approach. *Design Science Journal* 3: E8. https://doi.org/10.1017/dsj.2017.8.

Boy, G.A. (2020). *Human Systems Integration – From Virtual to Tangible*. Boca Raton, FL: CRC Press, Taylor & Francis.

Boy, G.A. and Grote, G. (2009). Authority in increasingly complex human and machine collaborative systems: application to the future air traffic management construction. In *Proceedings of the 2009 International Ergonomics Association World Congress*, Beijing, China.

Boy, G.A. and Narkevicius, J. (2013). Unifying human centered design and systems engineering for human systems integration. In: *Complex Systems Design and Management* (eds. M. Aiguier, F. Boulanger, D. Krob and C. Marchal). UK: Springer.

Boy, G.A., Salis, F., Figarol, S. et al. (2008). *Authority Sharing in the Airspace*. PAUSA: Final Technical Report. Paris, France: DGAC-DPAC.

Doyon-Poulin, P., Ouellette, B., and Robert, J.-M. (2012). Visual clutter – more than meets the eye. *Proceedings of HCI-Aero 2012: International Conference on Human-Computer Interaction in Aerospace*, Bruxelles, Belgium. Also in ACM Digital Library.

Endsley, M.R. (1996). Automation and situation awareness. In: *Automation and Human Performance: Theory and Applications* (eds. R. Parasuraman and M. Mouloua), 163–181. Mahwah, NJ: Lawrence Erlbaum.

Hilburn, B. (2004). *Cognitive Complexity in Air Traffic Control: A Literature Review*. Project COCA – COmplexity and CApacity. EEC Note No. 04/04.

Kaptelinin, V. and Nardi, B. (2006). *Acting with Technology: Activity Theory and Interaction Design*. Cambridge: MIT Press.

Lee, E.A. (2008). *Cyber Physical Systems: Design Challenges*. http://www.eecs.berkeley.edu/Pubs/TechRpts/2008/EECS-2008-8.html (accessed 10 May 2015).

Letondal, C., Vinot, J.L., Pauchet, S. et al. (2018). Being in the sky: framing tangible and embodied interaction for future airliner cockpits. *Proceedings of TEI 2018 12th International Conference on Tangible, Embedded and Embodied Interactions*, Stockholm, Sweden (March). doi:https://doi.org/10.1145/3173225.3173229. hal-01715317.

Minsky, M. (1985). *The Society of Mind*. New York: Simon and Schuster.

Pinet, J. (2015). *Facing the Unexpected in Flight – Human Limitations and Interaction with Technology in the Cockpit*. Boca Raton, FL: CRC Press, Taylor & Francis.

Russell, S. and Norvig, P. (2010). *Artificial Intelligence – A Modern Approach*, 3e. Boston, MA: Prentice Hall.

Sarter, N.B., Woods, D.D., and Billings, C.E. (1997). Automation surprises. In: *Handbook of Human Factors & Ergonomics*, 2e (ed. G. Salvendy). New York: Wiley.

Simon, H.A. (1996). *The Sciences of the Artificial*, 3e. Cambridge, MA, USA: The MIT Press.

Tan, W. (2015). Contribution to the Onboard Context-Sensitive Information System (OCSIS) of commercial aircraft. PhD dissertation, School of Human-Centered Design, Innovation and Arts, Florida Institute of Technology.

Tarnowski, E. (2006). The 4-loop approach to aeronautical automation evolution. *Keynote, HCI-Aero'06*, Seattle, USA.

Weyer, J. (2006). Modes of governance of hybrid systems. The mid-air collision at Überlingen and the impact of smart technology. *Science, Technology and Innovation Studies* 2: 127–149.

Wiener, E.L. (1989). *Human Factors of Advanced Technology ("Glass Cockpit") Transport Aircraft*. NASA Contractor Report 177528. CA, USA: NASA ARC.

Winograd, T. and Flores, F. (1986). *Understanding Computers and Cognition: A New Foundation for Design*. Boston, MA: Addison-Wesley.

Wolf, M. (2014). *High-Performance Embedded Computing: Applications in Cyber-Physical Systems and Mobile Computing*, 2e. Amsterdam: Morgan Kaufmann.

Section 3

Focus on Training and Skill Sets

8

Building a Socio-cognitive Evaluation Framework to Develop Enhanced Aviation Training Concepts for Gen Y and Gen Z Pilot Trainees

Alliya Anderson[1], Samuel F. Feng[1], Fabrizio Interlandi[2], Michael Melkonian[3], Vladimir Parezanović[1], M. Lynn Woolsey[3], Claudine Habak[3], and Nelson King[1]

[1] *Khalifa University of Science and Technology, Abu Dhabi, United Arab Emirates*
[2] *Capt, Etihad Aviation Training, Abu Dhabi, United Arab Emirates*
[3] *Emirates College for Advanced Education, Abu Dhabi, United Arab Emirates*

8.1　Introduction

The $5.7 billion flight simulator market (2019 estimate) is projected to grow by 5% per year. Airbus (2019) estimates that 550 000 new pilots will need to be trained in the next 20 years to offset retirement and attrition and to meet the demands of growing international and domestic airline markets. According to the 2018 "Pilot Demand Outlook" report, "Airlines and their training partners will need to produce an average of 70 new type-rated pilots per day globally to match the record high aircraft delivery rate and account for pilot attrition" (CAE 2016). With the recruitment rates and the increasing demand for pilots, airlines are already coping with pilot shortages (Garcia 2018; Okal and Kearns 2018; Prokopović 2019). While this demand has been significantly affected by the recent Covid-19 outbreak, it can be expected that pilot training requests will return to growth once the effects of the outbreak have subsided.

While there have been efforts to incentivize and recruit more trainees, the aviation industry is creating training tools that can support the development of modern pilot's competencies (e.g. Airbus VR Flight Trainer). This study focuses on the development of the novice pilots and trainees and the possible integration of the newly available technology with the training methodology. These low experienced pilot trainees are typically millennials (also known as Gen Y) and the younger Gen Z.

Low experienced pilots have less than 200 flying hours or with no previous flight experience on high-performance, complex aircraft. Depending on the definition, millennials are persons reaching young adulthood in the early twenty-first century. Therefore, the 1981–1996 birth cohort is a "widely accepted" definition. Generation Z, the demographic cohort after the millennials, typically use the mid-1990s to mid-2000s as starting birth years.

Evidence-based training (EBT), adopted by some airlines in the past decade, helps experienced pilots increase their resilience as they practice competencies that address actual errors and situations encountered by other pilots. Resilience represents the capability of the crew to deal effectively with unexpected occurrences that do not fit with standard operating procedures. In order to be resilient, the pilot necessarily has to be well prepared for anticipated failures but also to be "prepared to be unprepared" (Paries 2011). Therefore, modern pilot training focuses on the

A Framework of Human Systems Engineering: Applications and Case Studies, First Edition.
Edited by Holly A. H. Handley and Andreas Tolk.
© 2021 The Institute of Electrical and Electronics Engineers, Inc. Published 2021 by John Wiley & Sons, Inc.

development of core competencies, previously known as both technical and nontechnical knowledge, skills, and attitudes, aligning the training content with the actual competencies necessary in the context of contemporary aviation. EBT recognizes the need to develop and evaluate crew performance according to a set of competencies without necessarily distinguishing between the "nontechnical" (e.g. CRM) and the "technical" competencies needed to operate safely (International Civil Aviation Organization 2013).

United Arab Emirates (UAE)-based Etihad Aviation Training (EAT) was an early adopter of EBT for all its pilots. EAT approached the research team to see if their novice pilot training program could better support, in particular, younger pilot recruits, most of whom are from Generation Y and Z age groups. EAT is particularly interested in taking advantage of emerging digital technologies.

We take the perspective of other leaders in aviation like Nick Scarnato (director of global strategy – Collins Aerospace). Scarnato believes, "A training approach that includes time with an instructor in low- and high-fidelity simulators may produce pilots more efficiently, especially if there are appropriate ways to measure a student's progress" (Parsons et al. 2016). New digital technologies provide opportunities to develop low-fidelity simulators that historically have not been available to trainees. Digital technologies also allow gamification to measure progress.

Just adding new technologies does not address the underlying question of what support the trainees actually need. The goal of the project is to examine the social and cognitive differences in mastering an automated complex system (i.e. cockpit) with the current EBT approach. Generational and cultural differences are suspected of playing a role and will be addressed specifically. A gap analysis approach determines the tasks that these pilots have the most difficulty with; once these gaps have been identified, the appropriate training interventions can be proposed. We expect that gamification in learning and new technologies (e.g. virtual reality [VR] or augmented reality [AR]) will emerge from this initial gap analysis as a way to increase proficiency in the use of flight instruments early in the training program.

The challenges low experienced pilots face in an EBT-based training curriculum have yet to be documented in the academic or professional literature. However, it is discussed anecdotally in the profession. Thus, EAT must pioneer enhanced aviation training methods to satisfy its internal standards without the documented experiences of others in the industry. The team takes a human systems engineering (HSE) approach to the development of the training program such that the required interactions between trainees, instructors, and the training technology provide efficient data exchange and retention (Handley 2019). The training system design considers the limitations of the trainees and the mismatch of required competencies to take advantage of the advanced technologies available to trainees in later stages of training (e.g. Level D flight simulator).

This industry–academic partnership brings together visualization technologists, cognition and neuroscientists, and educational specialists. While there have been previous studies with similar goals for the cognitive analysis, to our knowledge, they do not fully utilize the biosensor data (Entzinger et al. 2018) or modern statistical methods particularly geared toward multimodal high-frequency data (Uenking 2000). Using an interdisciplinary approach will help us effectively answer the question: what are the system elements for a gamified learning environment to increase pilot trainee motor skill proficiency? Addressing this question could potentially allow millennial, Generation Z, and low experienced pilots (target group) to increase their functional competencies before engaging with EBT.

8.1.1 Gamification Coupled with Cognitive Neuroscience and Data Analysis

Gamified learning (Landers 2014), a theory that links the emerging fields of specific gaming and the gamification of learning, raises engagement with training manuals, the authoritative source of knowledge for operating an aircraft, and the training environment. The research proposes that a more important role of gamified learning will be in developing cognitive control (Kimura and Nakano 2019), automatic movement skills, and coordination to support motor learning (Pacheco et al. 2019). These skills are necessary for a pilot to translate procedural knowledge into action.

Anecdotal evidence suggests that millennials in US aviation respond well to the use of technology and specifically gamified learning (Niemczyk and Ulrich 2009). However, more realism in a simulated environment does not necessarily translate to better performance (van Lankveld et al. 2017; Buttussi and Chittaro 2018; Makransky et al. 2019). Gamified learning for pilot training takes the form of motor skill development to master the flight instruments and automated controls of a cockpit. Rather than trying to create a mental model of flight instruments from thousands of pages of training material, this research proposes developing physical and virtual training aids that use newer technologies such as AR and VR.

Psychological testing allows for the monitoring of workload to determine the impact of different technologies on attention, engagement, and performance. Visual interventions will be developed to address the deficiencies identified by the biosensor analysis. The analysis draws from recent advances in high-dimensional data analysis (e.g. deep neural networks) along with dynamic time-series prediction models widely used by researchers for analyzing neural signals (Ozaki 2012). Electroencephalography (EEG) data will be used to track the subjects' cognitive load, attentional focus, and resilience during specific control tasks during flight simulator training (Kumar and Kumar 2016).

AR continues to be underutilized in aviation (Anne et al. 2015). We anticipate that these interventions should reduce workload or improve performance. Cognitive factors, such as attention and memory, are especially crucial for motor learning when people are in demanding or complex environments (Warren and Warren 2006), such as a flight simulator. Accordingly, increased attention is linked to superior action control in pilot trainees (Yildiz et al. 2014).

Gamification in a flight training context provides measurable task repetition whose benefit can be measured by in-exercise and postexercise assessments. Post-task surveys or recall tests provide limited information (Macchiarella and Vincenzi 2004), so we will rely on cognitive measures and neurocognitive monitoring to determine training gaps as they occur. These biosensors potentially include EEG, eye tracking, galvanic skin response, and others as appropriate. The biosensor data will show the changes in cognitive workload as the trainee progresses through the gamified tasks or repeats the task (Ghasemian et al. 2017). These data will help predict breaks in attentional focus (as opposed to predicting errors) and recovery of focus, along with the resilience to distractors/stressors. Gamified exercises can then be developed with biosensor monitoring to determine improvements in task performance at a cognitive level. Merely flying more often in VR may not be the most effective training intervention, especially if trainees have limited time.

Gamification and VR take many different forms. In this context, rather than tracking gameplay, gamification will be used to track training activities and task performance. AR can be as traditional as interactive and smart textbooks or further augmented technology (e.g. smart glasses). For example, trainees may struggle with recalling a checklist in a specific set of flight conditions. Intelligent prompts may enable them to move forward with the rest of the checklist they do remember.

8.1.2 Generational Differences in Learning

The widespread consideration of generational differences between younger and older members of a given society is often popularized in terms of a "generation gap" (Bengtson 1970; Friedenberg 1969). Typically, the particular age-related cohorts placed into generational groupings may be characterized as being significantly distinct from each other in terms of intergenerational perceptions (Mannheim 1952; Mead 1970; Shapiro 2004), displayed attitudes and dispositions

(Fengler and Wood 1972; Arsenault 2004), and aspects of their lives such as their methods of social engagement and their career development (PricewaterhouseCoopers 2013).

In contrast to the popular and often stereotypical beliefs reported by the general media, the results of research investigating the presence of significant generational differences are largely inconsistent (Brink et al. 2015; Costanza and Finkelstein 2015, 2017). Some studies (Giancola 2006; Costanza and Finkelstein 2015), including a meta-analysis of 20 studies investigating generation-based differences (Costanza et al. 2012), report no significant generational differences, proposing that "mythical" differences are individual rather than group dependent and largely the application of generational stereotypes. However, other research is less clear about the absence of differences (Ng and Feldman 2010; Parry and Urwin 2011) and propose further examination to better understand the concepts involved (Inglehart 1981; Parry and Urwin 2011; Rudolph and Zacher 2017) and the methods of investigation (Costanza and Finkelstein 2017).

Despite the lack of construct clarity or research support for a generation gap within organizations, the perception of generational differences and organizational concerns regarding how to recruit and retain young talent to replace retiring older workers continues to prompt investigation (Smola and Sutton 2002; Cennamo and Gardner 2008; Twenge et al. 2010). The accounting firm PricewaterhouseCoopers was concerned by a high attrition rate among its millennial employees and conducted a global survey of 44 000 employees to establish what, if any, generational difference might be influencing the loss of young employees (PricewaterhouseCoopers 2013). The report concluded that despite some interesting generational preferences by millennials in terms of an improved work–life balance, a greater emphasis on teamwork, more need for recognition, and choice of work styles, overall, more similarities than differences were found to exist between the different generations (PricewaterhouseCoopers 2013).

The aviation sector, like other areas of industry, has also undergone substantial changes as the result of technological advances (Kantowitz and Campbell 1996; Schultz 2018), and these changes have been regarded as both beneficial and challenging by pilots (Wiener et al. 1991). Additionally, and in conjunction with the technological advances in aircraft manufacture and equipment, it is the growing ubiquity of technology that has not only been felt in the aviation industry but within all aspects of everyday life, including education.

As populations age, the demographic of employees in the general workplace also changes. This change has not gone unnoticed within the aviation industry. It is recognized that with this change is the attendant need to maintain an understanding of the evolving workforce. Specifically, the need to understand how any perceived generational differences may impact the education of incoming trainee pilots. Therefore, it is important to examine how the modern generation of trainee pilots may be perceived to differ from earlier generations. It is important to explore not only their attitudes, work ethic, and other considered generational differences but also their usage of, accepted reliance on, and attitudes toward technology. Irrespective of the presence of a "generation gap," it is of value to investigate student and instructor perceptions of current pilot training methods and procedures for the presence and/or absence of considered generational or age-related differences to establish a frame of understanding that may provide insight to improve pilot training within the twenty-first century.

8.2 Virtual Technologies in Aviation

The value of AR in aviation has long been recognized because of the ability to merge synthetic and real objects in an integrated scene (Macchiarella and Vincenzi 2004). Military combat pilots particularly benefit from VR as they always swivel their heads to look up and around for potential threats. However, recent naval aviation training approaches use VR to go beyond noncombat situations. They are using VR to research human performance, including simulator sickness, spatial disorientation countermeasures, fatigue/eye strain, and vergence–accommodation conflict issues (Biggs et al. 2018).

Virtual technologies are not the panacea, though AR is used in aviation maintenance training programs such as for assembly and maintenance training tasks (Wang et al. 2016). AR does have its limitations, especially with early generations of technology, in which inadequate frame rates cause simulator sickness. But beyond the technology, it is unclear how performance improvements are measured. As a result, AR continues to be underutilized in aviation (Anne et al. 2015). Hence, the team focuses on a scientific approach to first identify the gaps before considering different forms of virtual technology. Early advocates of VR in pilot training were for military aviators who need to scan the skies around them for threats for which VR is well suited. Commercial pilots spend most of their time looking forward at their displays or through the forward-facing cockpit windows. Different trade-offs come into play for commercial pilot training including the burden of wearing a headset for extended periods of time.

8.2.1 Potential Approaches for Incorporating Virtual Technologies

The project takes a more traditional educational intervention approach (e.g. training and learning approaches and instructor training) done in parallel with trying to understand the socio-cognitive impact of gamified technologies on the acquisition of procedural knowledge and the corresponding motor skills associated with the procedure. In essence, how do new technologies such as AR, mixed reality (MR), and VR help the trainee? Some of the procedures performed by pilots could be perceived as boring to learn and memorize or quite lengthy such as the "cold and dark" start of an aircraft (i.e. from all systems shut down to engine start). New training technologies could help change those perceptions or at least make them more interesting to learn.

Neurocognitive testing and monitoring are the underlying foundation of this research, as cognitive factors are strong predictors of student flight performance (Emery 2011). Test batteries are predictive for pilot performance, including acquired knowledge, perceptual processing, motor abilities, controlled attention, and general ability (ALMamari and Traynor 2019). Rather than relying on self-reports of satisfaction with technology or instructor perception of skill improvement, this research uses neurocognitive testing in conjunction with the gamification of teaching interventions to provide a quantitative basis for improvement.

An extension of interactive and smart manuals and their augmentation would be an immersive training manual for the problematic procedures. VR offers such an environment where the procedural knowledge is combined with the requisite motor learning; VR training of this type has been shown to transfer learning to real-world contexts (Kim et al. 2019).

Partial task simulators represent a class of training devices that allow for the repetition of spatial and procedural tasks with reasonable fidelity. For example, a dial can be turned digitally with the swipe of one finger. The spatial dimensions are preserved, and the procedural sequence can be programmed. These training devices, some of which can be 3D, represent physical mock-ups of a set of controls or an entirely digital representation of select portions of an instrument panel. These devices do not replicate the entire cockpit, as this can be done in hobby flight simulators to some extent already. Partial task simulators are not intended to replicate the flight training device or the Level D+ flight simulator that already faithfully reproduces not only the physical cockpit but also the associated motions of flight.

The bulk of expected learning probably still occurs in studying manuals on tablets or computers or through instructor-led training. Recording the computer-based training (CBT) sessions will allow us to observe how the students use these materials, along with eye tracking and other physiological measures (e.g. brain activity), while engaged in tasks.

8.3 Human Systems Engineering Challenges

The HSE approach taken in this project recognizes that the trainees work within a training system that prepares them to function effectively in a high-fidelity simulated training system (i.e. users operating a system). We are trying to engineer the training system to increase the integration with the existing simulation system. While some generational and/or cultural differences likely exist, the study seeks to identify gaps in trainee preparation, which probably results from at least some of the numerous factors alluded to in this chapter. The entire training system needs to be explored, not just the training that takes place in a high-fidelity simulator. Arguably, the front end of training has the most significant impact on the cognitive development of trainees.

Neurocognitive monitoring and analysis (e.g. deep neural network) allow engineers to determine the socio-cognitive demands placed upon operators performing tasks. Such knowledge allows engineers to redesign the controls and displays or develop appropriate training using state-of-the-art visualization technologies, including AR, VR, and 3D immersive technologies.

This research recognizes that system complexity, and the corresponding workload, takes different forms. For example, an airline cockpit presents a dense set of controls and displays within arm's reach of a pilot. The later phases of the research apply artificial intelligence (AI) analysis methods to human intelligence to commercial aviation pilot training. The plethora of biological responses collected from in situ experiments (e.g. hours of multichannel brain data) requires *big data* capture, storage, analysis, and retrieval methods.

Complex control systems, such as cockpits and control towers, put operators under high cognitive workload as they must quickly make decisions and execute sequences of control tasks based on a system's stimuli. System designers seek to automate many of these tasks, which trades off less complicated control panels with the need to retain a mental model of the underlying control system. The consumer electronics analogy is the Apple operating system. That system predetermines what can be done, whereas the Linux operating system makes available all functionality if the user knows what to do. Apple products are easy to use but offer limited access to functionality compared with other operating systems. In the case of complicated controls, the real challenge for pilots is "to be into the loop," maintaining a proper situational awareness, trying to understand the underlying logic of the systems and their interactions even if it is almost impossible to know thoroughly such a complex architecture (Chialastri 2011).

The complexity of control systems can result from the density of controls and displays, embeddedness of procedures due to automation, frequency of decisions, spatial richness (e.g. dimensions of visual input), and the required degree of vigilance. A cockpit consists of tightly packed physical controls in which the modern aircraft embeds automation. Complexity affects human performance and cognition, and workload tells us whether performance can be sustained through biosensor measurements, especially EEG. In addition, a recent study of trust in automated systems suggests that novice military pilots do not understand the full costs and potential benefits, which effectively puts their trust in a system not well understood (Lyons et al. 2017). The challenge faced by engineers designing the operation of control systems is to consider all the cognitive aspects of their design, so the design is more comfortable for human brains to use. Understanding the contributors to complexity (Habak et al. 2019) becomes increasingly important as AI plays an increasing role in technology. Numerous examples exist on the cognitive limitation of human systems integration, which often result in contradictory demands. For example, a driver of a self-driving car must remain vigilant despite the car driving itself except in those rare unanticipated moments when the control system fails to detect a hazard. However, humans do not remain vigilant in those settings. This research, in its later phases, will go deeper than usability to emphasize human systems integration that reflects the socio-cognitive aspects of operators.

8.4   Potential Applications Beyond Aviation Training

The potential benefits of the project extend beyond the training of pilots to a broader view of operator performance to enhance functional fidelity (i.e. simulated operational tasks) (Matthews et al. 2011). The findings from this research could also extend to other knowledge-intensive digital workplaces. While the role of cognition has been documented for physically active work (e.g. sports), little has been written about the role of cognition in the digital workplace for both training and resiliency in operations. The digital workplace typically requires the mastery of an automated control system, whether it be the cockpit or the control room of a process plant. All of these control systems require memorization and rapid recall of checklists and procedures, so we expect the principles gained from aviation training to be applicable to many other domains that require this particular combination of cognitive and fast procedure control, such as disaster control, emergency response teams, police force, energy plant control, security procedures, and even healthcare that borrows heavily from aviation safety protocols.

8.5   Looking Forward

The project, as described in this chapter, takes an HSE approach to the training of new commercial aviation pilots. Most of the attention gets focused on the realism of the flight simulator with the trainee expected to acquire the mental model to operate a complex system through manuals and CBTs. However, the training system has not explicitly accounted for the social and cognitive limitations of trainees. By taking a critical look at the training system, the project seeks out both pedagogical and technological interventions that, at the very least, increase the motivation of the trainee to acquire the skills necessary to take advantage of the Level D flight simulators.

With advances in computing and AI and the growing market demand for qualified commercial pilots, R2-D2 from *Star Wars* as a copilot is seeming less like science fiction and more like a not-so-distant possibility for commercial airlines. For example, the US Pentagon's Defense Advanced Research Projects Agency (DARPA) has already started testing Aircrew Labor In-Cockpit Automation System (ALIAS) in the UH-60 Black Hawk, the largest fleet in the US Army, for that exact purpose – "'offloading pilots' cognitive burden [to free] them to focus on mission execution" (DARPA 2019). This project could provide the foundation for identifying which tasks are suited for automation and which tasks may build pilot resiliency.

Acknowledgement

A portion of this publication is based upon work supported by the Khalifa University of Science and Technology under Award No. RC2 DSO.

References

Airbus (2019). *Global Services Forecast 2019–2038*. https://services.airbus.com/content/dam/corporate-topics/strategy/global-market-forecast/GMF-2019-2038-Airbus-Commercial-Aircraft-book.pdf (accessed 15 January 2020).

ALMamari, K. and Traynor, A. (2019). Multiple test batteries as predictors for pilot performance: a meta-analytic investigation. *International Journal of Selection and Assessment* 27 (4): 337–356. https://doi.org/10.1111/ijsa.12258.

Anne, A., Wang, Y., and Ropp, T. (2015). Using augmented reality and computer-generated three-dimensional models to improve training and technical tasks in aviation. In: *18th International Symposium on Aviation Psychology*, 452. Wright State University – Conferences & Events.

Arsenault, P.M. (2004). Validating generational differences: a legitimate diversity and leadership Issue. *Leadership and Organization Development Journal* 25 (2): 124–141. https://doi.org/10.1108/01437730410521813.

Bengtson, V.L. (1970). The generation gap: a review and typology of social-psychological perspectives. *Youth & Society* 2 (1): 7–32. https://doi.org/10.1177/0044118X7000200102.

Biggs, L., Geyer, D., Schroeder, V. et al. (2018). *Adapting Virtual Reality and Augmented Reality Systems for Naval Aviation Training*. No. NAMRU-D-19-13. Naval Medical Research Unit Dayton, Wright-Patterson AFB, United States.

Brink, K.E., Zondag, M.M., and Crenshaw, J.L. (2015). Generation is a culture construct. *Industrial and Organizational Psychology* 8: 335–340. https://doi.org/10.1017/iop.2015.45.

Buttussi, F. and Chittaro, L. (2018). Effects of different types of virtual reality display on presence and learning in a safety training scenario. *IEEE Transactions on Visualization and Computer Graphics* 24 (2): 1063–1076. https://doi.org/10.1109/TVCG.2017.2653117.

CAE (2016). *Airline Pilot Demand, 10-year Outlook at a Glance*, p. 36. https://www.cae.com/media/documents/Civil_Aviation/CAE-Airline-Pilot-Demand-Outlook-Spread.pdf (accessed 29 January 2020).

Cennamo, L. and Gardner, D. (2008). Generational differences in work values, outcomes and person-organisation values fit. *Journal of Managerial Psychology* 23 (8): 891–906. https://doi.org/10.1108/02683940810904385.

Chialastri, A. (2011). Resilience and ergonomics in aviation. In: *Resilience Engineering in Practice: A Guidebook* (eds. E. Hollnagel, É. Rigaud and D. Besnard), 65–71. Farnham, UK: Ashgate Publishing Ltd.

Costanza, D.P. and Finkelstein, L.M. (2015). Generationally based differences in the workplace: is there a there there? *Industrial and Organizational Psychology* 8 (3): 308–323. https://doi.org/10.1017/iop.2015.15.

Costanza, D.P. and Finkelstein, L.M. (2017). Generations, age, and the space between: introduction to the special issue. *Work, Aging and Retirement* 3: 109–112. https://doi.org/10.1093/workar/wax003.

Costanza, D., Badger, J., Fraser R., Severt, J. and Gade, P. (2012) 'Generational differences in work-related attitudes: a meta-analysis', *Journal of Business and Psychology*, 27, pp. 375–394. http://about.jstor.org/terms (accessed 8 April 2020).

DARPA (2019). US army tests DARPA autonomous flight system, pursuing integration with Black Hawk. *Defense Advanced Research Projects Agency*. https://www.darpa.mil/news-events/2018-10-29 (accessed 30 January 2020).

Emery, B. (2011). Neurocognitive predictors of flight performance of successful solo flight students. Doctoral dissertation. Available from PQDT Open (3489209).

Entzinger, J.O., Uemura, T., and Suzuki, S. (2018). Individualizing flight skill training using simulator data analysis and biofeedback. *31st Congress of the International Council of the Aeronautical Sciences, ICAS 2018*, Boca Raton.

Fengler, A.P. and Wood, V. (1972). The generation gap: an analysis of attitudes on contemporary issues. *Gerontologist* 12 (2): 124–128. https://doi.org/10.1093/geront/12.2_Part_1.124.

Friedenberg, E.Z. (1969). The generation gap. *The Annals of the American Academy of Political and Social Science* 382 (1): 32–42. https://doi.org/10.1177/000271626938200105.

Garcia, M. (2018) A "perfect storm" pilot shortage threatens global aviation, *Forbes*, p. 3. https://www.forbes.com/sites/marisagarcia/2018/07/27/a-perfect-storm-pilot-shortage-threatens-global-aviation-even-private-jets/#2738d9c71549 (accessed 29 January 2020).

Ghasemian, M., Taheri, H., Saberi Kakhki, A., and Ghoshuni, M. (2017). Electroencephalography pattern variations during motor skill acquisition. *Perceptual and Motor Skills* 124 (6): 1069–1084. https://doi.org/10.1177/0031512517727404.

Giancola, F. (2006). The generation gap: more myth than reality. *Human Resource Planning* 29 (4): 32.

Habak, C., Seghier, M.L., Brûlé, J. et al. (2019). Age affects how task difficulty and complexity modulate perceptual decision-making. *Frontiers in Aging Neuroscience* 11 https://doi.org/10.3389/fnagi.2019.00028.

Handley, H. (2019). Human system engineering. In: *Topics in Safety, Risk, Reliability and Quality*, 7–15. Netherlands: Springer https://doi.org/10.1007/978-3-030-11629-3_2.

Inglehart, R. (1981). Post-materialism in an environment of insecurity. *American Political Science Review* 75 (4): 880–900. https://doi.org/10.2307/1962290.

International Civil Aviation Organization (2013). *Manual of evidence-based training* (1). Quebec: IATA, https://www.iata.org/contentassets/c0f61fc821dc4f62bb6441d7abedb076/ebt-implementation-guide.pdf (accessed 8 April 2020).

Kantowitz, B.H. and Campbell, J.L. (1996). Pilot workload and flightdeck automation. In: *Automation and Human Performance: Theory and Applications*, 3–17. Boca Raton, FL: CRC Press https://doi.org/10.1201/9781315137957.

Kim, A., Schweighofer, N., and Finley, J.M. (2019). Locomotor skill acquisition in virtual reality shows sustained transfer to the real world. *Journal of Neuroengineering and Rehabilitation* 16 (1): 113. https://doi.org/10.1186/s12984-019-0584-y.

Kimura, T. and Nakano, W. (2019). Repetition of a cognitive task promotes motor learning. *Human Movement Science* 66: 109–116. https://doi.org/10.1016/j.humov.2019.04.005.

Kumar, N. and Kumar, J. (2016). Measurement of cognitive load in HCI systems using EEG power spectrum: an experimental study. *Procedia Computer Science* 84: 70–78. https://doi.org/10.1016/j.procs.2016.04.068.

Landers, R.N. (2014). Developing a theory of gamified learning: linking serious games and gamification of learning. *Simulation and Gaming* 45 (6): 752–768. https://doi.org/10.1177/1046878114563660.

van Lankveld, G., Sehic, E., Lo, J.C. et al. (2017). Assessing gaming simulation validity for training traffic controllers. *Simulation and Gaming* 48 (2): 219–235. https://doi.org/10.1177/1046878116683578.

Lyons, J.B., Ho, N.T., Van Abel, A.L. et al. (2017). Comparing trust in auto-GCAS between experienced and novice air force pilots. *Ergonomics in Design* 25 (4): 4–9.

Macchiarella, N.D. and Vincenzi, D.A. (2004). Augmented reality in a learning paradigm for flight and aerospace maintenance training. *AIAA/IEEE Digital Avionics Systems Conference – Proceedings*. doi: https://doi.org/10.1109/dasc.2004.1391342.

Makransky, G., Terkildsen, T.S., and Mayer, R.E. (2019). Adding immersive virtual reality to a science lab simulation causes more presence but less learning. *Learning and Instruction* 60: 225–236. https://doi.org/10.1016/j.learninstruc.2017.12.007.

Mannheim, K. (1952). The Sociological Problem of Generation. In: *Essays on the Sociology of Knowledge*, 163–195. New York: Harcourt, Brace and World.

Matthews, G., Warm, J., Reinerman-Jones, L. et al. (2011). The functional fidelity of individual differences research: the case for context-matching. *Theoretical Issues in Ergonomics Science* 12 (5): 435–450. https://doi.org/10.1080/1463922X.2010.549247.

Mead, M. (1970). *Culture and Commitment: A Study of the Generation Gap*. The Bodley Head.

Ng, T.W.H. and Feldman, D.C. (2010). The relationships of age with job attitudes: a meta-analysis. *Personnel Psychology* 63 (3): 677–718. https://doi.org/10.1111/j.1744-6570.2010.01184.x.

Niemczyk, M. and Ulrich, J.W. (2009). Workplace preferences of millennials in the aviation industry. *International Journal of Applied Aviation Studies* 9 (2): 207–219.

Okal, A. and Kearns, S. (2018). Pilot shortage: sustainability perspective. *National Training Aircraft Symposium (NTAS)*. https://commons.erau.edu/ntas/2018/presentations/5 (accessed 29 January 2020).

Ozaki, T. (2012). *Time Series Modeling of Neuroscience Data* (eds. N. Keiding et al.). Boca Raton, FL: CRC Press https://doi.org/10.1201/b11527.

Pacheco, M.M., Lafe, C.W., and Newell, K.M. (2019). Search strategies in the perceptual-motor workspace and the acquisition of coordination, control, and skill. *Frontiers in Psychology* 10 (August) https://doi.org/10.3389/fpsyg.2019.01874.

Paries, J. (2011). Lessons from the Hudson. In: *Resilience Engineering in Practice* (eds. E. Hollnagel, J. Paries, D.D. Woods and J. Wreathall), 9–26. Ashgate Publishing Limited.

Parry, E. and Urwin, P. (2011). Generational differences in work values: a review of theory and evidence. *International Journal of Management Reviews* 13 (1): 79–96. https://doi.org/10.1111/j.1468-2370.2010.00285.x.

Parsons, B., Magill, T., Boucher, A. et al. (2016). Enhancing cognitive function using perceptual-cognitive training. *Clinical EEG and Neuroscience* 47 (1): 37–47. https://doi.org/10.1177/1550059414563746.

PricewaterhouseCoopers (2013) *PwC's NextGen: A Global Generational Study*. PriceWaterhouseCoopers, the University of Southern California and the London Business School, p. 16. www.pwc.com/structure (accessed 8 April 2020).

Prokopovič, K. (2019) Pilot-hungry airlines struggle with global pilot shortage, *Aviation Voice*. https://aviationvoice.com/pilot-hungry-airlines-struggle-with-global-pilot-shortage-201906101447 (accessed 29 January 2020).

Rudolph, C.W. and Zacher, H. (2017). Considering generations from a lifespan developmental perspective. *Work, Aging and Retirement* 3 (2): 113–129. https://doi.org/10.1093/WORKAR/WAW019.

Schultz, T.P. (2018). *The Problem with Pilots: How Physicians, Engineers, and Airpower Enthusiasts Redefined Flight*. Johns Hopkins University Press.

Shapiro, A. (2004). Revisiting the generation gap: exploring the relationships of parent/adult-child dyads. *International Journal of Aging & Human Development* 58 (2): 127–146. https://doi.org/10.2190/EVFK-7F2X-KQNV-DH58.

Smola, K.W. and Sutton, C.D. (2002). Generational differences: revisiting generational work values for the new millennium. *Journal of Organizational Behavior* 23 (special issue): 363–382. https://doi.org/10.1002/job.147.

Twenge, J., Campbell, S., Hoffman, B., and Lance, C. (2010). Generational differences in work values: leisure and extrinsic values increasing, social and intrinsic values decreasing. *Journal of Management* 36 (5): 1117–1142. https://doi.org/10.1177/0149206309352246.

Uenking, M.D. (2000). Pilot biofeedback training in the cognitive awareness training study (CATS). *Modeling and Simulation Technologies Conference*. doi: https://doi.org/10.2514/6.2000-4074.

Wang, Y., Anne, A., and Ropp, T. (2016). Applying the technology acceptance model to understand aviation students' perceptions toward augmented reality maintenance training instruction. *International Journal of Aviation, Aeronautics, and Aerospace* 3 (4) https://doi.org/10.15394/ijaaa.2016.1144.

Warren, W.H. and Warren, H. (2006). The dynamics of perception and action. *Psychological Association* 113 (2): 358–389. https://doi.org/10.1037/0033-295X.113.2.358.

Wiener, E.L., Chidester, T.R., Kanki, B.G. et al. (1991). *The Impact of Cockpit Automation on Crew Coordination and Communication. Volume 1: Overview, LOFT Evaluations, Error Severity, and Questionnaire Data*. NASA archives.

Yildiz, A., Quetscher, C., Dharmadhikari, S. et al. (2014). Feeling safe in the plane: neural mechanisms underlying superior action control in airplane pilot trainees – a combined EEG/MRS study. *Human Brain Mapping* 35 (10): 5040–5051. https://doi.org/10.1002/hbm.22530.

9

Improving Enterprise Resilience by Evaluating Training System Architecture

Method Selection for Australian Defense

Victoria Jnitova[1], Mahmoud Efatmaneshnik[2], Keith F. Joiner[3], and Elizabeth Chang[4]

[1] *School of Engineering and Information Technology, University of New South Wales at Australian Defence Force Academy, Canberra, ACT, Australia*
[2] *Defence Systems Engineering at the University of South Australia (UNISA), Adelaide, SA, Australia*
[3] *Capability Systems Center, School of Engineering and Information Technology, University of New South Wales at Australian Defence Force Academy, Canberra, ACT, Australia*
[4] *University of New South Wales at Australian Defence Force Academy, Canberra, ACT, Australia*

9.1 Introduction

Military organizations are undergoing a dramatic change in their operational environment and capabilities (Bitzinger 2016). Defense posture has shifted from the Cold War to the global anti-terrorism war (Jordan et al. 2016), and its capability systems are struggling to stay on top of rapid technological progress and future uncertainty while remaining secure and closed from potential probing by adversaries and competitors (Alberts et al. 2000; Joiner and Tutty 2018; U.S. Government Accountability Office 2018). Department of Defense (DoD) capability systems are required to be trusted that they are able to effectively respond to new or changing conditions in a wide range of operational contexts (Joiner et al. 2019). To effectively manage these challenges and uncertainties while addressing the capability gaps and anticipating future requirements, DoD increasingly relies on systems thinking from the perspective of systems engineering (SE) discipline. The Australian Defense Force (ADF) headquarters and individual service commands are leading the systems thinking implementation in the integrated defense capability life cycle (DCLC) context. They focus on future-proofing the systems' design and overcoming traditional "stovepiping" and separations between defense departments, processes, and artifacts (Schuel 2018; Unewisse 2016). The challenge is to extend the SE practices to nontraditional SE areas such as human resources and training and to "engineer" human factors as described in Chapter 1 into the organizational design.

We propose to meet this challenge by applying systems thinking in the defense training organization context to engineer a resilient Defense Training System (DTS). The "As Is" and "To Be" DTS design is determined in the phased whole Defense Survey, followed by the system architecture development using the DoD Architecture Framework (DoDAF) and modular design methodologies. The DTS resilience attributes and their associated measures are derived from the reviewed academic literature and embedded into the DTS architectural framework to support model-based measurement of the system resilience.

A Framework of Human Systems Engineering: Applications and Case Studies, First Edition.
Edited by Holly A. H. Handley and Andreas Tolk.
© 2021 The Institute of Electrical and Electronics Engineers, Inc. Published 2021 by John Wiley & Sons, Inc.

This chapter presents an initial phase of this project, where the DTS is defined and conceptualized, followed by the resilience-focused literature review to support our effort of formulating a model-based conceptual resilience framework and measurement methodology. Basic factors and measures for DTS resilience are proposed in the DTS case study section of the chapter to guide development of the transformation road map from "As Is" to "To Be" DTS design targeting improved system's resilience.

9.2 Defense Training System

9.2.1 DTS Conceptualization

Defense Learning Manual (DLM) and Individual Service Training Policies mandate use of the Systems Approach to Defense Learning underpinned by analyze/design/develop/implement/evaluate (ADDIE) methodology across the DoD for training planning and management. Fakult et al. (1988) argue that instructional design models such as ADDIE are derivatives of the systems thinking approach, grounded in the SE discipline, but developed in isolation from SE common tools and innovations. Indeed, DoD Systems Approach to Defense Learning does neither describe numerous DTS artifacts and processes in SE terms nor use standard SE practices. Moreover, with its academic underpinning almost 45 years old and strong accretional focus on individual training, the ADDIE model is likely not adequately equipped to prepare the DoD workforce to operate in the modern workplace and to deal with new emergent challenges of the global anti-terrorism war, complexities of collective training and rapidly evolving technologies, and uncertainties associated with a changing strategic environment, which could potentially impact the DTS fitness for purpose and resilience.

We argue DTS is a complex sociotechnical system with its boundaries extending far beyond SADL and ADDIE. DTS require a more context-specific suite of tools and processes to address the combination of its human and system components. As stated in Chapter 1, human factors are traditionally employed in systems' approach and concerned with manpower, personnel, training, human factors engineering, health and safety, habitability, and survivability – all of which need to be integrated into the DTS architecture together with traditional "technical" training resources and services. The emphasis of the DTS engineering effort is therefore on the trade-offs within and across these domains. With relation to human factors integration into the system design, one of the SE traditional design challenges is a trade-off between the design usability and cost effectiveness. In commercial setting, the SE needs to ensure the system design is competitive; therefore commercial systems are traditionally designed for usability to be achieved at a lowest possible cost. In contrast, DoD is a noncommercial enterpriser with its capability systems that are not commonly designed for usability but rely heavily on training the individuals to use them. As such, the DoD integrates DTS in its architecture and places a much higher importance on the DTS enabling service compared with commercial extended enterprisers. In this research, DTS is conceptualized as a sociotechnical system, setting a foundation for our future work to formulate DTS design alternatives integrating DTS workforce and comparing these alternatives in terms of their resilience.

9.2.2 DTS as an Extended Enterprise Systems

DTS is best described as an extended enterprise system as defined in the reviewed academic literature. Boardman and Sauser (2008) define *enterprise* as a collection of parts and relationships assembled to form a whole with properties, attributes, and purpose that make it essentially different from its components. An *enterprise* is also defined as a complex system consisting of people, processes, information systems, and technology infrastructure to produce goods or services using physical, financial, and human resources (Fiksel 2015). What lies outside the enterprise is as important as what lies inside it, because for a resilient system it is necessary to observe how an enterprise can most effectively adapt when facing both external and internal disruptions. Ignatiadis and Nandhakumar (2007) use the term *enterprise systems* to refer to the enterprise-wide information systems and applications that support business functions, processes, operations, and services of the enterprise. The effective use of enterprise systems can provide timely information and faster decision-making abilities that result in increased system's flexibility, agility, and adaptability.

Enterprise systems are required to have the ability to access resources and information quickly and to change their processes rapidly in order to satisfy emerging business needs (Boardman and Clegg 2001). This need for agility requires a shift to more integrated, open, and collaborative enterprise structures, where the enterprise is viewed as a whole along with its environment and stakeholders. Applications, people, and data are connected and integrated in an extended inter- and intra-enterprise processes in order to improve their efficiency and effectiveness (Kosanke et al. 1999). The concept of *extended enterprise* emerged from the need to collaborate, interact, and share information and resources among the distributed parts of the enterprise and across its boundaries. This need is primarily driven by market globalization and a competitive environment that requires increased efficiency; reduced cycle times, cost, and effort; and increased customer satisfaction (Gold-Bernstein and Ruh 2004). An *extended enterprise* is viewed as a system of enterprises that assumes particular characteristics not necessarily present in any of its components (Boardman and Sauser 2008).

9.2.3 Example: Navy Training System

The DTS key design considerations and associated measures of its success are illustrated using the Navy Training System example, considered in DTS and DoD contexts.

9.2.3.1 Navy Training System as a Part of DTS
Navy Training System is an integral part of the DTS extended enterpriser. Success of the Navy Training System stated intent is continually assessed through measuring the achievement of the associated actions as illustrated in Table 9.1.

9.2.3.2 Navy Training System as a Part of DoD
Navy Training System is also considered in a wider context of the DoD extended enterprise systems. It is one of the enablers of the four Navy strategic objectives as stated in the Royal Australian Navy (RAN) Plan Pelorus, namely, (1) warfighting, (2) capability, (3) workforce, and (4) reputation and reform. According to the Navy Training Force Plan 2018, the Navy Training System support for each of the objectives is as follows:

1) Sea Training Group provides collective training against prescribed competencies in direct support to the Fleet Commander.
2) Training Force is a critical stakeholder in projects delivering new Navy capability.
3) Training authorities provide career and position prerequisite training.
4) Contribution to RAN reputation and reform: training is delivered to equip each sailor and officer to achieve their full potential as ethical leaders throughout their careers.

Table 9.1 RAN training Force intent, associated actions, and measures of achievement.

Intent	Associated actions	Measures of achievement
Training delivery		
Exploit existing and explore new training technologies	• Identify opportunities to use simulation • Implement learner-centric training technique • Identify opportunities for e-learning • Implement coaching focused in sea training and assessment • Replace paper-based sea training assessment with the digital technologies • Participate in acquisition projects to continuously support meeting training requirements	• Simulation-based training and assessment become a norm not an exception • PowerPoint presentations are challenged and replaced by innovative learner-centric methods • Use of distance learning methods, ADELE and Google apps • Positive engagement and coaching by the Sea Training Group become a norm not an exception • Sea training assessment is digitalized, and feedback is implemented • New capabilities introduction into service are supported by simulators, new technologies, infrastructure, and modern information management means
Training pipeline efficiency		
Optimize the efficiency and effectiveness of the training pipelines	• Manage resources to meet training requirements within time and cost constraints • Continuously review schedules and processes to minimize wastage • Evaluate training to seek the most efficient delivery approach	• Courses are delivered at maximum capacity and at a minimum cost feasible • Delivering category/PQ training blocks to enable cohort postings and 100% job readiness • Courses are continuously internally evaluated to identify gaps, with external evaluation is conducted to confirm and address the gaps
Professional development (PD)		
PD opportunities for TS community to learn new TS skills	• PD training and coaching is provided where required • Individual development and succession planning • Annual instructor evaluation and tracking gaps if identified • Identify instructional exemplars and internally develop them • Provide coaching training to all in the Sea Training Group • Award exceptional instructors	• TS Officers are seen as SADL and National Training Framework professional experts • Training delivery is learner centric, innovative, and in a learning conducive environment • All instructors are evaluated, and the identified gaps are addressed • Provision of TS advice and PD workshops on a regular basis • Certification of the sea trainers includes coaching training • Instructional talents are acknowledged through the honors and award system
Training governance		
Ensure implementation and ongoing scrutiny of SADL governance	• Course documentation is reviewed to meet the standard • Senior TS are accountable for training development, quality control, and training governance • The findings of navy training audits and Defense Seaworthiness Boards are addressed	• Version controlled, accessible via defense information management system and meets the standard • Senior TS manage and report training development, quality control, and training governance in their respective areas of responsibility • Achieving a measurable reduction in deficiencies identification by the navy training audits and Defense Seaworthiness Boards

Source: Extract from the RAN Training Force Plan 2018.

9.3 Concept of Resilience in the Academic Literature

Resilience emerges as a desirable system attribute demanded at all levels within the DoD due to uncertainties and the continuous need to change and adapt quickly in turbulent periods while retaining a desired level of performance to consistently satisfy stakeholder needs and requirements. Resilience is often discussed in various disciplines, but the plethora of the concept interpretations present a challenge for defining a meaningful set of resilience measures for a given system. There is little consensus regarding what resilience is, what it means for organizations, and, more importantly, how they may achieve greater resilience in the face of increasing threats. Although it is not our intent to provide an in-depth review of such diverse literature, we conduct a multidisciplinary literature review to form a holistic view of resilience and to gather insights to support our study of model-based DTS resilience framework.

More than 90 papers were selected for this review, of which about half are related to each part. The papers for this review were selected using combinations of the following keywords: military training, workforce planning, system, resilience, measuring resilience, future-proofing, resilience engineering, measuring resilience, enterpriser, architecture, and others.

9.3.1 Definition of Resilience: A Multidisciplinary and Historical View

The reviews emphasize multidisciplinary nature and role of resilience and provide examples of definitions from the disciplines of ecology, psychology, engineering, organizational or enterprise supply chain management, economics, SE, computer science, materials science, disaster management, organizational theory, risk management, sociology, and defense, to name a few. There are several comprehensive literature reviews of the resilience concept, for example, in Kamalahmadi and Parast (2016), McManus (2008), Burnard and Bhamra (2011), and Bhamra et al. (2011). Several other papers offer tabulated summaries of the multidisciplinary resilience definitions (Downes et al. 2013; Fraccascia et al. 2017; Francis and Bekera 2014), while others provide a historical perspective (Bouaziz 2018; De Sanctis et al. 2018; Erol et al. 2010c) or tailor their reviews to their projects but highlight the multidisciplinary origin of the selected definitions (Conz and Magnani 2019; Dalziell and McManus 2004).

9.3.2 Definition of Resilience: Key Aspects

This great variety of resilience definitions in the academic literature calls for a critical analysis aimed at clarifying the main aspects of the concept and identifying commonalities and differences in its interpretations. For this literature review, we will seek to critically analyze the literary responses to the following three key questions:

- **What?** (resilience is/is not)
- **Why?** (resilience triggers)
- **How?** (resilience mechanisms and how resilience response can be measured)

The result of this research effort is presented below.

9.3.2.1 What? (Resilience Is and Is Not)

Due to a vast variation in the resilience definitions in the reviewed academic literature, we commenced our effort by identifying the "like terms" and different terminology perspectives to facilitate our comparative analysis. To achieve this, we divided the resilience definitions into their

Table 9.2 Key elements of resilience definition and their associated vocabulary.

Resilience is	Resilience of what?	To do what?	To achieve what?
• Ability	• System	• Absorb	• Reduce vulnerability
• (Dynamic/adaptive/ potential) capacity	• Complex system	• Adapt	• Resist disorder
• Capability	• Ecosystem	• Anticipate	• Restore original state and evolve post-disturbance
• Attribute	• Social system	• Avoid	• Reduce the probabilities of disruptions
• Collection of attributes	• Socio-ecological system	• Cope	
• Property	• Engineered system	• Evolve	• Cope with unexpected disturbances
• Emergent property	• System of systems	• Learn	• Recognize and adapt to handle unanticipated perturbations
• Characteristic	• Organization	• Predict	
• Factor	• Enterpriser	• Prepare	
• Response	• Business	• React	• Recover from irregular variations, disruptions, and degradation of expected working conditions. Respond and develop new features in the face of change
• Reaction	• DoD	• Recover	
• Form of control	• (System/network/enterpriser/ organizational/operational/ service) architecture	• Reduce	
	• People (individuals, teams, employees, workers, human resources)	• Resist	
		• Respond	
		• Restore	
	• Supply chain	• Retain	• Prepare for unexpected events
		• Survive	
		• Tolerate	

common conceptual elements and identified the variations in terminology associated with those elements with examples provided in Table 9.2.

We select the "attribute" and "system" from Table 9.2 to represent the like terms for "resilience" and the entity that possesses it to use in this paper.

Our analysis of the reviewed literature revealed two common trends in defining resilience, namely:

- Resilience as a singular attribute of the entity.
- Resilience as a collection of attributes.

9.3.2.1.1 *Resilience as a Singular System's Attribute*

Resilience is defined as a singular system attribute in 36 out of 91 reviewed papers. When comparing these definitions, we noticed they were closely aligned with each other, with the main difference being adding new components to simpler versions of the definition. Based on our literature review, we propose six "increments" of resilience definitions that take the system from a simple response triggered by a change to a system's creative continuous learning and adjustment to adapt and evolve in an uncertain and changing environment. The six increments and their respective examples of the resilience definitions from the reviewed academic literature are in Table 9.3:

Several reviewed papers also categorize the resilience concepts using similar criteria to ours, for example, resilience perspectives proposed in Giannoccaro et al. (2018) and Lengnick-Hall et al. (2011). They, however, focus not only on analysis of resilience definitions but also on resilience mechanism alternatives. These considerations are further detailed in the resilience mechanisms and measurements section of this chapter.

9.3.2.1.2 *Resilience as a Collection of System's Attributes*

Resilience is described as a collection of attributes in 48 of 91 reviewed papers. Our analysis of the "resilience attributes" presented a

Table 9.3 Six "increments" of resilience concept.

Paper	Element 3: To do what?	Element 4: To achieve what?
Increment 1: Deal with the changes as they occur while retaining original system design		
Holling (1973)	To absorb change and disturbance	To maintain the same relationships between populations or state variables still
Uday and Marais (2015)	To combine survivability and recoverability, to bounce back	Reduce the likelihood of failure and to recover from unexpected disturbances in the operating environment
Walker et al. (2004)	To absorb disturbances and reorganize while changing	To retain essentially still the same functions, structures, identity, and feedbacks
Increment 2: Prepare and deal with the changes while retaining original system design		
Ayyub (2014)	To prepare for and adapt to changing conditions	To withstand and recover rapidly from disruptions
Increment 3: Anticipate, prepare, and deal with the changes while retaining original system design		
Tran et al. (2017)	To apply classical reliability methods, such as redundancy at the component level and use of preventative maintenance at the system level	To anticipate and resist disruptions
Hollnagel and Woods (2006)	To create foresight, recognize, and anticipate	To defend against the changing shape of risk before adverse consequences occur
Small et al. (2017)	To create through design choices an inherent system quality	To maintain system performance objectives in the face of diverse operational challenges, in either a preparative or recovery sense, within acceptable time and cost parameters
Increment 4: Anticipate, prepare, deal, and learn from the changes (might alter original system design as part of lessons learnt)		
Lengnick-Hall et al. (2011)	To effectively absorb, develop situation-specific responses to, and ultimately engage in transformative activities	To capitalize on disruptive surprises that potentially threaten organizational survival
Menéndez Blanco and Montes Botella (2016)	To take both proactive and reactive measures including knowledge, learning, and innovation to balance efficiency and adaptability	To survive in response to adversity and achieve competitive advantage
Erol et al. (2010c)	To maintain positive adjustment under challenging conditions. The term "challenging conditions" is used to describe disruptive events such as discrete errors, scandals, crises, shocks, disruptions of routines, and ongoing risks (e.g. competition) that cause stress and strain. Accordingly, creating organizational resilience is associated with personnel and management concerns	Such that the organization emerges from those conditions strengthened and more resourceful

(Continued)

Table 9.3 (Continued)

Paper	Element 3: To do what?	Element 4: To achieve what?
Increment 5: Anticipate, prepare, deal, and learn to live with uncertain changes (continuous system adaptation to new conditions)		
Berkes (2007)	(1) Learn to live with change and uncertainty (2) Nurture various types of ecological, social, and political diversity for increasing options and reducing risks (3) Increase the range of knowledge for learning and problem solving (4) Create opportunities for self-organization, including cross-scale linkages and problem-solving networks	To deal with uncertainty and future change and to cope with change characterized by surprises and unknowable risks
Increment 6: Resilience is an emergent property of a continuously evolving system		
Erol et al. (2010a)	Interaction of the characteristics and capacities of a system, which will eventually evolve in the case of a disruptive event	To adapt to an uncertain and changing environment and recover from the impacts of the disruptive event
Park et al. (2013)	The outcome of a recursive process that includes sensing, anticipation, learning, and adaptation where resilience is the emergent property of a system undergoing adaptive process – that is, not as something a system has, but a characteristic of the way it behaves	Preparing for and living with the unexpected, where hazards cannot be identified with certainty

biggest challenge to us due to a plethora of different interpretations of what resilience attributes are and are not and the lack of uniformed terminology and agreed standard to represent resilience concept. For example, Erol et al. (2009) state that extended enterprise resilience and flexibility is a function of agility, efficiency, and adaptability, while the same authors in their 2010 paper formulate resilience as a function of vulnerability, flexibility, adaptability, and agility. Ponomarov and Holcomb (2009) select readiness and preparedness, response and adaption, and recovery or adjustment as resilience attributes. Tierney and Bruneau (2007) propose their "R4" model of resilience consisting of robustness, redundancy, rapidity, and resourcefulness, while Ma et al. (2018) argue that the attributes such as robustness, redundancy, and recovery are concepts overlapping with, but not resilience. In our literature review, we noticed that some of the attributes of resilience where used frequently and in multiple papers, while others feature in a single source only. Examples of the common and frequently used resilience attributes from the reviewed literature are illustrated in Table 9.4, followed by examples of resilience attributes from a single source in Table 9.5.

Further synthesis and analysis of the reviewed definitions are required to reconcile the differences in the attributes' interpretations and to tighten the pool of the available attribute propositions. However, despite the differences in terminology and lack of agreement between the academics on the matter of resilience attributes, we were able to identify nine attributes that featured most prominently in the reviewed publications, namely, (1) flexibility, (2) agility, (3) adaptive capacity, (4) adaptability, (5) robustness, (6) vulnerability reduction, (7) redundancy, (8) recovery/

Table 9.4 Attributes commonly associated with the concept of resilience.

Attribute	Is an attribute of resilience	Is not resilience
Flexibility	<ul><li>An attribute of a resilient system (Erol et al. 2010c)</li><li>Resilience can be achieved by increasing flexibility (Gomes 2015)</li><li>Implicit in the proposed supply chain resilience definition (Kamalahmadi and Parast 2016)</li><li>Integrated into the supply chain resilience factors and measures (Falasca et al. 2008)</li><li>Flexibility versus stiffness: the system's ability to restructure itself in response to external changes or pressures (Woods 2017)</li><li>System ability linked to the static perspective of resilience (Giannoccaro et al. 2018)</li><li>One of four traits of resilience in small enterprisers (Bhamra et al. 2011)</li><li>Identified as one of the top-level attributes, resulting from the application of one or more resilience architecture principles (Nuss et al. 2017)</li></ul>	<ul><li>System attribute that closely relates to resilience in the system's design (Uday and Marais 2015)</li><li>Overlapping but different from resilience (Xiao and Cao 2017)</li><li>Might overlap but insufficient to describe resilience (Lengnick-Hall et al. 2011)</li></ul>
Agility	<ul><li>An attribute of a resilient system (Erol et al. 2010c)</li><li>Implicit in the proposed supply chain resilience definition (Kamalahmadi and Parast 2016)</li><li>One of the supply chain resilience enablers (Soni et al. 2014)</li><li>A dimension of organizational resilience (Bouaziz 2018)</li></ul>	<ul><li>Overlapping but different from resilience (Xiao and Cao 2017)</li><li>Might overlap but insufficient to describe resilience (Lengnick-Hall et al. 2011)</li></ul>
Adaptive capacity	<ul><li>An attribute of a resilient system (Erol et al. 2010c)</li><li>One of three resilience capacities (Francis and Bekera 2014)</li><li>Adaptation ability is one of the dimensions of organizational resilience (Chen 2016)</li><li>Adaptive capability is one of the supply chain resilience enablers (Soni et al. 2014)</li><li>Resilience is a function of McManus (2008)</li><li>Lee et al. (2013) agree with and quote McManus (2008)</li><li>Resilience is a function of Dalziell and McManus (2004)</li><li>A dimension of system resilience (Fraccascia et al. 2017)</li><li>Resilience metric: the ability to change itself and adapt to the changing environment (Erol et al. 2010a)</li><li>System ability linked to a dynamic perspective of resilience (Giannoccaro et al. 2018)</li><li>Often interchangeable with resilience (Limnios et al. 2014)</li></ul>	
Adaptability	<ul><li>Implicit in the proposed supply chain resilience definition (Kamalahmadi and Parast 2016)</li><li>Is considered one of the main capabilities of the supply chain resilience (Carvalho 2011)</li></ul>	<ul><li>Adaptability overlapping but different from resilience (Xiao and Cao 2017)</li><li>Adaptability might overlap but insufficient to describe resilience (Lengnick-Hall et al. 2011)</li></ul>

(Continued)

Table 9.4 (Continued)

Attribute	Is an attribute of resilience	Is not resilience
Robustness	• An attribute of a resilient system (Erol et al. 2010c) • A part of R4 resilience framework (Tierney and Bruneau 2007) • A dimension of organizational resilience (Bouaziz 2018) • System ability linked to the static perspective of resilience (Giannoccaro et al. 2018)	• System attribute that closely relates to resilience in the system's design (Uday and Marais 2015) • Overlapping but different from resilience (Xiao and Cao 2017) • Influence resilience but narrow: recognize a single stable state where the system can return after perturbation and are usually applied to improve systems efficiency (Limnios et al. 2014)
Vulnerability reduction	• Resilience can be achieved by reducing vulnerability (Gomes 2015) • (Keystone vulnerabilities) resilience is a function of (McManus 2008) • Lee et al. (2013) agree with and quote McManus (2008) • Resilience is a function of Dalziell and McManus (2004) • Resilience metric (a system's capability to decrease its level of vulnerability to expected and unexpected events) (Erol et al. 2010a)	• A concept to define or measure the resilience of systems (Erol et al. 2010c)
Redundancy	• Resilience can be achieved by creating redundancy (Gomes 2015) • A part of R4 resilience framework (Tierney and Bruneau 2007) • Integrated into the supply chain resilience factors and measures (Falasca et al. 2008)	• Overlapping but different from resilience (Xiao and Cao 2017)
Recovery/ restorative capacity	• One of three resilience capacities (Francis and Bekera 2014) • A dimension of system resilience (Fraccascia et al. 2017) • Resilience metric: system's ability to recover in the least possible time in case of a disruptive event (Erol et al. 2010a) • System ability linked to the static perspective of resilience (Giannoccaro et al. 2018)	• (Recovery) overlapping but different from resilience (Xiao and Cao 2017)
Form of control	• System resilience is a form of control. A system is in control when "it is able to minimise or eliminate unwanted variability" (Hollnagel and Woods 2006)	• Reduction in control may serve in some circumstances as an enabler to organizational resilience (Ignatiadis and Nandhakumar 2007)
Efficiency	• Focus of engineering resilience (Dalziell and McManus 2004) • Savings from reusable services; is related to optimal use of resources (Erol et al. 2009)	

Table 9.4 (Continued)

Attribute	Is an attribute of resilience	Is not resilience
Reliability	• System attribute that closely relates to resilience in the system's design (Uday and Marais 2015) • Resilience replaced reliability as a richer metric (Han et al. 2012) • Not resilience; the ability of a system and its components to perform required functions under stated conditions for a specified period (Uday and Marais 2015)	
Tolerance	• How a system behaves near a boundary – whether the system gracefully degrades as stress/pressure increase or collapses when pressure exceeds adaptive capacity (Woods 2017) • Identified as one of the top-level attributes, resulting from the application of one or more resilience architecture principles (Nuss et al. 2017)	
Absorptive capacity	• One of three resilience capacities: the degree to which a system can absorb the impacts of system perturbations and minimize consequences with little effort; attained through the system's practice of adverse event mitigation (Francis and Bekera 2014) • (Buffering capacity) the size or kinds of disruptions the system can absorb or adapt to without a fundamental breakdown in performance or the system's structure (Woods 2017)	
Capacity	• Identified as one of the top-level attributes, resulting from the application of one or more resilience architecture principles (Nuss et al. 2017) • Availability of assets to enable sustained production levels (Fiksel 2015)	
Cohesion	• Is considered one of the main capabilities of the supply chain resilience (Carvalho 2011) • Identified as one of the top-level attributes, resulting from the application of one or more resilience architecture principles (Nuss et al. 2017)	

restorative capacity, and (9) efficiency. Examples of the nine attributes' definitions as reported in the reviewed academic literature are in Table 9.6.

Analysis of proposed definitions of the selected nine resilience attributes revealed, again, the lack of consistency and uniformity. For example, according to Erol et al. (2010c), efficiency, agility, and adaptability are sub-attributes of flexibility; Golden and Powell (2000) define flexibility as "the capacity to adapt through the use of efficiency, responsiveness, versatility and robustness," while Bhamra et al. (2011) consider flexibility to be one of the traits of resilience of a small enterprise. We propose our interpretation of the nine selected resilience attribute definitions elicited from the reviewed academic literature:

1) *Flexibility* is a system's ability to adapt to change with minimum time and effort. It is a complex attribute of resilience consisting of multiple sub-attributes, such as agility, efficiency, and adaptability. It is also enabled by system's robustness, redundancy, and restorative capacity.

Table 9.5 Resilience attributes as proposed in a single source.

Attribute	Definition
Improvisation	Overlapping but different from resilience (Xiao and Cao 2017)
Diversity	Is considered one of the main capabilities of the supply chain resilience (Carvalho 2011)
Shared vision	One of the dimensions of organizational resilience (Chen 2016)
Willingness to learn	One of the dimensions of organizational resilience (Chen 2016)
Cooperative awareness	One of the dimensions of organizational resilience (Chen 2016)
Work enthusiasm	One of the dimensions of organizational resilience (Chen 2016)
The organizational capability to change in responses to unexpected events such as shocks or disasters	One of the two resilience capabilities (Xiao and Cao 2017)
The organizational capability to anticipate the occurrence of such unexpected events	One of the two resilience capabilities (Xiao and Cao 2017)
Nodes density (centralization vs. dispersion)	One of the supply chain resilience factors and measures (Falasca et al. 2008)
Nodes criticality (cost vs. resilience)	One of the supply chain resilience factors and measures (Falasca et al. 2008)
Nodes complexity (flexibility and redundancy vs. efficiency)	One of the supply chain resilience factors and measures (Falasca et al. 2008)
Collaboration among players	One of the supply chain resilience enablers (Soni et al. 2014)
Information sharing	One of the supply chain resilience enablers (Soni et al. 2014)
Sustainability in supply chain	One of the supply chain resilience enablers (Soni et al. 2014)
Risk and revenue sharing	One of the supply chain resilience enablers (Soni et al. 2014)
Supply chain visibility	One of the supply chain resilience enablers (Soni et al. 2014)
Supply chain structure	One of the supply chain resilience enablers (Soni et al. 2014)
Overall situational awareness	• Resilience is a function of McManus (2008) • Lee et al. (2013) agree with and quote McManus (2008)
Resourcefulness	A part of R4 resilience framework (Tierney and Bruneau 2007)
Rapidity plus technology	A part of R4 resilience framework (Tierney and Bruneau 2007)
Integrity	A dimension of organizational resilience (Bouaziz 2018)
Margin	How closely or how precarious the system is currently operating relative to one or another kind of performance boundary (Woods 2017)
Motivation	One of four traits of resilience in small enterprisers (Bhamra et al. 2011)
Perseverance	One of four traits of resilience in small enterprisers (Bhamra et al. 2011)
Optimism	One of four traits of resilience in small enterprisers (Bhamra et al. 2011)
Risk	The concept that helps to define or measure the resilience of systems (Erol et al. 2010c)
Disruptions	The concept that helps to define or measure the resilience of systems (Erol et al. 2010c)
Innovation	Critical for organizational resilience (Menéndez Blanco and Montes Botella 2016)
Efficiency	• The focus of engineering resilience (Dalziell and McManus 2004) • Savings from reusable services; is related to optimal use of resources (Erol et al. 2009)

Table 9.6 Examples of the nine resilience attributes' definitions.

Attribute	Description and comments	Reporting paper
1) Flexibility	The ability of a system to adapt to the changing requirements of its environment and its stakeholders with minimum time and effort	De Leeuw and Volberda (1996)
	To rapidly adapt to its changing environment	Helaakoski et al. (2007)
	The capacity to adapt through the use of efficiency, responsiveness, versatility, and robustness	Golden and Powell (2000)
	Relates to agility and adaptability	Helaakoski et al. (2007)
	The ability of an enterprise to adapt to changing business and stakeholder requirements more efficiently, easily, and quickly	Erol et al. (2009)
	Real-time adjustment of actions in response to actual events	Ma et al. (2018)
	• Flexibility in sourcing: ability to quickly change inputs or the mode of receiving inputs (subfactors: common product platforms, supply contract flexibility, supplier capacity, supplier expediting, alternate suppliers)	Fiksel (2015)
	• Flexibility in manufacturing: the ability to quickly and efficiently change the quantity and type of outputs (subfactors: product/service modularity, multiple pathways, multiple skills, manufacturing postponement, changeover speed, batch size, manufacturing expediting, reconfigurability, scalability, rerouting of requirements	
	• Flexibility in fulfillment: ability to quickly change the method of delivering outputs (subfactors: multisourcing, demand pooling, inventory management, alternate distribution modes, multiple service centers, transportation capacity, transportation expediting)	
	Resilience is a system's ability to bounce back from disruptions and disasters by building in redundancy and flexibility	Hu et al. (2008)
	• Emphasize efficiency, agility, and adaptability	Erol et al. (2010c)
	• Requires the integration of business processes and systems within the enterprise and across the partners of the enterprise. Alignment of business processes and information technology is also an enabling factor for enterprise flexibility that requires a simple and manageable enterprise architecture	
2) Agility	The ability to change rapidly	Fricke and Schulz (2005)
	Agility has been used in conjunction with flexibility as a defining attribute of resilience (Carpenter et al. 2001; Christopher and Peck 2004; Walker et al. 2004)	Erol et al. (2010c)
	Agility characterizes a system's ability to change rapidly	Fricke and Schulz (2005)
	Agility may introduce new risks and vulnerabilities that result in lower resilience	Morello (2002)

(Continued)

Table 9.6 (Continued)

Attribute	Description and comments	Reporting paper
	The system's ability to respond to changes in an uncertain and quickly evolving environment	Helaakoski et al. (2007)
	Resilience involves agility, and it helps a system to reorganize itself rapidly	Christopher and Peck (2004)
	Related to more timely and faster response to rapidly changing business requirements	Erol et al. (2009)
	Agility dimension includes items assessing how easily and rapidly firms adapt to changing circumstances	Kantur and Say (2015)
	The ability of a supply chain to rapidly respond to change by adapting its initial stable configuration	Kamalahmadi and Parast (2016)
	The reactive dimension of resilience, connected with the responsiveness of supply chains in case of disruptions and emergencies	Wieland and Marcus Wallenburg (2013)
	Responding rapidly to changes in demand, in terms of both volume and variety	Cabral et al. (2012)
	The capability to change and move in and out of different domains	Ma et al. (2018)
3) Adaptive capacity	An aspect of resilience that reflects learning; the flexibility to experiment and adopt novel solutions and development of generalized responses to broad classes of challenges	Walker et al. (2004)
	Indicates the ability of systems to revert to their initial state after partial damage	Zhang (2008)
	In order to enhance resilience, adaptive capacity should be increased even after a disruption. Adaptive capacity has also been related to concepts of robustness, agility, and adaptability. The adaptive capacity of a system in the case of disruption can be increased by designing, planning, and building flexibility in systems (Carpenter et al. 2001; Walker et al. 2004; Sheffi and Rice 2005)	Erol et al. (2010c)
	Functional redundancy is a way to increase a system's adaptive capacity	He (2008)
	The ability of an enterprise to alter its strategy, operations, management systems, governance structure, and decision support capabilities to withstand perturbations and disruptions	Starr et al. (2003)
	Connect resilience to redundancy and adaptive capacity	Dalziell and McManus (2004)
	• The ability of an enterprise to alter its "strategy, operations, management systems, governance structure, and decision-support capabilities" to withstand perturbations and disruptions • Organizations that focus on their resilience in the face of disruption generally adopt adaptive qualities and proactive responses. Furthermore, they emphasize positive behavior within the enterprise and employees and look at disruptions as being opportunities for advancement	McManus (2008)
	• The adaptive capacity of a system also reflects the learning aspect of system behavior in response to a disruption. Within organizations, adaptive capacity refers to the ability to cope with unknown future circumstances • Flexibility to change in response to new pressures	Bhamra et al. (2011)

4) Adaptability	• The ability to adapt to changing environments while delivering the intended functionality under varying operating conditions Fricke and Schulz (2005)	Erol et al. (2010c)
	• The ability of an enterprise to respond to the changing business requirements and to change its business processes	
	• Is associated with the integration of new and existing systems and processes	
	The capability to reestablish a state of fit with a changing environment	Ma et al. (2018)
	Ability to modify operations in response to challenges or opportunities (subfactors: seizing advantage from disruptions, alternative technology development, learning from experience, strategic gaming and simulation, environmental sustainability)	Fiksel (2015)
5) Robustness	Requires that systems do not get any damage; characterizes the ability to forego hyper-responses to changing environments	Fricke & Schulz (2005)
	Redundancy and robustness as the two primary attributes of resilience	Haimes et al. (2008)
	• The property of a system that allows it to satisfy a fixed set of requirements, despite changes in the environment or within the system	Uday and Marais (2015)
	• No performance loss is allowed in the case of robustness; a resilient system may permit a (sometimes temporary) performance loss in "bouncing back" from the adverse event	
	The robustness dimensions included items aiming to measure the resistance capacity of the firms	Kantur and Say (2015)
	Not resilience, to take hits with minimal damage to functional capability, which sometimes can capitalize on environmental disruptions to create new options and capabilities	Ma et al. (2018)
	Proactive dimension of resilience	Wieland and Marcus Wallenburg (2013)
6) Vulnerability reduction	• Vulnerability is being at risk and the likelihood of having disruptions	Christopher and Peck (2004)
	• The level of its vulnerability can measure the resilience of a system to a specific risk	
	Vulnerability assessment should be a part of strategic planning for resilience. The level of a vulnerability is the probability of the occurrence of disruption and the extent of the ensuing consequences. As the probability of the occurrence of any disruption and the severity of the consequences increases, the system's vulnerability quotient will be higher	Sheffi and Rice (2005)
	Is a measure of the criticality, preparedness, and susceptibility of the components of an organizational system and that vulnerability is but one of the components of resilience, if a given system is highly resilient then it has a low vulnerability	McManus (2008)

(Continued)

Table 9.6 (Continued)

Attribute	Description and comments	Reporting paper
7) Redundancy	• To keep extra capacity or resources kept in reserve to be used in case of a disruption • Redundancy and flexibility are keys to producing resilience	Sheffi and Rice (2005)
	• The ability of certain components of a system to support the functions of failed components without any considerable effect on the performance of the system itself • Resilience is a system's ability to bounce back from disruptions and disasters by building in redundancy and flexibility	Haimes et al. (2008)
	• The extra capacity of an enterprise to withstand potentially high impact disruptive events • Redundancy and robustness as the two primary attributes of resilience	Hu et al. (2008)
	Redundancy relates to the adaptive capacity of the system. Researchers suggest that resilience can be enhanced by increasing the adaptive capacity of the system, either by ensuring that the system design includes enough redundancy to provide continuity of a function or through increasing the ability and speed of the system to evolve and adapt to new situations as they arise	Dalziell and McManus (2004)
	Functional redundancy as a way to increase a system's adaptive capacity	He (2008)
	• The ability of a system to persist even when some parts of it are compromised • Not resilience • Adding extra functions in a system that can play a role when other functions fail	Ma et al. (2018)
8) Recovery (restorative capacity)	Returns to its baseline conditions	Ma et al. (2018)
	Ability to return to normal operational state rapidly (subfactors: equipment repairability, resource mobilization, communications strategy, crisis management, consequence mitigation)	Fiksel (2015)
	A time interval or period from the time the system failed to function to the time the system gets back with its function; the input to the model is a failure event; the output is the recovery time	Raj et al. (2014)
9) Efficiency	Performance with modest resource consumption	Bhamra et al. (2011)
	Is related to the optimal use of resources	Erol et al. (2009)
	Capability to produce outputs with minimum resource requirements (subfactors: labor productivity, asset utilization, quality management, preventive maintenance, process standardization, resource productivity)	Fiksel (2015)
	Organizational resilience focus management could be understood as a strategy of continuous anticipation and adjustment to disturbances by balancing efficiency and adaptability and, accordingly, an organization's ability to continuously create competitive advantages based on innovations	Kayes (2015)

2) *Agility* is a sub-attribute of flexibility, which describes a system's capacity to change easily, rapidly, and timely in response to changes in an uncertain and quickly evolving environment; if not balanced with other resilience attributes, it may introduce new risks and vulnerabilities and result in lower resilience.

3) *Adaptability* is a sub-attribute of flexibility, which describes a system's capacity to reestablish fit with the environment and presume a new, externally determined equilibrium in the desired state.

4) *Efficiency* is a sub-attribute of flexibility, which describes a system's capacity to achieve desired outcomes with minimum effort and resource consumption.

5) *Robustness* is a system's ability to withstand stresses and demands without suffering any damage, degradation, or loss of function. Robustness enables a system's flexibility to respond to changes in the environment or within the system to ensure that the set of system's requirements continues to be satisfied.

6) *Recovery (restorative capacity)* is a system's capacity to rapidly return to equilibrium in the system's desired state in response to changes in the environment or within the system. It enables the system's flexibility and is measured as a time interval from the moment the system failed to function to the time the system gets back to perform its function.

7) *Redundancy* is the extra capacity of a system, or resources kept in reserve to be used in case of a disruption to prevent system's failure or considerable effect on the performance and continuity of its functions where certain components of a system can support the functions of failed components, or through increasing the ability and speed of a system to evolve and adapt to new situations as they arise. Redundancy is an enabler of the system's flexibility and adaptive capacity.

8) *Adaptive capacity* is an aspect of resilience that reflects learning, the flexibility to experiment and adopt novel solutions and develop generalized responses to broad classes of challenges enabling an organization's capacity to alter its "strategy, operations, management systems, governance structure and decision-support capabilities" to withstand perturbations and to evolve and adapt to new situations as they arise.

9) *Vulnerability level* is achieved through a system's resilience and is a measure of the criticality, preparedness, and susceptibility of the components of an organizational system to risk and the likelihood of having disruptions. If a given system is highly resilient, then it has a low vulnerability level.

Based on the above, we propose our original interpretation of the interrelationships between the nine attributes in Figure 9.1, as elicited from the reviewed academic literature. It is intended to use Figure 9.1 to derive resilience measures for our DTS case study. The system resilience is described via the level of its vulnerability. The higher the system resilience is, the lower is its vulnerability to threats or changes in the system's environment. The decrease in the system's vulnerability can be achieved through the increase in the system's flexibility, which, according to Erol et al. (2010c), is a function of agility, efficiency, and adaptability. Flexibility, however, is insufficient for a comprehensive system resilience evaluation as it does not include the system's adaptive capacity that captures system's ability to evolve and settle in a new equilibrium rather than simply returning to the same equilibrium as experienced before the impact. The higher the system's adaptive capacity is, the more resilient it is. The system's flexibility and adaptive capacity are enabled by the system's physical characteristics such as robustness, restorative capacity, and redundancy.

9.3.2.2 Why? (Resilience Triggers)

The notion of resilience requiring some form of a trigger to manifest itself is featuring in all 91 papers selected for the review. There are, however, variations in interpretations of what the "trigger" is, with the area of application and trigger types summarized in Table 9.7.

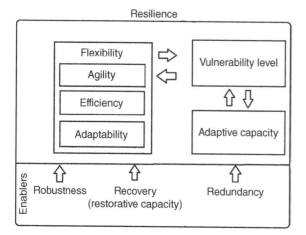

Figure 9.1 Resilience attributes' interrelationships.

As depicted in Table 9.7, resilience is triggered by changes, either in circumstances, processes, or environment. These changes are predominantly negative but could be any that are significant enough to trigger the system's reaction. The uncertain nature of these changes is featuring prominently in the resilience-focused academic literature with more than a third of the papers reporting the challenges of building resilience against "unexpected"/"potential" events and "unforeseeable"/"unknown"/"unpredictable" hazards/changes/disturbances.

Some papers take the uncertainty argument further by firmly linking resilience to uncertainty. For example, Park et al. (2013) state that resilience thinking requires consideration of all types of inherent uncertainties – including internal variability and dynamics, external drivers, and shocks. Hollnagel and Woods (2006) add that due to uncertainty, the only potential for resilience could be measured but not resilience itself. They compare the measurement of reliability as the probability that a given function or component would fail under specific circumstances to measuring resilience that includes recovery from irregular variations, disruptions, and degradation of expected working conditions.

9.3.2.3 How? (Resilience Mechanisms and Measures)

System design must incorporate resilience to provide stakeholders with the solution most appropriate for their life cycle needs (Ferris 2019). To incorporate resilience, the resilience mechanisms and parameters need to be determined, followed by establishing metrics for the system's resilience. Jackson et al. (2015b) highlight the challenge of measuring system resilience as it means different things to different people and in different contexts and concern with different types of systems and different aspects of resilience. The lack of common terminology and agreed resilience framework in a vast and fast-growing resilience literature exacerbates this problem even further. In response to this challenge, several reviewed papers propose their conceptual models and perspectives of resilience to support analysis of the system performance and behavior in response to a threatening situation or a change, as well as their solutions to measure them. Below is our selection of the resilience mechanisms and measures, as reported in the reviewed literature for consideration in our research.

Table 9.7 Resilience triggers.

Source	Area of application	Terms used
Trigger type: negative change		
Akgün and Keskin (2014)	Organizations (business)	Disruptions, environmental turbulence that involves technological discontinuities
Azvedo et al. (2011)	Supply chain	Disturbance occurrence
Berkes (2007)	Organizational learning (HRM)	Previous crises
Bhamra et al. (2011)	Multidisciplinary (lit review)	The diverse and ever-changing environment
Bukowski (2016)	System of systems	The impact of a hazard (technical) or disaster (social)
Carpenter et al. (2001)	Socioecological system	Disturbance
Carthey et al. (2001)	Organizations (healthcare)	Adversity
Carvalho and Machado (2007)	Supply chain	Disturbance topology: natural events, accidents, human-made events
Cooper et al. (2014)	Organizations (HRM)	Failure
Crichton et al. (2009)	Organizations (emergency response)	Accident or incident
Dalziell and McManus (2004)	Systems	Disaster event, failure
Kamalahmadi and Parast (2016)	Supply chain	Disruptions
Kantur and Say (2015)	Organizations (HRM)	Natural disasters and crises situations including economic, political, and social events
Downes et al. (2013)	Multidisciplinary	Undesirable changes
Erol et al. (2009, 2010c)	Enterpriser (system)	Turbulent change, disruption
Falasca et al. (2008)	Supply chain	Disasters, disturbances due to internal and external risks
Ferris (2019)	Engineering (system)	Threat encounters, adversity
Gilbert et al. (2012)	Business transformation	Changes in business environment turning business model into disruptive
Gomes (2015)	Enterprise architecture (system)	External and internal disruptions
Han et al. (2012)	System of systems	Different levels of failure
Henry and Ramirez-Marquez (2010)	Enterpriser (systems)	Man-made and natural disasters and calamities
Kayes (2015)	Organizations (learning)	Disturbances and organizational failures
Kim et al. (2015)	Supply network	Disruption
Lee et al. (2013)	Organizations (emergency response)	Emergencies, crisis, rapid change
Limnios et al. (2014)	Organizations (resilience architecture)	External and internal disturbances
Lucero (2011)	Organizations (DoD)	New CONOPs, external threats, late discovery of defects, suboptimal engineering tools that are not standardized

(Continued)

Table 9.7 (Continued)

Source	Area of application	Terms used
Mafabi et al. (2012)	Organizations (knowledge management)	Environmental demands and threats
Mallak (1999)	Organizations (employee resilience)	Rapid amounts of change from many sources
McManus (2008)	Organizations (business)	Hazard event
Munoz and Dunbar (2015)	Supply chain	Operational disruptions
Pettit et al. (2016)	Operational resilience (military aviation)	Changing environment, environmental exposure, disruptive conditions
Ponomarov and Holcomb (2009)	Supply chain	Disruptions
Raj et al. (2014)	Supply chain	Consequences of unavoidable risk events or disruptions
Rosenkrantz et al. (2009)	Service-oriented networks (system)	Failures and attacks
Tierney and Bruneau (2007)	Disaster resilience	Environmental shocks and disruptive events
Uday and Marais (2014)	System of systems	Change over their lifetimes to recover from
Uday and Marais (2015)	System of systems	Disturbances and disruptions
Walker et al. (2004)	Socio-ecological systems	Disturbance
Ahlquist et al. (2003)	Enterpriser	Disruption/disaster
Trigger type: any changes		
Alliger et al. (2015)	Team resilience	Stressors and chronic and acute challenges, team challenges (resources, assignments, conflicts etc.); normalization of deviance
Bardoel et al. (2014)	HRM (employee resilience)	Individuals' exposure to subjectively significant threat, risk, or harm
Chen (2016)	Organizations (R&D teams and businesses)	Incremental change and sudden disruptions; unexpected, stressful, adverse situations
Cumming et al. (2005)	Complex systems	A potential and specific change in the system
De Sanctis et al. (2018)	Organizations (HRM context)	Unexpected (i.e. failures or other disruptions) and expected events (i.e. cross-training, personnel relocation)
Erol et al. (2010a, 2010b)	Enterpriser (system)	Expected and unexpected threats/events; changing environment, disruptive event
Fraccascia et al. (2017)	Complex systems	Operational management of internal and external resources, disruptions, and unexpected events
Hoffman and Hancock (2017)	Organizations (HRM)	Circumstances that push the system beyond the boundaries
Kativhu et al. (2018)	Organizations (business)	Disruptive events are weighted according to their predictability, potential to disrupt a system, and whether the disruption is internal or external

Table 9.7 (Continued)

Source	Area of application	Terms used
Menéndez Blanco and Montes Botella (2016)	Organizations (business)	Unexpected disruptions and change measures to achieve competitive advantage
Sitterle et al. (2015)	Resilience engineering (DoD systems)	External and internal changes
Trigger type: unexpected change (uncertainty)		
Annarelli and Nonino (2016)	Organizations (supply chain)	Disruptions and unexpected events, e.g. natural disasters, pandemic disease, terrorist attacks, economic recession, equipment failure, and human errors
Ayyub (2014)	Disaster risk reduction	Changing conditions and disruptions, natural disasters, includes uncertainty context
Baran (2016)	Organizations (HRM)	Catastrophic failures, calamities
Bouaziz (2018)	Organizations (HRM)	Terrorist attacks, economic downturns, global financial crises, uncertainties in the competitive market, and political and social conditions
Buchanan et al. (2015)	Resilience engineering (DoD systems)	Disruption, adversity in the context of uncertainty
Burnard and Bhamra (2011)	Organizations (complex systems)	Systematic discontinuities; potentially devastating implications of disruptions; unexpected complexity and disorder
Cabral et al. (2012)	Supply chain	Unexpected disturbance, in conjugation with environmental responsibilities
Conz and Magnani (2019)	Organizations (business)	An unexpected critical event, shocks
Goerger et al. (2014)	Resilience engineering (DoD systems)	Rapid changes in missions and mission requirements, as well as the emergence of new asymmetric threats in the operational environment, create uncertainties and surprises
Ho et al. (2014)	Organizations (HRM)	Disruptions to organizational life through natural disasters or economic crises
Hollnagel and Woods (2006)	Systems (resilience engineering)	Complexity, unexpected events
Jackson et al. (2015b)	Engineering (systems)	Threat conditions and events
Jackson et al. (2015a)	Engineering (systems)	Threat event
Fiksel (2006)	Enterpriser (systems)	Disturbances and turbulent changes, perturbations and unexpected changes
Fiksel (2015)	Complex supply chains	Unexpected disruptions, turbulent change, and catastrophic disruptions
Francis and Bekera (2014)	Engineering (system)	Realized risk events in the context of uncertainty, unforeseeable hazards in design and management activities
Giannoccaro et al. (2018)	Hard (industrial) and soft (social) infrastructure systems	Disturbance
Lengnick-Hall et al. (2011)	Organizations (HRM)	Unexpected, stressful, adverse situations
Ma et al. (2018)	Organizations (emergency response)	Disruptions, unforeseen events

(Continued)

Table 9.7 (Continued)

Source	Area of application	Terms used
Mansouri et al. (2009)	Transportation system of systems	Illness, change, or misfortune, potential disruption
Nuss et al. (2017)	System architectures	Changing conditions, failures or losses before their failures are investigated to identify root causes
Park et al. (2013)	Systems (emergency response)	Changing conditions, failure, unknown hazards
Sawalha (2015)	Organizations (business)	Disasters and crises, potential risks
Sharma and Sharma (2016)	Team resilience	Setbacks, significant changes, the potential negative effect of stressors
Shirali et al. (2016)	Process industry (systems)	Regular and irregular disruptions and disturbances
Small et al. (2017)	Resilience engineering (systems)	Environmental and adversarial threats and future evolving threats
Soni et al. (2014)	Supply chain	Unpredicted disruptions
Tran et al. (2017)	Complex systems	Disruptions, potential threats, or failures
Woods (2015)	Organizations (resilience engineering)	Trauma, unanticipated side effects, and sudden dramatic failures
Woods (2017)	Organizations (resilience engineering)	Disruptions, external changes, or pressures
Xiao and Cao (2017)	Organizations (business)	Crises and disasters, emergency, and disruption in an ever-changing and uncertain environment

9.3.2.3.1 Resilience Mechanisms

The literature reports different perspectives on how resilience works, with the most common listed below followed by an explication of each:

a) Engineering and ecological resilience
b) Conditional and total resilience
c) Structural resilience
d) Static and dynamic resilience
e) System equilibrium and state transitions resilience models

Engineering and ecological resilience. Pritchard and Gunderson (2002) and Dalziell and McManus (2004) distinguish two types of resilience: ecological and engineering, with the former focusing on the magnitude of disturbance the system can withstand before it is modified and the latter on the speed of system's return to equilibrium. High engineering resilience implies maximizing the efficiency of systems and processes, whereas ecological resilience is concerned with the system's flexibility to continue functioning in the face of adversity and often at an efficiency expense.

Conditional and total resilience. Han et al. (2012) develop their original propositions to facilitate resilience measurement in a system-of-systems (SoS) context. Conditional resilience is a percentage of the SoS's performance that remains after a failure of a specific set of systems within that SoS. The total resilience is how the SoS's performance gradually degrades as more and more of its components fail until the SoS performance breaches the "minimum required performance" threshold constituting a total failure of the SoS.

Structural resilience. We propose the term "structural resilience" as a superficial construct to represent the resilience perspectives from the academic literature that focuses on the resilience of system architectures. For example, Gomes (2015) proposes dimensions and activities for the enterprise architecture (EA) development focused on supporting the system resilience requirements. Kim et al. (2015) formulate a structural perspective of the supply chain network resilience. Their study conceptualizes supply network disruption and resilience by examining the structural relationships among entities in the network architecture. They propose that resilience can change with the change in the network architecture configuration. The network resilience is measured by the number of system failures recorded for different system configurations, each achieved by manipulating the network structural factors, namely, density, length, connectivity, and criticality.

Static and dynamic resilience perspective. Giannoccaro et al. (2018) propose to divide the resilience concept into static and dynamic perspectives. Static resilience focuses on the system's ability to absorb disturbance and bounce back to the original equilibrium state maintaining its core functions when shocked, with its focus on the recovery of the original shape and features once stretched (robustness) and the capacity of the system to take alternative positions to respond better to change (flexibility). Dynamic resilience is defined as the system's ability to evolve and to move to a new, more favorable equilibrium state from which to react to the disturbances experienced. Dynamic resilience concerns the adaptive capacity of the system, which can react to disturbance by changing its structure, processes, and functions in order to increase its ability to persist or in an organismic context to evolve. A similar proposition is featured in Lengnick-Hall et al. (2011) that presents a small literature review of two resilience perspectives: one focused on returning to original state post-disruption, while the other adds development of new capabilities and system modifications to keep pace with and create new opportunities. Park et al. (2013) add to the discussion a system view of resilience, where it is represented as the outcome of the dynamic system behavior that includes sensing, anticipation, learning, and adaptation. In this approach, resilience displays characteristics of the system's emergent property that enables the system to prepare for and to live with uncertainty, where hazards are not fully known. Haimes et al. (2008) highlight the importance of understanding the organizational emergent properties, which they define as those system features that are not designed in advance but evolve based on sequences of collected events that create the motivation and responses for properties that ultimately emerge into system features. System's resilience response is an example of the system's emergent properties where it cannot be directly measured within the "as-is" state of a system; it can only be predicted. In other words, a reliable measure of enterprise resilience cannot be deduced from the normal operation of the enterprise. Enterprise resilience is the sum of those characteristics and abilities that will eventually evolve in the case of disruptive events, and so in order to measure resilience, it is important to identify both the inherent and adaptive attributes of the enterprise to confront such events, as will be defined in the later sections of this chapter.

The concept of dynamic resilience is also reflected in the organismic evolution of a complex system through governance mechanisms by Keating and Bradley (2015), as further engineered by Keating et al. (2019) into enterprises using a pathological approach to the enterprise governance. Another interpretation of the static and dynamic resilience perspectives is proposed by Erol et al. (2010c): wherefrom one perspective, resilience relies upon the anticipating unexpected disruptive events and designing solutions to eliminate errors, and the other suggests that resilience is more about detecting unexpected events sooner when they can be more easily corrected and building the capacity to recover from such events. Rose and Liao (2005) also distinguish two characteristics of resilience that they title inherent and adaptive. They define inherent resilience as the ability under normal circumstances and adaptive resilience as the ability in crises to draw upon ingenuity and extra effort.

Small et al. (2017) also consider a similar approach to resilience perspectives but in the context of defense organizations: (1) system response to known threats using existing system characteristics and (2) developing new system features to deal with new threats. The first perspective is titled "**Mission resilience**," defined as "the ability of a system to repel, resist, absorb, and recover from environments and threats that occur on planned missions." The second perspective, titled "**Platform resilience**," is defined as "the ability of a system to adapt to new missions and new threats."

System equilibrium and state transitions resilience models. Resilience representation as a transition between the system states in response to change is featured prominently in the reviewed literature, for example, in Bhamra et al. (2011), Buchanan et al. (2015), Burnard and Bhamra (2011), Carpenter et al. (2001), Holling et al. (1995), Limnios et al. (2014), Mafabi et al. (2012), Park et al. (2013), Sawalha (2015), Tran et al. (2017), and others. We selected six examples of this perspective from the reviewed literature for consideration in our research as follows:

- Adaptive cycle
- Four qualities model
- Resilience Architecture Framework
- Operational resilience
- Thermodynamic system states model
- Five-state model
- Generic state-machine model

Holling et al. (1995) introduce a concept of the "adaptive cycle," where "the dynamical systems such as ecosystems, societies, corporations, economies, nations and socio-ecological systems do not commonly function in some stable or equilibrium condition, but pass through the following four phases [system states]: rapid growth and exploitation (r), conservation (K), collapse or release ("creative destruction" [Ω]), or/and renewal or reorganization (α)." Resilience changes throughout the adaptive cycle, and different aspects of resilience assume prominence at different times during the cycle.

Hollnagel and Woods (2006) presented a four qualities model that considers system resilience to be a form of control. A system is in control when "it is able to minimise or eliminate unwanted variability, either in its own performance, in the environment, or in both." The authors propose three qualities a system must have to be in control and therefore resilient, with time as a fourth, dependent quality. The three main qualities are (1) anticipation (based on the knowledge what to expect), (2) attention (based on the competence of what to look for), and (3) response (based on resources and knowledge of what is required to be done). These qualities must be exercised continuously and are not sequential. The transition between them could occur in any direction and between any of the pairs of these qualities. The system must constantly be watchful and prepared to respond and constantly update its knowledge, competence, and resources by learning from successes and failures – its own as well as those of others. This concept of the importance of continuous adaptive control is also evident in the meta-functions of the complex system governance model (Keating and Bradley 2015).

Limnios et al. (2014) propose a Resilience Architecture Framework that examines the relationship between the magnitude of system resilience and desirability of system state. In this model, the relationships are presented as four possible scenarios:

1) "Adaptability quadrant" (high magnitude and high desirability)
2) "Vulnerability quadrant" (low magnitude and high desirability)
3) "Rigidity quadrant" (low magnitude and low desirability)
4) "Transient quadrant" (high magnitude and low desirability)

The Resilience Architecture Framework is used to differentiate between two opposing manifestations of resilience as either offense (adaptation) or defense (resistance) to internal or external disturbance and then to recognize when resilience is desirable or undesirable depending on the system state.

Pettit et al. (2016) propose an operational resilience model using their variation of state transition model in a defense context. They define operational resilience as "the ability of an organisation to avoid disruption, and if unsuccessful in avoidance, to recover as quickly as possible," based on a priori assumption that organizations change their modes of operation (states in the Jackson et al., 2015b, model) to achieve this. The operational resilience relies on policies and procedures associated with the transition between operational modes. These authors are working on identifying generalizable principles and practices that may enhance resilience in all types of operations and organizations concerning effective transition between the modes.

Fiksel (2003) presents his simplified conceptual model of thermodynamic system states to explain the state transition-based mechanism of resilience. According to this model, each system has a stable state representing the lowest potential energy at which the system maintains its order and function. When triggered by a change event, the stable state might transition to the higher potential energy adjustment state. The system operating within a narrow band of possible states constitutes low resilience. These systems can resist small disturbances from its equilibrium state but have low resilience to larger-scale or high-impact events. The broader the system's operational band of possible states is, the larger-scale or higher-impact events the system can cope with. Some systems have multiple equilibrium states and can shift to different states under certain conditions, enabling the system to tolerate larger perturbations. However, the shift to a different equilibrium point represents a fundamental change in the system's structure and function.

Henry and Ramirez-Marquez (2010) formulate a five-state conceptual model of system resilience and associated quantitative approach to measure it. The five states of resilience are as follows: stable original state, disruptive event, impact/disrupted state, resilience response, and stable recovered state. A system passes through five states starting from the original stable state. A disruptive event, which could be a combination of internal and external factors, triggers the system's disrupted state. Resilience action is then taken in response, enabling the system to bounce back to a recovered state.

Jackson et al. (2015a, 2015b) propose a system resilience model consisting of seven states and their associated transitions. The states range from a fully operational system to the system's decommissioning. The transitions between the states are in essence transition scenarios that contain a description of the transition trigger, such as threat encounter, system response to mitigate the threat, systems' post-threat restoration, decommissioning event, and to what state the system is transitioning as a result. These authors present an initial 28 of these scenarios with more likely to be identified as the work progresses.

9.3.2.3.2 *Measuring Resilience* According to Ferris (2019), the assessment of the system resilience enables the system stakeholders to determine if the system resilience level is appropriate to achieve its purpose and how the changes in the system design and operation affect this level. The measurement method should incorporate the effect of the uncertainty on the system and account for interdependencies between the resilience and other system attributes. Ideally, the resilience measure should also be useful to guide management choices during the system life cycle. There is a growing pool of academic literature devoted to the development and application of resilience measurement methodologies. Some useful examples follow that build on earlier reviewed resilience concepts.

Carpenter et al. (2001) propose to measure resilience based on the adaptive cycle model by Holling et al. (1995) in the context of their two original case studies of socio-ecological systems. The authors emphasize the importance of a careful selection of the key system state variables associated with the adaptive cycle phases, and determination of the preferred state, the timescale, and period within which the system resilience is to be measured. Walker et al. (2004) add the notion of the "cross-scale effects" to the system resilience consideration. According to the authors, the adaptive cycle model is based on observed system changes and does not imply fixed or regular cycling. Systems can move back from K toward r, or from r directly into Ω, or back from α to Ω, and the cycles occur at several scales and interact with other cycles across multiple scales. Cumming et al. (2005) also use adaptive cycle model in combination with potential future scenarios to distinguish different future states of the system in question. The authors propose to determine a priori notion of system identity that constitutes the system's "healthy" resilient state and a priori set of alternative future system states that would indicate a loss of identity for the purpose of operational assessment of the system's resilience.

The five-state model by Henry and Ramirez-Marquez (2010) is used to develop measures of resilience. To measure the resilience of the system of interest, the functions performed by system S are captured as performance variables "PVi." These should be quantifiable and measurable. While the system could perform many functions, only the functions that are relevant to resilience are selected. For example, the system would require resources to perform its functions, so these are captured as resource variables "RVi." Commonly, one or more resources would be required to perform a function. Also, one resource could be required by more than one function. Hence a many-to-many relationship could exist between performance variables and resource variables.

Erol et al. (2009) then propose the attributes to base the resilience metrics upon and define enterprise resilience and enterprise flexibility as a function of agility, efficiency, and adaptability and list the attributes of SOA, the business outcome of each attribute, and each attribute's impact on agility, efficiency, and adaptability.

The (Ferris 2019) system resilience measurement and analysis approach lie in recognizing the purpose of engineering of systems to provide a service to one or more stakeholders whose primary interest in the system is the service it provides rather than the system itself. Ferris approach is based on a variation of the generic state-machine model by Jackson et al. (2015a, 2015b). These are tailored to a specific system context to measure resilience by enumerating the events that could cause state transitions (E), then selecting the number of the system's functional capability levels (F), followed by determination of the number of system states (Q). The values are then allocated to the (E), (F), and (Q) variables for input into the following two functions to enable discrete automation of the resilience of the system in question:

a) Function N that relates every pair of elements (e_t, q_t) from E and Q to the next state q_{t+1}, i.e. the state transitions

b) Function W that relates every pair of elements (e_t, q_t) to an element in F, i.e. f_{t+1}.

The conventional system design approach to decide between multiple design options is to establish a trade-space model to find a figure of merit (FOM) for each contending choice. To develop the system resilience measures for the purpose of finding FOM of the realized and proposed system designs, the authors first select the set of attributes or measures that are instrumental for their system's resilience. The number of attributes needs to be sufficient to account for all key influences but carefully selected to ensure the impact of each of the attributes on the decision is not diluted. Then they determine relative importance weights for each attribute in the set followed by determination of value-for-scale functions for each attribute in the set. The measure of each attribute likely to be achieved under normal operating conditions is estimated for each design proposal in the analysis, followed by determining the value of each proposal, usually as the sum of the product of the attribute weights and the value-for-scale functions of the achievable quantity of that attribute.

9.4 DTS Case Study Methodology

Our approach uses our interpretation of Ferris (2019) system resilience measurement and analysis methodology combined with the DoDAF methodology and our original resilience attributes and their interrelationships concept in Figure 9.1, tailored to our DTS case study to develop a model-based framework for DTS design for resilience. We also use an original survey to determine the current and desired DTS architectures. These methodologies are further outlined below.

9.4.1 DTS Resilience Measurement Methodology

We use Ferris (2019) system resilience measurement and analysis approach outlined in Section 9.3.2.3.2. The authors propose five generic resilience attributes manifestation to base the system's resilience measures' development upon. The DTS resilience is measured using the main resilience manifestations. These are the failure probability of the system elements and the probability of occurrence of the external threats to the system and the nature of the impairments they cause. Restorative transition events occur as a result of maintenance of the system, in whole or in part, and can be incorporated into the model as distributions of the time to affect a repair. We propose the DTS resilience driving factors in Figure 9.2 to be further refined based on results of our DTS Resilience Survey outlined in Section 9.4.3.

We map our set of nine resilience attributes in Figure 9.1, with the system resilience manifestations proposed in Ferris (2019) in Table 9.8 to form the basis for development of the DTS Resilience Survey, which is further described in Section 9.4.3.

9.4.2 DTS Architecture

We choose to use DoDAF methodology to conceptualize the DTS design. This methodology provides a simplified and well-defined view of the system's key elements, enables the alignment of technology and business, and supports system's integration. We also propose to enhance the DoDAF views with the human systems engineering framework presented in Chapter 1 of this book to ensure DTS workforce is placed on more equal footing with design elements such as hardware and software. When utilized together, systems integration and system architecture result in connectivity within the extended enterprise and help in achieving agility, adaptability, and flexibility that should support the resilience at the extended enterprise level. EA also helps to identify the interrelationships and interdependencies between the elements of the extended enterprise and therefore can be used as a tool to assess the vulnerabilities. EA thus has a significant role in creating enterprise resilience by addressing design parameters that can lead to increased system's resilience. EA has a role in facilitating the integration and alignment of enterprise information systems, further leading to timely response and adaptation to change (Hoogervorst 2009).

Erol et al. (2010c) propose a framework with two key enablers of enterprise resilience. The first enabler is the capability of an enterprise to connect people, processes, and information in a way that allows the enterprise to become more flexible and responsive to the dynamics of its environment, stakeholders, and competitors. Such connectivity requires integration within the enterprise

External threats to the training system:
- Fluctuations in demand: too low/too high for the TS capacity
- Resource availability
  - Instructor availability
  - Sessions availability
  - Equipment availability
  - Course capacity (min/max numbers)
- Resource quality and flexibility
  - Impacted by the lacking policy
  - Lack of training composability and modularity
  - Poor preparation of TS specialists
  - Poor quality control practices (feedback loops)
  - Lack of OQE
- Project issues (FMS repatriation; non-compatible formats)
- Lack of multidimensionality (multi-factorial influences)
- Disconnect of training products across the lifecycle; services; training continuation

Probability of occurrence of the external threats to the system and the nature of the impairments they cause

Training system elements:
- Training courses (content; format; theory and practice balance; versatility to address different learning styles)
- Training pipelines (workgroups, specializations, equipment application-career-general skills; from individual to collective, sequence, hierarchy (from simple to complex or from career to equipment application); composability and modularity)
- Training delivery (mode face-to-face; instructor driven or facilitated student centered – e-learning – blended; duration; location)
- Training lifecycle: from acquisition to disposal (e.g. project delivered training; transition training, repatriation of training from Foreign Military Sales);
- Training resources:
  - Workforce (Training Specialists; Instructors; Support Personnel (administrators, maintainer/operators of training equipment; planners etc.)
  - Technology (simulators, emulators; e-learning packages; various training support equipment, e.g. computer, stationary)
- Target audience (trainees)
- Workplace:
  - Job role/tasks (core and auxiliary)
  - Supervisory arrangements
  - Capability (needs and requirements)

The failure probability of the system elements

Distributions of the time to effect repair (restorative transition events)

Nature of the impairments to the elements that could be caused by external threats
- Ineffective training for the work role:
  - Undertraining
  - Overtraining
  - Irrelevant training
  - Lack of appropriate practice opportunities
- Ineffective training delivery
  - Different Learner styles not addressed
  - Disrupted/disjointed (instructor availability, venue availability; poor schedule planning with competing priorities)
  - Unbalanced supply/demand relationship (lack of availability; insufficient demand)
  - Poor transition planning
- Disjointed training continuum:
  - Training products not logically progress throught the training continuum from individual to collective
  - Disjointed governance across the training continuum
Stakeholder competing interests and priorities are not reconciled

Figure 9.2 DTS resilience driving factors.

Table 9.8 Resilience attributes and their manifestations.

Resilience attributes	Ferris (2019) resilience manifestations
Flexibility (a function of): • Agility • Adaptability • Efficiency	• Agnostic of threat type • Generalizable to any engineered system • Agnostic of the outcome, both short and long term, following a threat encounter
Enablers (robustness, recovery, and redundancy)	Cognizant that performance is multidimensional
Adaptive capacity	Cognizant that threat events occur at any time during the life cycle, with statistically predictable frequency

Source: Based on Ferris (2019).

and across the partners, suppliers, and customers of the enterprise. The second enabler is the alignment of information technology and business goals. Attaining this alignment requires simplification of the underlying technology infrastructure and creation of a consolidated view of, and access to, all available resources in the enterprise. A well-defined EA provides a simple and consolidated view of the enterprise and supports the integration and connectivity at inter- and intra-enterprise levels.

Our DTS architecture development is guided by the Cumming et al. (2005) five-step approach as follows:

Step 1: Define current system: First conceptualize the four essential system attributes (structural components, functional relationships, innovation, and continuity) most relevant to our research hypothesis. Second, for each of these, select specific variables that are most likely to change in response to changes in the intensity or extent of the drivers we selected. The degree of identity change that occurs in multivariate space (and in particular, whether the system crosses any key identity thresholds) then becomes the response variable of interest that best represents whether or not the systems are resilient to the changes brought about by increased connectivity.

Step 2: Define possible future systems – same and different identities. The second element of our research design is to define a small number of plausible future identities for each study system using a scenario-building approach. We attempt to define a set of systems that our system might conceivably become, including entirely new systems as well as systems that retain the same identity while having experienced growth and reorganization. If we see no scope for the maintenance of system identity, the system lacks resilience; the focus of the exercise then shifts to the management of change, both to mitigate negative outcomes and enhance positive outcomes. An important objective is to specify quantitatively the amount of change in the key variables comprising identity that would constitute a new system.

Step 3: Clarify change trajectories. This step is undertaken interactively with the second step. It involves defining the main causes of change in the system, with particular relevance to their impacts on properties of interest. At this stage, we also identify the kinds of perturbation and disturbance against which resilience will be assessed.

Step 4: Assess likelihoods of alternate futures. Are we likely to lose the system properties of interest, or not? If system identity is likely to be lost, our system is not very resilient. By contrast, if our system is likely to maintain its identity across a broad range of scenarios, it is resilient. If both identity and the likelihood of identity change in alternate scenarios are rigorously and quantitatively determined, the likelihood of a change in identity (and its magnitude) will provide a quantitative measure of resilience.

Step 5: Identify mechanisms and levers for change. These mechanisms are used to suggest manipulation points, or levers, at which changes in the system trajectory could be brought about. Ultimately, if we (that is, the community of stakeholders) are working with an agenda in which we desire to facilitate or prevent certain kinds of system change, this step will identify some of the key issues that would need to be addressed by planners and policy makers (policies often constitute mechanisms).

Source: Cumming et al. (2005). © Springer Nature.

9.4.3   DTS Resilience Survey

We use the DTS survey to determine the relative weights and the value-for-scale functions for each attribute in accordance with the (Ferris 2019) system resilience measurement approach. The survey's purpose is to determine resilience of the current and desired DTS architecture as perceived by the DTS stakeholders. Our approach to the survey design and conduct is outlined below.

9.4.3.1   DTS Resilience Survey Design

The survey questionnaire in Annex A is formulated for the five key stakeholder groups, namely:

- Training specialist
- Defense training instructor
- Trainee
- Supervisor
- Capability stakeholders

The survey contains demographic section followed by the DTS resilience questions informed by the resilience attributes and their manifestations in Table 9.8. An example of the resilience questions developed for training specialist is provided in Table 9.9. The survey questionnaires for the remaining four target audiences contain the same value questions but formulated from the perspective of their respective groups. For example, Question 2.1 from Table 9.9 reads for different target groups as follows:

- Training specialist and instructor:
  "In your experience, does defense individual training adequately prepare defense personnel to perform their collective [unit, task group and joint operation] tasks?"
- Trainee:
  "In your experience, does defense individual training adequately prepare you to perform their collective [unit, task group and joint operation] tasks?"
- Supervisor:
  "In your experience, has defense individual training adequately prepared your subordinates to perform their collective [unit, task group and joint operation] tasks?"
- Capability stakeholder:
  "In your experience, has defense individual training adequately prepared your capability users to perform their collective [unit, task group and joint operation] tasks?"

9.4.3.2   DTS Resilience Survey Conduct

The ADF ethics committee has approved our survey design. The survey is implemented in a phased approach, with the phase 1 that is currently under way. The survey implementation phases are as follows:

Phase 1: Inaugural distribution. The RAAF training establishment has been selected for the inaugural survey conduct. Trip to Melbourne to meet inaugural RAAF stakeholders. Survey results

Table 9.9 DTS resilience questions for training system specialist.

#	Main question	Supporting questions	Possible values
1	Is DTS and its products agnostic of threat (change) type (agility)?		
1.1	To your knowledge, is the following statement about defense training resources true?	• Insufficient • Inconsistent across defense • Some gaps due to scheduling and sharing issues • Sufficient	Not sure Never Sometimes Often Always
1.2	To your knowledge, is the following statement about defense training true?	• There is often a long wait to attend courses • Inconsistent across defense • Often under-subscribed • Available and fully subscribed	Not sure Never Sometimes Often Always
1.3	In your experience, are the following statements about defense use of training technology true?	• Obsolescent (legacy) and hard to replace • Different across training pipelines making it hard to learn • Meets serving members' expectations • Meets new recruits' expectations • State of the art	Not sure Never Sometimes Often Always
1.4	Does defense training develop students' learning skills, e.g. by focusing on study techniques and best learning practices?		Not sure Never Sometimes Often Always
2	Is DTS and its products generalizable to any capability systems engineering (adaptability)?		
2.1	In your experience, does defense individual training adequately prepare defense personnel to perform their collective (unit, task group, and joint operation) tasks?		Not sure Never Sometimes Often Always
2.2	To your knowledge, can same defense courses be delivered to different audiences?	• Different ranks of the same workgroup within one service • Same workgroup across two or three services • More than one workgroup within one service • More than one workgroup across two or three services • Cross-defense	Not sure Never Sometimes Often Always
2.3	To your knowledge, can be a defense course delivered as a whole to one group and partially (as a selection of some of its modules) to another group?	• Different ranks of the same workgroup within one service • Same workgroup across two or three services • More than one workgroup within one service • More than one workgroup across two or three services • Cross-defense	Not sure Never Sometimes Often Always

(*Continued*)

Table 9.9 (Continued)

#	Main question	Supporting questions	Possible values
2.4	To your knowledge, can defense training be delivered at different levels?	• As basic/intermediate/advanced skill levels • If different depth is required by different workgroups • Cross-service collaboration • For professional development	Not sure Never Sometimes Often Always
2.5	To your knowledge, is defense training delivered in the following formats?	• Instructor led (e.g. PowerPoint presentation) • Instructor facilitated (e.g. facilitator answers your questions) • Apprenticeship • Online course • Reading package • Combination of the above	Not sure Never Sometimes Often Always
3	Is DTS and its products agnostic of the outcome, both short and long term, following a threat encounter (efficiency)?		
3.1	Have you experienced the following issues during defense training?	• Over-training • Under-training • Irrelevant training • Negative training	Not sure Never Sometimes Often Always
3.2	To your knowledge, does the following occur during defense training?	• Training session not reaching minimum trainee numbers • Training bottleneck (many waiting to start) • Training backlog (many waiting to enroll) • Training failures • Non-release of nominees for training	Not sure Never Sometimes Often Always
3.3	To your knowledge, how often on job training (OJT) is used to consolidate defense training?		Not sure Never Sometimes Often Always
3.4	To your knowledge, are the following approaches used in the workplace to consolidate defense training?	• Competency log • Task book • Competency log • Task book • Authorization matrix • Platform endorsement • Work under supervision	Not sure Never Sometimes Often Always
4	Is DTS and its products cognizant that performance is multidimensional (enabler: robustness)?		
4.1	Have you experienced a diminished defense training service similar to examples listed below?	• Total loss of training service • Partial loss of service (limited availability or reduced content) • Delivered but at a lower quality • Delivered but at a higher cost • Service is of the lowest cost and at the required standard	Not sure Never Sometimes Often Always

Table 9.9 (Continued)

#	Main question	Supporting questions	Possible values
4.2	Have you experienced the following restoration scenarios of defense training conduct after disruption?	• Remain disrupted with no restoration plans • Next reporting period • For the next session of the course • On the same day • Immediately and comprehensively	Not sure Never Sometimes Often Always
5	Is DTS and its products cognizant that performance is multidimensional (enablers: recovery and redundancy)?		
5.1	To your knowledge, is defense training affected by the following factors?	• Unavailability of instructors • Unavailability of appropriate training equipment and facilities • Unavailability of "on location" training • Unavailability of suitable schedule options	Not sure Never Sometimes Often Always
5.2	In your opinion, are the following statements about defense personnel transitioning between platforms and/or equipment true?	• Same platform and equipment for the career duration • Transition between platforms and equipment is rare and is conducted ad hoc • It is possible to transition, but the process is difficult • The transition process is well defined and well implemented	Not sure Never Sometimes Often Always
5.3	Please select responses that best describe your experience of ad hoc training in defense	• Used in emergency only; stressful and strains resources • Used primarily to fill the planned training gaps • Easily available; supports training scheduling and PD	Not sure Never Sometimes Often Always
5.4	To your knowledge, are the following training delivery options available in defense?	• More than one delivery location • Many schedule options throughout a year • Options to attend as intensive delivery, or over a longer period • Same training could be attended in class, as a computer package, and/or a mix of both • Can adjust to different learning styles • Can adjust to different learner abilities and prior knowledge	Not sure Never Sometimes Often Always
6	Is DTS and its products cognizant that threat events occur at any time during the life cycle, with statistically predictable frequency (adaptive capacity)?		
6.1	In your opinion, does defense adapt its training well to the following changes in its internal environment?	• New or updated capabilities • New or updated equipment • New or updated functionality • New or updated technology • New specializations • New positions • Commencement of a retiring and transitioning processes	Not sure Never Sometimes Often Always

(Continued)

Table 9.9 (Continued)

#	Main question	Supporting questions	Possible values
6.2	In your opinion, does defense adapt its training well to the following changes in its external environment?	• Rapid escalation of anti-terrorism warfare • Rapid escalation of cyber and information warfare • Rapid increase of unmanned warfare • Unforeseeable circumstances (e.g. fire, flood, epidemics) • Rapidly evolving technology • Extensive use of modeling, simulation, and AI • Complex problem solving assisted by technology • Globalization • Preparing for future uncertainties	Not sure Never Sometimes Often Always
6.3	Have you experienced the following defense training scenarios due to changes in defense capability requirements?	• Training fails to adapt to the change • Training adapts well, but introduced with a significant delay • Training is well planned and well adapted to the change	Not sure Never Sometimes Often Always
6.4	Have you experienced the following methods to increase defense training capacity?	• Increase in max numbers of students per course session • Increase in number of course sessions per year • Changes to the course and personnel rotation schedules • Increasing the training specialist workforce • Training policy change	Not sure Never Sometimes Often Always

Free responses

7   Are there any key Defense Training System performance measures that you use (should use)? Please specify

8   In your opinion, on what learning needs should Defense Training System focus more?

9   In your opinion, what could be done to improve defense training?

to form the basis for the DTS architecture and resilience measurement. The survey population is asked for the feedback for the questions and resilience measurement scale to be incorporated into the future phases of the survey.

Phase 2: Cross-DoD distribution. The postinaugural version of the survey is to be distributed to a randomly selected sample of targeted populations from the three services.

Phase 3: Mitigation distribution. If any of the data is insufficient, the phase 3 of the survey will be conducted to address any information gaps from the previous phases.

Phase 4: Publishing the survey methodology and results. It is intended to publish a separate article in a peer-reviewed journal outlining the results of the survey effort.

9.5   Research Findings and Future Directions

Our research established that DTS is worthy of an enterprise resilience approach, especially in the Australian context where SE input and processes appear not to have kept pace, and therefore there is a good prospect of significant improvements from such an approach. A substantial literature review of enterprise resilience found differences in terminology and lack of agreement between the academics on the matter of resilience attributes. However, we were able to identify nine attributes for our research use that featured most prominently in the reviewed publications, namely, (1) flexibility, (2) agility, (3) adaptive capacity, (4) adaptability, (5) robustness, (6) vulnerability reduction, (7) redundancy, (8) recovery/restorative capacity, and (9) efficiency.

The literature review also found different perspectives on resilience mechanism, wherefrom the most common reviewed, we selected a resilience model with system equilibrium, and state transitions called the generic state-machine model by Jackson et al. (2015a, 2015b). The preferred approach to implementing this model we chose is that of Ferris (2019). His approach to developing a system resilience measurement and analysis methodology was tailored to our DTS case study and overviewed here. Basic factors and measures for DTS resilience were proposed and supported by DTS Resilience Survey to determine the current DTS design and to guide the DTS transformation road map development from "As Is" to "To Be" architecture.

Our future work is to implement the training system surveys to measure these resilience measures and build the resilience model to research what limitations can be addressed to improve the resilience of the DTS. Having established the attributes of interest, we then proceed to establish the relevant causes of change in these attributes and how they might respond to both gradual and abrupt kinds of change. This process, in turn, leads to the elaboration of a small number of plausible futures, designed to cover a range of uncertainty rather than produce any single "correct" prediction. By assessing the likelihood that the system will change in certain specified ways in the future, we obtain a surrogate measure of resilience that will tell us how likely it is that the system properties that we are interested in will be maintained at a specified time interval in the future.

The research used here to establish a defense enterprise resilience method to improve training systems has wide applicability for any enterprise seeking to be more resilient, especially through better engineering of its training systems and personnel.

References

Ahlquist, G., Irwin, G., Knott, D., and Allen, K. (2003). Enterprise resilience. *Best's Review* 104 (3): 88–88.

Akgün, A.E. and Keskin, H. (2014). Organisational resilience capacity and firm product innovativeness and performance. *International Journal of Production Research* 52 (23): 6918–6937.

Alberts, D.S., Garstka, J.J., and Stein, F.P. (2000). *Network Centric Warfare Developing and Leveraging Information Superiority: Command and Control Research Program (CCRP)*. Washington, DC: US Office of the Secretary of Defence.

Alliger, G.M., Cerasoli, C.P., Tannenbaum, S.I., and Vessey, W.B. (2015). Team resilience. *Organizational Dynamics* 44 (3): 176–184.

Annarelli, A. and Nonino, F. (2016). Strategic and operational management of organizational resilience: current state of research and future directions. *Omega* 62: 1–18.

Ayyub, B.M. (2014). Systems resilience for multihazard environments: definition, metrics, and valuation for decision making. *Risk Analysis* 34 (2): 340–355.

Azvedo, S.G., Govindan, K., Carvalho, H. et al. (2011). GResilient index to assess the greenness and resilience of the automotive supply chain. Discussion Papers on Business and Economics (August). University of Southern Denmark, Department of Business and Economics.

Baran, B.E. (2016). High-reliability HR: preparing the enterprise for catastrophes. *People & Strategy* 39 (1): 34–38.

Bardoel, E.A., Pettit, T.M., De Cieri, H., and McMillan, L. (2014). Employee resilience: an emerging challenge for HRM. *Asia Pacific Journal of Human Resources* 52 (3): 279–297.

Berkes, F. (2007). Understanding uncertainty and reducing vulnerability: lessons from resilience thinking. *Natural Hazards Review* 41 (2): 283–295.

Bhamra, R., Dani, S., and Burnard, K. (2011). Resilience: the concept, a literature review and future directions. *International Journal of Production Research* 49 (18): 5375–5393.

Bitzinger, R.A.E. (2016). *Emerging Critical Technologies and Security in the Asia-Pacific*. Hampshire, UK: Palgrave Macmillan.

Boardman, J.T. and Clegg, B.T. (2001). Structured engagement in the extended enterprise. *International Journal of Operations & Production Management* 21: 795–811.

Boardman, J. and Sauser, B. (2008). *Systems Thinking: Coping with 21st Century Problems*. New York: CRC Press.

Bouaziz, F. (2018). Strategic human resource management practices and organizational resilience. *Journal of Management Development* 37 (7): 537–551. https://doi.org/10.1108/JMD-11-2017-0358.

Buchanan, R.K., Goerger, S.R., Rinaudo, C.H. et al. (2015). Resilience in engineered resilient systems. *The Journal of Defense Modeling Simulation*: 1548512918777901. https://doi.org/10.1177/1548512918777901.

Bukowski, L. (2016). System of systems dependability – theoretical models and applications examples. *Reliability Engineering System Safety* 151: 76–92.

Burnard, K. and Bhamra, R. (2011). Organisational resilience: development of a conceptual framework for organisational responses. *International Journal of Production Research* 49 (18): 5581–5599.

Cabral, I., Grilo, A., and Cruz-Machado, V. (2012). A decision-making model for lean, agile, resilient and green supply chain management. *International Journal of Production Research* 50 (17): 4830–4845.

Carpenter, S., Walker, B., Anderies, J.M., and Abel, N. (2001). From metaphor to measurement: resilience of what to what? *Ecosystems* 4 (8): 765–781.

Carthey, J., de Leval, M., and Reason, J. (2001). Institutional resilience in healthcare systems. *BMJ Quality Safety* 10 (1): 29–32.

Carvalho, H. (2011). Resilience index: proposal and application in the automotive supply chain. Paper presented at the Proceedings of the 18th EUROMA conference, Cambridge, UK.

Carvalho, H. and Machado, V.C. (2007). Designing principles to create resilient supply chains. Paper presented at the IIE Annual Conference Proceedings.

Chen, S.-H. (2016). Construction of an early risk warning model of organizational resilience: an empirical study based on samples of R&D teams. *Discrete Dynamics in Nature Society* 2016: 4602870. https://doi.org/10.1155/2016/4602870.

Christopher, M. and Peck, H. (2004). Building the resilient supply chain. *The International Journal of Logistics Management* 15 (2): 1–14.

Conz, E. and Magnani, G. (2019). A dynamic perspective on the resilience of firms: a systematic literature review and a framework for future research. *European Management Journal* 38: 400–412.

Cooper, C.L., Liu, Y., and Tarba, S.Y. (2014). Resilience, HRM practices and impact on organizational performance and employee well-being. *International Journal of Human Resource Management* 25 (17): 2466–2471.

Crichton, M.T., Ramsay, C.G., and Kelly, T. (2009). Enhancing organizational resilience through emergency planning: learnings from cross-sectoral lessons. *Journal of Contingencies Crisis Management* 17 (1): 24–37.

Cumming, G.S., Barnes, G., Perz, S. et al. (2005). An exploratory framework for the empirical measurement of resilience. *Ecosystems* 8 (8): 975–987.

Dalziell, E.P. and McManus, S.T. (2004). Resilience, vulnerability, and adaptive capacity: implications for system performance. *First International Forum for Engineering Decision Making (IFED)*, Stoos, Switzerland (5–8 December 2004).

De Leeuw, A.C. and Volberda, H.W. (1996). On the concept of flexibility: a dual control perspective. *Omega* 24 (2): 121–139.

De Sanctis, I., Ordieres Meré, J., and Ciarapica, F.E. (2018). Resilience for lean organisational network. *International Journal of Production Research* 56 (21): 6917–6936.

Downes, B.J., Miller, F., Barnett, J. et al. (2013). How do we know about resilience? An analysis of empirical research on resilience, and implications for interdisciplinary praxis. *Environmental Research Letters* 8 (1): 014041.

Erol, O., Mansouri, M., and Sauser, B. (2009). A framework for enterprise resilience using service oriented Architecture approach. Paper presented at the 2009 3rd Annual IEEE Systems Conference.

Erol, O., Henry, D., and Sauser, B. (2010a). 3.1.2 Exploring resilience measurement methodologies. Paper presented at the INCOSE International Symposium.

Erol, O., Henry, D., Sauser, B., and Mansouri, M. (2010b). Perspectives on measuring enterprise resilience. Paper presented at the 2010 IEEE International Systems Conference.

Erol, O., Sauser, B., and Mansouri, M. (2010c). A framework for investigation into extended enterprise resilience. *Enterprise Information Systems* 4 (2): 111–136.

Fakult, N., Pfledderer, J.A., and Bills, C.G. (1988). Comparison of instructional system development (ISD) with systems engineering from the training system perspective. Paper presented at the Proceedings of the IEEE 1988 National Aerospace and Electronics Conference, Dayton, OH, USA (23–27 May 1988).

Falasca, M., Zobel, C.W., and Cook, D. (2008). A decision support framework to assess supply chain resilience. Paper presented at the Proceedings of the 5th International ISCRAM Conference.

Ferris, T.L. (2019). A resilience measure to guide system design and management. *IEEE Systems Journal* 13 (4): 3708–3715.

Fiksel, J. (2003). Designing resilient, sustainable systems. *Environmental Science Technology* 37 (23): 5330–5339.

Fiksel, J. (2006). Sustainability and resilience: toward a systems approach. *Sustainability: Science, Practice* 2 (2): 14–21.

Fiksel, J. (2015). From risk to resilience: learning to deal with disruption. In: *Resilient by Design*, 19–34. Springer.

Fraccascia, L., Giannoccaro, I., and Albino, V. (2017). Rethinking resilience in industrial symbiosis: conceptualization and measurements. *Ecological Economics* 137: 148–162.

Francis, R. and Bekera, B. (2014). A metric and frameworks for resilience analysis of engineered and infrastructure systems. *Reliability Engineering System Safety* 121: 90–103.

Fricke, E. and Schulz, A.P. (2005). Design for changeability (DfC): principles to enable changes in systems throughout their entire lifecycle. *Systems Engineering* 8 (4): 342–359.

Giannoccaro, I., Albino, V., and Nair, A. (2018). Advances on the resilience of complex networks. *Complexity* 2018: 8756418. https://doi.org/10.1155/2018/8756418.

Gilbert, C., Eyring, M., and Foster, R.N. (2012). Two routes to resilience. *Harvard Business Review* 90 (12): 65–73.

Goerger, S.R., Madni, A.M., and Eslinger, O.J. (2014). *Engineered Resilient Systems: A DoD Perspective*. Army Corps of Engineers Vicksburg Ms Engineer Research and Development Center.

Gold-Bernstein, B. and Ruh, W. (2004). *Enterprise Integration: The Essential Guide to Integration Solutions*. Addison Wesley Longman Publishing Co., Inc.

Golden, W. and Powell, P. (2000). Towards a definition of flexibility: in search of the Holy Grail? *Omega* 28 (4): 373–384.

Gomes, R. (2015). Resilience and enterprise architecture in SMES. *ISTEM-Journal of Information Systems* 12 (3): 525–540.

Haimes, Y.Y., Crowther, K., and Horowitz, B.M. (2008). Homeland security preparedness: balancing protection with resilience in emergent systems. *Systems Engineering* 11 (4): 287–308.

Han, S.Y., Marais, K., and DeLaurentis, D. (2012). Evaluating system of systems resilience using interdependency analysis. Paper presented at the 2012 IEEE International Conference on Systems, Man, and Cybernetics (SMC).

He, Y. (2008). A novel approach to emergency management of wireless telecommunication system. Master dissertation, Department of Mechanical Engineering, University of Saskatchewan Saskatoon, Saskatchewan, Canada

Helaakoski, H., Iskanius, P., and Peltomaa, I. (2007). Agent-based architecture for virtual enterprises to support agility. Paper presented at the Working Conference on Virtual Enterprises.

Henry, D. and Ramirez-Marquez, J.E. (2010). 3.1.1 A generic quantitative approach to resilience: a proposal. Paper presented at the INCOSE International Symposium.

Ho, M., Teo, S.T., Bentley, T. et al. (2014). Organizational resilience and the challenge for human resource management: conceptualizations and frameworks for theory and practice. Paper presented at the International Conference on Human Resource Management and Professional Development for the Digital Age (HRM&PD). Proceedings.

Hoffman, R.R. and Hancock, P.A. (2017). Measuring resilience. *Human Factors* 59 (4): 564–581.

Holling, C. (1973). Resilience and stability of ecological systems. *Annual Review of Ecology* 4 (1): 1–23.

Holling, C.S., Gunderson, L., and Light, S. (1995). *Barriers and Bridges to the Renewal of Ecosystems*. New York: Columbia University Press.

Hollnagel, E. and Woods, D.D. (2006). Epilogue: resilience engineering precepts. In: *Resilience engineering: Concepts* (eds. E. Hollnagel, D.D. Woods and N. Leveson), 347–358. Aldershot: Ashgate.

Hoogervorst, J.A. (2009). *Enterprise Governance and Enterprise Engineering*. The Netherlands: Springer Science & Business Media.

Hu, Y., Li, J., and Holloway, L.E. (2008). Towards modeling of resilience dynamics in manufacturing enterprises: literature review and problem formulation. Paper presented at the 2008 IEEE International Conference on Automation Science and Engineering.

Ignatiadis, I. and Nandhakumar, J. (2007). The impact of enterprise systems on organizational resilience. *Journal of Information Technology* 22 (1): 36–43.

Jackson, S., Cook, S., and Ferris, T.L. (2015a). A generic state-machine model of system resilience. *Insight* 18 (1): 14–18.

Jackson, S., Cook, S.C., and Ferris, T.L. (2015b). Towards a method to describe resilience to assist system specification. Paper presented at the INCOSE International Symposium.

Joiner, K.F. and Tutty, M.G. (2018). A tale of two allied defence departments: new assurance initiatives for managing increasing system complexity, interconnectedness and vulnerability. *Australian Journal of Multi-Disciplinary Engineering* 14 (1): 4–25.

Joiner, K.F., Atkinson, S.R., Castelle, K., and Bradley, J. (2019). Near-term Asian war: impacts and options for Australia's submarine program. Paper presented at the 5th Submarine Science, Technology and Engineering Conference (SubSTEC5), Fremantle, Western Australia (18–22 November 2019).

Jordan, D., Kiras, J.D., Lonsdale, D.J. et al. (2016). *Understanding Modern Warfare*. Cambridge, U.K.: Cambridge University Press.

Kamalahmadi, M. and Parast, M.M. (2016). A review of the literature on the principles of enterprise and supply chain resilience: major findings and directions for future research. *International Journal of Production Economics* 171: 116–133.

Kantur, D. and Say, A.I. (2015). Measuring organizational resilience: a scale development. *Journal of Business Economics Finance* 4 (3): 456–472.

Kativhu, S., Mwale, M., and Francis, J. (2018). Approaches to measuring resilience and their applicability to small retail business resilience. *Problems Perspectives in Management* 16 (4): 275–284.

Kayes, D.C. (2015). *Organizational Resilience: How Learning Sustains Organizations in Crisis, Disaster, and Breakdown*. USA: Oxford University Press.

Keating, C.B. and Bradley, J.M. (2015). Complex system governance reference model. *International Journal of System of Systems Engineering* 6 (1–2): 33–52. https://doi.org/10.1504/ijsse.2015.068811.

Keating, C.B., Katina, P.F., Jaradat, R.M. et al. (2019). Framework for improving complex system performance. *INCOSE International Symposium* 29 (1): 1218–1232. https://doi.org/10.1002/j.2334-5837.2019.00664.x.

Kim, Y., Chen, Y.-S., and Linderman, K. (2015). Supply network disruption and resilience: a network structural perspective. *Journal of Operations Management* 33: 43–59.

Kosanke, K., Vernadat, F., and Zelm, M. (1999). CIMOSA: enterprise engineering and integration. *Computers in Industry* 40 (2–3): 83–97.

Lee, A.V., Vargo, J., and Seville, E. (2013). Developing a tool to measure and compare organizations' resilience. *Natural Hazards Review* 14 (1): 29–41.

Lengnick-Hall, C.A., Beck, T.E., and Lengnick-Hall, M.L. (2011). Developing a capacity for organizational resilience through strategic human resource management. *Human Resource Management Review* 21 (3): 243–255.

Limnios, E.A.M., Mazzarol, T., Ghadouani, A., and Schilizzi, S.G. (2014). The resilience architecture framework: four organizational archetypes. *European Management Journal* 32 (1): 104–116.

Lucero, S. (2011). *Engineered Resilient Systems-DoD Science and Technology Priority*. Washington, DC: Office of the Deputy Assistant Secretary of Defense (Systems Engineering) https://apps.dtic.mil/dtic/tr/fulltext/u2/a586459.pdf (accessed 28 August 2020).

Ma, Z., Xiao, L., and Yin, J. (2018). Toward a dynamic model of organizational resilience. *Nankai Business Review International* 9 (3): 246–263.

Mafabi, S., Munene, J., and Ntayi, J. (2012). Knowledge management and organisational resilience: organisational innovation as a mediator in Uganda parastatals. *Journal of Strategy Management Research Review* 5 (1): 57–80.

Mallak, L.A. (1999). Toward a theory of organizational resilience. Paper presented at the PICMET'99: Portland International Conference on Management of Engineering and Technology. Proceedings Vol-1: Book of Summaries (IEEE Cat. No. 99CH36310).

Mansouri, M., Sauser, B., and Boardman, J. (2009). Applications of systems thinking for resilience study in maritime transportation system of systems. Paper presented at the 2009 3rd Annual IEEE Systems Conference.

McManus, S.T. (2008). Organisational resilience in new zealand. PhD thesis, University of Canterbury. https://ir.canterbury.ac.nz/bitstream/handle/10092/1574/?sequence=1 (accessed 28 August 2020).

Menéndez Blanco, J.M. and Montes Botella, J.L. (2016). What contributes to adaptive company resilience? A conceptual and practical approach. *Development Learning in Organizations: An International Journal* 30 (4): 17–20.

Morello, D.T. (2002). *The Blueprint for the Resilient Virtual Organization Spurred by the Needs for Security, Protection and Recovery, Enterprises Are Taking on the New Challenge of Deliberately Designing Resilience into Their Management of People, Places, Infrastructure and Work Processes*. Gartner, Inc.

Munoz, A. and Dunbar, M. (2015). On the quantification of operational supply chain resilience. *International Journal of Production Research* 53 (22): 6736–6751.

Nuss, A.J., Blackburn, T.D., and Garstenauer, A. (2017). Toward resilience as a tradable parameter during conceptual trade studies. *IEEE Systems Journal* 12 (4): 3393–3403.

Park, J., Seager, T.P., Rao, P.S.C. et al. (2013). Integrating risk and resilience approaches to catastrophe management in engineering systems. *Risk Analysis* 33 (3): 356–367.

Pettit, T.J., Simpson, N.C., Hancock, P.G. et al. (2016). Exploring operational resilience in the context of military aviation: finding the right mode at the right time. *Journal of Business Behavioral Sciences* 28 (2): 24.

Ponomarov, S.Y. and Holcomb, M.C. (2009). Understanding the concept of supply chain resilience. *The International Journal of Logistics Management* 20 (1): 124–143.

Pritchard, L. and Gunderson, L.H. (2002). *Resilience and the Behavior of Large Scale Systems*. Island Press.

Raj, R., Wang, J., Nayak, A. et al. (2014). Measuring the resilience of supply chain systems using a survival model. *IEEE Systems Journal* 9 (2): 377–381.

Rose, A. and Liao, S.Y. (2005). Modeling regional economic resilience to disasters: a computable general equilibrium analysis of water service disruptions. *Journal of Regional Science* 45 (1): 75–112.

Rosenkrantz, D.J., Goel, S., Ravi, S., and Gangolly, J. (2009). Resilience metrics for service-oriented networks: a service allocation approach. *IEEE Transactions on Services Computing* 2 (3): 183–196.

Sawalha, I.H.S. (2015). Managing adversity: understanding some dimensions of organizational resilience. *Management Research Review* 38 (4): 346–366.

Schuel, D. (2018). Force integration – integrated capability realisation for the ADF. Paper presented at the Systems Engineering Test & Evaluation Conference, Sydney (2 May 2018).

Sharma, S. and Sharma, S.K. (2016). Team resilience: scale development and validation. *Vision* 20 (1): 37–53.

Sheffi, Y. and Rice, J.B. Jr. (2005). A supply chain view of the resilient enterprise. *MIT Sloan Management Review* 47 (1): 41.

Shirali, G.A., Motamedzade, M., Mohammadfam, I. et al. (2016). Assessment of resilience engineering factors based on system properties in a process industry. *Cognition, Technology Work* 18 (1): 19–31.

Sitterle, V.B., Freeman, D.F., Goerger, S.R., and Ender, T.R. (2015). Systems engineering resiliency: guiding tradespace exploration within an engineered resilient systems context. *Procedia Computer Science* 44: 649–658.

Small, C., Parnell, G., Pohl, E. et al. (2017). Engineered resilient systems with value focused thinking. Paper presented at the INCOSE international symposium.

Soni, U., Jain, V., and Kumar, S. (2014). Measuring supply chain resilience using a deterministic modeling approach. *Computers Industrial Engineering* 74: 11–25.

Starr, R., Newfrock, J., and Delurey, M. (2003). Enterprise resilience: managing risk in the networked economy. *Strategy and Business* 30: 70–79.

Tierney, K., & Bruneau, M. (2007). Conceptualizing and measuring resilience: a key to disaster loss reduction. *TR News*(250). http://www.trb.org/Publications/Blurbs/158982.aspx (accessed 28 August 2020).

Tran, H.T., Balchanos, M., Domerçant, J.C., and Mavris, D.N. (2017). A framework for the quantitative assessment of performance-based system resilience. *Reliability Engineering System Safety* 158: 73–84.

U.S. Government Accountability Office (2018). *Weapon Systems Cybersecurity: DOD Just Beginning to Grapple with Scale of Vulnerabilities*. Washington, DC.

Uday, P. and Marais, K.B. (2014). Resilience-based system importance measures for system-of-systems. *Procedia Computer Science* 28: 257–264.

Uday, P. and Marais, K. (2015). Designing resilient systems-of-systems: a survey of metrics, methods, and challenges. *Systems Engineering* 18 (5): 491–510.

Unewisse, M. (2016). The new Defence Capability Lifecycle: a conceptual overview. Paper presented at the Systems Engineering, Test and Evaluation, Melbourne.

Walker, B., Holling, C.S., Carpenter, S., and Kinzig, A. (2004). Resilience, adaptability and transformability in social–ecological systems. *Ecology Society* 9 (2) https://www.gbcma.vic.gov.au/downloads/wshop_resilience_reading/resilience_2.pdf (accessed 28 August 2020).

Wieland, A. and Marcus Wallenburg, C. (2013). The influence of relational competencies on supply chain resilience: a relational view. *International Journal of Physical Distribution* 43 (4): 300–320.

Woods, D.D. (2015). Four concepts for resilience and the implications for the future of resilience engineering. *Reliability Engineering System Safety* 141: 5–9.

Woods, D.D. (2017). Essential characteristics of resilience. In: *Resilience Engineering*, 21–34. CRC Press.

Xiao, L. and Cao, H. (2017). Organizational resilience: the theoretical model and research implication. Paper presented at the ITM Web of Conferences.

Zhang, W. (2008). *Resilience Engineering – A New Paradigm and Technology for Systems*. Canada: University of Saskatchewan.

10

Integrating New Technology into the Complex System of Air Combat Training*

Sarah M. Sherwood[1], Kelly J. Neville[2], Angus L. M. T. McLean, III[3], Melissa M. Walwanis[4], and Amy E. Bolton[5]

[1] Naval Medical Research Unit Dayton, Wright-Patterson AFB, OH, USA
[2] The MITRE Corporation, Orlando, FL, USA
[3] Collins Aerospace, Cedar Rapids, IA, USA
[4] Naval Air Warfare Center Training Systems Division, Orlando, FL, USA
[5] Office of Naval Research, Arlington, VA, USA

10.1 Introduction

Training fighter aircrew is a time-consuming, expensive, and challenging process. A critical part of ensuring their readiness is live air combat training, which is especially resource intensive, requiring large amounts of airspace, fuel, and personnel and contributing to the wear and tear of valuable assets. Training resources are stretched thin to provide as much live training as possible, and recent advances in air combat technology have introduced new training requirements with new resource demands. These demands include the need for larger range spaces to support long-range tactics, opportunities to practice against larger numbers of adversary aircraft, and opportunities to practice against faster and more capable adversary aircraft.

For many years, the military has strategically employed simulation-based training to reduce the cost and risk of air combat training and increase training opportunities (Ausink et al. 2011; Rickard et al. 2013). However, though simulation training is valuable, it has its limits. For example, simulation training does not give aircrew the opportunity to practice while experiencing the stressors of live combat, including G-forces, cockpit noise and vibration, and potential high-severity consequences. Simulation training does not allow aircrew to become attuned to many of the perceptual details of the live environment, such as vibration associated with aircraft performance or visual artifacts such as the smoke trails of a missile. Simulation training also exposes aircrew to more artificialities than they experience in live training due to its relatively limited fidelity. These artificialities can lead to negative transfer of training, i.e. the development of habits that work well in the simulated environment but not in live combat.

Simulation-based training is valuable, but live training remains a critical last stop prior to deployment. In an effort to maintain the effectiveness of live training in the face of rising fuel and maintenance costs and next-generation threats, the US military is considering the augmentation of

---

*Approved for Public Release; Distribution Unlimited. Public Release Case Number 20-0871.

A Framework of Human Systems Engineering: Applications and Case Studies, First Edition.
Edited by Holly A. H. Handley and Andreas Tolk.
© 2021 The Institute of Electrical and Electronics Engineers, Inc. Published 2021 by John Wiley & Sons, Inc.

live training with virtual and constructive aircraft. The virtual and constructive aircraft will be controlled by aircrew in simulators and software agents, respectively. These non-live assets will enable the virtual extension of training ranges and amplification of adversary capabilities in live training events. The proposed variant of *Live, Virtual, Constructive (LVC)* training will differ from previous LVC variants in the way live operational aircraft and their crews participate. In past LVC exercises, live aircraft populated the environments of virtual and constructive participants, but the relationship was not reciprocal. The non-live entities did not populate the live aircraft's training environment. In the proposed new variant, the aircrew of live aircraft will see non-live simulated entities on the radar and sensor system displays of their aircraft. To the aircrews in the air, it will appear as though these non-live aircraft are flying around in the sky with them. When appropriate for training, the non-live aircraft may also operate beyond the physical boundaries of the range and employ advanced tactics that live aircraft would be restricted or prevented from executing.

Despite the potential for LVC technology to overcome key live training limitations, many unknowns exist regarding its effect on the sociotechnical system of air combat training. In live air combat training, there is an ever-present risk of costly mishap with aircrew lives at stake. Flying fighter jets will never be perfectly safe; there will always be an assumed level of danger. Nevertheless, the Navy air combat training system has evolved over decades to continuously optimize both training effectiveness and the safety of operating in a complex and dynamic high-stakes environment. In other words, the sociotechnical training system has developed a resilience that allows it to operate effectively and safely across a wide range of challenging conditions. This system consists of, for example, the Navy's rigorous briefing, debriefing, and instruction practices; training rules; training exercise oversight practices; and the Naval Air Training and Operating Procedures Standardization (NATOPS) program (COMNAVAIRFOR, 2016), as well as personnel, aircraft, and other technologies.

For the most part, this training system is not designed to prevent specific mishaps. Instead, it is designed to be resilient to safety threat events in general, both foreseeable and unpredictable. The system has evolved over time to protect itself, its mission, and the well-being and safety of the elements and actors within it. It has evolved to be able to detect a wide range of threats ranging from routine to rare and to respond to threats so they do not precipitate into an actual mishap. In the case that a mishap does occur, the system is designed to limit the damage and recover.

This view of the air combat training system is consistent with principles of resilience engineering (Hollnagel et al. 2006). It contrasts with the more static and linear view taken by safety management systems, such as Heinrich's (1931) domino model, which tend to focus on the nominally emergent resilience (NER) features of an organization (i.e. the resilience features that can be reduced and/or predicted from a combination of individual agent properties). The resilience engineering approach draws from complexity science and systems theory to improve the robustness of real-world, complex organizations against threats. Typically, in addition to NER, such organizations also exhibit weakly emergent resilience (WER) (Pariès 2006). WER refers to the resilience features of an organization that cannot be predicted on the macro level based on individual agent properties without performing a one-to-one simulation. In essence, the whole is not always equal to the sum of its parts, which presents an interesting challenge to test and evaluation (T&E) and safety management professionals alike: how does one evaluate the robustness of an organization against threats that one cannot foresee?

Safety is an emergent property of a complex system. The interaction of individual agent components, such as the Navy's training rules and its briefing and debriefing practices, produces an aggregate air combat safety system that is more flexible and adaptive than its component parts. Such systems should be engineered with a certain amount of residual disorder. That is, the system processes should not be optimized to protect against specific safety threats. Although this engineering choice will, of course, result in system features that do not produce the best results on the agent level, these same features will enable the system to better maintain stability and resilience at the macro level despite variability in the environment (Hollnagel et al. 2006; Pariès 2006).

The stability and resilience of the Navy's air combat training system additionally depends on a balance between two high-priority system objectives – safety during training exercises and adequate fidelity to prepare aviators for combat. The introduction of non-live entities into this system has the potential to disrupt this balance as well as the mechanisms that provide each individual objective its stability and resilience.

The introduction of non-live entities into live air combat training will involve the incorporation of a complex simulation paradigm into the already complex sociotechnical naval flight training system. In its most basic form – a single exercise in isolation – the new LVC training system will involve interactions by virtual and constructive entities with one another, technical system components, live exercise participants, and exercise management personnel. However, if one digs a little deeper, the interactions among these system elements take place in the context of the Navy's methods of exercise planning, briefing, and debriefing; training rules; standard operating procedures; emergency procedures; and the air combat community culture. These are also part of the training system. In a system of such complexity, unanticipated emergent behavior – including new threats to aircrew safety or training and readiness – can arise. In addition to these intra-system interactions, external financial and environmental pressures on the system could lead to further emergent behavior, some of which could be hazardous.

The remainder of this chapter describes evaluation work conducted to determine whether the new LVC training paradigm will introduce new safety risks, increase the likelihood or severity of existing risks, or compromise training effectiveness and aircrew readiness (which we treat as a risk to post-training safety). The methods used are adapted from and influenced by cognitive engineering, resilience engineering, safety management, and action research paradigms. Results were used to guide the development of recommended requirements for a subsequent training system acquisition program.

10.2 Method

10.2.1 Data Collection

Participants. Twelve active duty Navy pilots, six former Navy and Air Force pilots, one active duty Naval Flight Officer (NFO), one National Air Guard pilot, and two military modeling-and-simulation (M&S) experts with experience coordinating large-scale training exercises participated in interviews of approximately one hour.

All but one of the interviewed pilots had experience flying high-performance jets (Table 10.1). Active Navy pilot participants reported an average of 1180.8 flight hours (range: 340–1800 hours) in their primary aircraft. Nine of the active Navy pilots and the NFO were TOPGUN or Strike Fighter Advanced Readiness Program (SFARP) instructors. Four former military pilot participants reported an average of 2612.0 flight hours in their primary aircraft (range: 2150–3600 hours; flight hours of the fifth pilot are missing). To protect participant anonymity, flight hours are not reported for single representatives of a given experience category. Participants other than the M&S experts, who considered the interview to be a discussion among colleagues, were given an informed consent document to read and sign. All participants, including the M&S experts, were asked for permission to audiotape the interview.

Table 10.1 Aircraft flown by participants.

Platform	Participants	Service	Tactical aircraft	Status
F/A-18 Hornet/Super Hornet	13	USN	Yes	In service
F-14 Tomcat	2	USN	Yes	Retired
F-16 Fighting Falcon	1	ANG	Yes	In service
F-15 Eagle/Strike Eagle	1	USAF	Yes	In service
F-111 Aardvark	1	USAF	Yes	Retired
Tornado	1	GAF	Yes	In service
E-2 Hawkeye	1	USN	No	In service

Notes. USN, US Navy; ANG, Air National Guard; USAF, US Air Force; GAF, German Air Force.

Two research team members conducted event-centered interviews with individuals ($n = 12$) and groups ($n = 9$; two groups of five sharing one common participant). Six of the interview sessions with individuals were conducted over the telephone. All other interviews were conducted in person.

Procedure. At the start of each interview session, participants were briefed on the LVC training concept and the Navy's goal of using it to improve training efficiency and meet training requirements of next-generation aircraft. They were told that the purpose of the interview was to identify potential safety issues, training concerns, and training benefits associated with introducing non-live entities into live air combat training.

To conduct the interviews, we used an adaptation of the Critical Decision Method (CDM) (Flanagan 1954; Klein et al. 1989), a technique used to elicit expert knowledge. Participants were asked to describe past training exercises and scenarios and to envision and describe potential impacts of including non-live entities in them. We posed *what-if questions* such as "What if the first package of blue-force fighters had been constructive?" to prompt interviewees to consider further impacts of LVC technology on training. Some interviewees brought pre-identified concerns and objections with them to the interview session. We listened to their concerns and objections and, for each, asked them to describe a past event or other type of experience in which something similar occurred.

Interviews were captured by audio recording. When permission to audio record was not obtained, the research team took detailed notes that were reviewed and elaborated by the interviewees.

10.2.2 Data Analysis

Data analysis consisted of a thematic analysis performed to identify specific hazards and hazard categories, and an evaluation of the level of risk associated with each hazard and hazard category. Figure 10.1 depicts a visual overview of the analysis process.

Low-Level Thematic Analysis. We performed a *thematic analysis* (Braun and Clarke 2006) of the interview data. Interview notes and transcripts were reviewed to identify statements made by interviewees about potential LVC hazards, training concerns, and training benefits. Each statement was reviewed in the context of the interview narrative and assigned a code corresponding to the hazard or training issue being described. As statements, referred to as *data extracts*, were identified and coded, we iteratively expanded the set of codes and reassessed previously coded data in light of new codes.

Once this iterative coding process was complete, data extracts describing potential LVC hazards – the focus of this paper – were reorganized by their associated code, hereafter referred to simply as a *hazard*, although at this stage each was considered to be a hypothesized hazard. For each hazard, we organized the associated data extracts into contrasting positions, representing ways LVC training should versus should not be implemented with respect to the potential hazard.

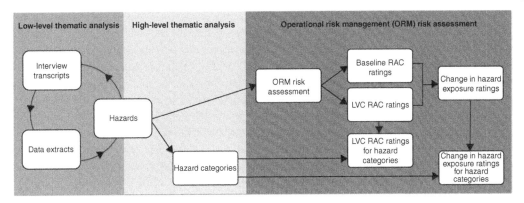

Figure 10.1 Overview of the data analysis procedure. *Notes.* ORM, operational risk management; RAC, risk assessment category; LVC, Live, Virtual, Constructive.

Table 10.2 Naval air combat expert profiles.

Expert 1	Expert 2
7600 total flight hours	8000 total flight hours
2400 F/A-18 flight hours	2800 F/A-18 flight hours
Three deployments	Three deployments
Navy adversary pilot (3 yr)	Navy adversary pilot (13 yr)
Navy range training officer (6 yr)	Navy range training officer (10 yr)

Expert Review. Two naval air combat experts (see expert demographics in Table 10.2) reviewed and critiqued hazards. We presented a subset of the hazards, together with summaries of the associated interview data extracts to both experts during a 2.5-hour meeting. During two subsequent four-hour meetings, we presented each remaining hazard and associated interview data to Expert 1. Each expert commented and elaborated on each presented hazard. Expert responses during the first meeting were captured in researcher notes. The second and third meetings were audio recorded and captured in researcher notes.

High-Level Thematic Analysis. A second thematic analysis was conducted on the identified hazards. Three researchers independently categorized the hazards according to the underlying interviewee concerns they represented using a bottom-up assessment approach. The researchers then met to reconcile their categorizations of the hazards. The result of this process was the identification of hypothesized *hazard categories*.

Risk Level Evaluation. Two researchers independently assessed the safety risk level associated with each hazard using the severity and probability scales of the Navy's operational risk management (ORM) system (Department of the Navy 2010). On the basis of related interview data, the researchers assessed the severity and probability of the consequences of each hazard if the non-live entities and LVC training system were to be implemented without any changes to the Navy's air combat training system (i.e. in the *future unmitigated LVC environment*). The researchers then made a second set of independent severity and probability assessments of the consequences of each hazard in the current air combat training environment (i.e. in the *baseline environment* in which non-live entities do not participate).

Hazard severity was assessed on a four-point scale ranging from I to IV, with category I representing hazards with the most severe consequences. Severity is defined by the ORM system as the "potential consequence that can occur as a result of a hazard and is defined by the degree of injury, illness, property damage, loss of assets (time, money, personnel), or effect on the mission or task" (Department of the Navy 2010, p. 8). Probability was assessed on a four-point scale ranging from A to D, with A indicating a high degree of probability that a hazard will occur. Probability is defined as "an assessment of the likelihood that a potential consequence may occur as a result of a hazard and. . .by assessment of such factors as location, exposure (cycles or hours of operation), affected populations, experience, or previously established statistical information" (Department of the Navy 2010, p. 9).

The ORM risk assessment system specifies mappings between risk levels and combinations of probability and severity ratings. For example, a probability rating of A (likely) and a severity rating of III map to a Risk Assessment Code (RAC) of 2. The RAC is the level of risk for each hazard, expressed as an Arabic number ranging from 1 to 5. A RAC of 2 indicates a serious level of risk (Department of the Navy 2010).

Researcher agreement on initial hazard RAC assessments was assessed using a linear weighted kappa, which indicated a good level of agreement ($\kappa = 0.68$, 95% CI: 0.52, 0.83). Once their independent ratings were completed, the researchers met to reconcile their ratings.

We assigned RACs to each overarching hazard category based on the RACs of the individual hazards within it. If a hazard category contained hazards of differing RACs, it was assigned the highest of those hazards' RACs. The hazard and hazard-category RACs for baseline and unmitigated LVC environments were compared to determine each hazard and hazard category's potential *change in hazard exposure.*

10.3 Results and Discussion

The thematic analysis produced 26 hazards that fall into the following seven hazard categories:

- Unseen tracks flying within visual range (WVR) of live aircraft.
- Unexpected virtual and constructive (non-live) aircraft behavior.
- Complacency and increased risk taking.
- Inadequate support for human–machine interaction.
- Inadequate support for exercise management.
- Reduced big picture awareness.
- Negative transfer of training to the operational environment.

It is important to keep in mind that these hazard categories represent sets of *potential, hypothesized* hazards.

The assessed RACs of the hazard categories ranged from 1 (critical) to 4 (minor), as shown in Column 2 of Table 10.3. Five of the hazard categories exist in both the baseline and unmitigated LVC environments. This means they do not represent new hazards and so should be mitigated by existing safety and resilience mechanisms of the air combat training system. However, one of these five hazard categories, unseen tracks flying within visual range, was rated as more probable in an LVC environment. This indicates a potential for increased hazard exposure. Two additional hazard categories represent a new type of hazard and thus a potential increase in hazard exposure during air combat training.

Each hypothesized hazard category is described below. The description of each hazard category is followed by an overview of how similar hazards are dealt with in current naval air combat

Table 10.3 Results of the assessment of potential LVC training hazards using the Navy operational risk management system.

Hazard category	Assessed risk level	Comparison with hazards in baseline environment	Projected change in hazard exposure
Unseen aircraft within visual range	1	Similar	Increase
Unexpected virtual and constructive (non-live) aircraft behavior	2	Similar	No change
Complacency/risk taking	2	Similar	No change
Inadequate support for human–machine interaction	2	New	Increase
Inadequate support for exercise management	3	Similar	No change
Reduced big picture awareness	3	New	Increase
Negative transfer of training to the operational environment	4	Similar	No change

training exercises (i.e. in the baseline environment) and a discussion of whether the given hazard category could introduce new or unmitigated risk to live training exercises (i.e. whether it represents a change in hazard exposure).

10.3.1 Unseen Aircraft Within Visual Range

Within-visual-range (WVR) training exercises involve fighter jets flying within visual proximity of each other. Once within visual range, pilots head to "the merge," a head-on neutral pass within close proximity after which air combat maneuvering (i.e. dogfighting) may begin. When entering the merge, pilots must focus their visual attention on the environment outside their aircraft while monitoring their own cockpit systems. The pilot's cognitive workload increases substantially during this period. "Tremendous workload, channelized attention, and misprioritization are very common [in WVR engagements]," said one interviewee.

The introduction of non-live aircraft during this period of intense workload was seen as a potential source of confusion and distraction that could lead to a fatal misjudgment. One pilot gave the following example: "[You expect] to look left [and see] a virtual entity that appears on your radar screen when you've actually got a real world aircraft on your right – something in your attention is drawing you to the left side of the cockpit when the actual physical threat is to the right." In addition, because non-live aircraft would only appear on cockpit displays, pilots might miss critical visual information outside their cockpits while looking down at their sensor displays. Two examples of hazards in this category include the following:

- The pilot could become distracted by trying to keep track of live versus non-live aircraft.
- When a pilot goes to the merge with a mixed group of live and non-live aircraft, his or her radar could erroneously lock onto a non-live aircraft instead of the live aircraft the pilot was trying to target. The pilot, seeing nothing in their target designator (TD) box when they expect to see a live aircraft, could then end up searching the sky for the missing live aircraft. They meanwhile would be neglecting to attend to the other aircraft in the sky for longer than would be safe, risking a collision.

Comparison with Hazards in Baseline Environment. Confusion, distraction, and looking down at cockpit sensor system displays instead of out the window are all existing hazards in the WVR environment. In fact, the radar display is known as the "drool cup" due to pilots' tendency to fixate on it. Another existing source of tunneled attention is loss of sight or failure to obtain sight ("a visual") of adversary aircraft. This situation is similar to what a pilot in an LVC exercise would experience if he or she erroneously tried to obtain sight of a non-live aircraft out the canopy. Measures that are currently in place to help pilots stay safe include codes they can use at any time to end an engagement (by rocking their aircraft's wings or transmitting "terminate") or an exercise (by transmitting "knock-it-off") if they become confused or distracted (Naval Strike and Air Warfare Center n.d.).

Change in Hazard Exposure. This hazard category was assessed as increasing hazard exposure because non-live entities are an additional source of pilot confusion and distraction. Since non-live aircraft do not exist in present-day training, they could increase the likelihood of collision with other live aircraft. In the words of one interviewee, "Mixing non-live targets with real airplanes risks channelized attention on something that cannot hurt you at the expense of something that can." Furthermore, since non-live entities are only visible on cockpit displays and not out the window, pilot head-down time – and therefore risk of collision – is increased.

Existent safety measures, in particular the wing-rock, terminate, and knock it off codes, increase the acceptability of pilot exposure to additional sources of distraction introduced by LVC training. However, these safety measures, although robust, are not infallible even in the current live training environment. For instance, in live training, communication lines are limited, and exercise participants frequently "step on" one another's communications. Consequently, as a number of interviewed pilots have pointed out, terminate and knock-it-off calls can be missed. Moreover, this problem is not isolated to verbal cues; visual cues, most notably wing rocking, are only effective if the engaging aircraft has a visual on the terminating aircraft.

A pilot who calls terminate or knock-it-off is usually experiencing elevated workload and confusion. Without mitigations, LVC training has the potential to further increase this confusion. Consider the following scenario: A live aircraft is engaged by a non-live aircraft and, for whatever reason, the pilot of the live aircraft signals to terminate the exercise by rocking their wings. The engaging non-live aircraft does not detect the rocking wings of the terminating live aircraft and fails to disengage. Since live pilots sometimes fail to detect wing rocking, the terminating pilot must assume that the engaging aircraft could be live. If this were a typical training engagement with a live adversary, the live pilot would still have his or her radar and separation tactics as a fallback. However, in this LVC scenario, while the engaging non-live aircraft is visible on the live pilot's radar, it is not visible outside his or her window. The live pilot, assuming that the engaging aircraft on their radar could be live, searches for it in the sky. This futile search, driven by the inconsistency between the live pilot's radar and out-the-window view, could add to the confusion of the situation. In short, non-live aircraft could make it more complicated for pilots to terminate or knock it off in a moment of crisis.

That being said, the ability of a pilot to end an engagement or exercise when the situation becomes potentially hazardous is an important safety measure that is largely effective and, despite the additional risk exposure potentially posed by the introduction of non-live aircraft, is expected to transfer well to the LVC training environment once appropriate mitigations are developed. One potential solution to prevent pilot confusion due to radar and out-the-window view mismatch is to provide pilots with a control that allows them to "turn off" all LVC data on their displays. Another solution, advocated by all pilot interviewees, is to prevent non-live aircraft from entering the WVR arena. The prevention of non-live aircraft WVR could be accomplished procedurally through training rules or technically, e.g. by automatically turning all non-live aircraft away when they approach within 10 nautical miles of a live aircraft.

10.3.2 Unexpected Virtual and Constructive Aircraft Behavior

The LVC training paradigm involves the implementation of state-of-the-art incipient technology. Injected constructive and virtual tracks will need to be responsive to myriad factors and forces in the fast-moving complexity of live air combat. In addition, this responsiveness needs to be manifested in both their behavior and in the sensor system displays of their behavior. Non-live tracks that respond in unexpected, unnatural ways could interfere with pilot performance and impact safety. Unexpected or inappropriate behavior by constructive entities, such as "continuing to trundle down range" after someone ends the exercise with a "knock-it-off" call, could lead to live pilot reactive maneuvering, attentional tunneling, or distraction. "You see something behaving not right," says one interviewee. "You find yourself glommed into trying to solve it – figure out why it is acting weird – and that becomes a situation awareness (SA) suck hole." Unexpected or inappropriate behavior by constructive entities could also confuse the E-2 Hawkeye controllers who serve airborne early warning, airborne battle management, and command and control functions. In essence, they are airborne air traffic control and the "eyes of the fleet," responsible for tagging and directing aircraft on radar, identifying friend or foe, and conducting threat analysis. Unnatural non-live aircraft behavior could draw their eyes away from the big picture and undermine tactical decision making (NASC 2018).

Of particular concern is the use of "magic moves" by non-live entities and the possibility of a delay or absence of constructive entity response to dynamic training events. "Magic moves," such as the sudden disappearance, appearance, unrealistic exercise physics, or violation of physical airspace limits by non-live tracks, could result in both E-2 controller and pilot confusion. If non-live tracks disappear due to an exercise reset or after being shot down, pilots and controllers might question whether the disappeared track was actually a live aircraft that suffered a mishap or lost its transponder. "If I'm going into the merge," says one pilot, "and I lose radar SA. . .if they just like disappear, that's going to make me very nervous." If this uncertainty distracts a pilot during this cognitively demanding period of time, a collision could occur. In addition, non-live tracks are not bound by physical airspace limits. Less experienced pilots could follow non-live entities performing magic moves that involve flying into off-limits airspace or unsafe maneuvering. "Watch one of our new dudes chase him out there," says one pilot with a laugh. Critically, the use of non-live aircraft in place of live adversary aircraft would mean a loss of the exercise-safety oversight that adversary pilot role-players traditionally provide.

Comparison with Hazards in Baseline Environment. At present, it is not uncommon to witness inappropriate behavior by live tracks on cockpit sensor systems. Radar is not infallible, and the expert reviewers reported that pilots sometimes see tracks disappear, reappear, or even fly underground at Mach 2. Likewise, E-2 Hawkeye radar is known for producing less than perfect information, a fact mentioned by many of the interviewees. Another point made by the expert reviewers is that unexpected events occur in real-world operations and one goal of training is to learn to expect the unexpected.

Change in Hazard Exposure. This hazard category was assessed as posing no change in hazard level because "magic moves" already exist in the training system in the form of radar anomalies. Current training rules accept the risks associated with imperfect radar and seek to mitigate them by stipulating that a pilot can call "terminate" or "knock-it-off" if he or she loses SA. "The way we deal with it is we brief that right up front," says one expert, referring to the many ways an exercise or mission can deviate from the plan. "We brief if you have a failure. . .. It's the real world. . .your system goes out, your sensors are jammed. . .it's a contested environment out there." It is also possible that computer-generated tracks will be less likely to deviate from the intended ground truth than real radar tracks. Further, ongoing work in modeling and simulation is expected to result in computer-generated track behavior that is increasingly realistic in its responsiveness to live pilot tactics.

10.3.3 Complacency and Increased Risk Taking

While Navy pilots are highly skilled professionals, flying high-performance jets is not an easy task, and pilots will take advantage of any factors that will increase their performance. In a training scenario, the upper hand could be gained through gaming the system and exploiting training artificialities. This is not to suggest that pilots would deliberately break training rules in order to gain the upper hand; rather, during periods of exceptionally high workload and stress, pilots will make use of any advantages they have in order to increase their performance and/or decrease their workload. "If you're not cheating, you're not trying," stated one pilot.

Thus, it is possible that LVC training could increase the possibility of pilot complacency and risk-taking behavior. For instance, pilots might be tempted to marginalize tracks that they know to be non-live (if non-live tracks are represented using unique symbols) or suspect to be non-live (if non-live tracks are not given unique symbols). In the presence of non-live entities, pilots might be more likely to cross altitude block boundaries or break other training rules, to cut corners in the heat of an engagement if they are being outmaneuvered, or to continue flying an engagement despite degraded SA. They may take risks rather than calling "terminate" or "knock-it-off" because there is "less metal" in the sky. Even if live and non-live radar tracks are represented using the same symbol set, it is possible that pilots could implicitly categorize tracks as live, virtual, or constructive based on their characteristics and behavior and then treat them differently. If presumed non-live tracks turn out to be live aircraft tracks, a collision could occur. In addition, aircrew would be practicing a way of flying that differs from how they need to fly with live aircraft.

Comparison with Hazards in Baseline Environment. Complacency and risk taking are threats to safety in the current air combat training system. The mindfulness and discipline required to follow training rules and safety practices add to pilots' high mental workload, and the potential for relaxing vigilance is ever present. Because relaxed vigilance and training rule violations are a risk in the current training environment, reinforcement methods are already in place. Training rule violations are kept in check by consistent enforcement. "We are brutal with people who break training rules," said one instructor pilot, "and then we shake hands and it will never happen again." This community-wide reinforcement and provision of direct feedback helps aircrew learn and make it a priority to practice diligence and stay within boundaries.

Change in Hazard Exposure. This hazard category was assessed as posing no change in hazard level because the Navy air combat training system already includes multiple mechanisms for detecting unacceptable risk taking and training rule violations. Pilots will be expected to treat all entities as if they are live aircraft and to adhere to all training rules despite the presence of non-live aircraft.

However, the use of non-live entities in training exercises may weaken a key mechanism for detecting training rule violations: the exercise staff playing the role of adversary pilots. Staff in the role of adversary pilot have multiple responsibilities. One responsibility is to monitor the exercise and exercise participants in order to detect if anyone loses SA or violates training rules. If training exercises replace a high proportion of adversary pilots with non-live aircraft as planned, the ability to effectively monitor exercise participants might be compromised. Additional research is needed to determine the adaptations required to maintain the safety system's integrity when fewer exercise staff members are in a position to monitor exercise safety.

10.3.4 Human–Machine Interaction

LVC training technology will introduce new capabilities and sources of information to the aircrew's work environment. It is therefore likely that system controls and displays will need to be updated so that aircrew have access to these new capabilities. However, changes to the interface design must be well considered; poor integration of LVC training capabilities with existing system components could result in new hazards. Examples of interface design changes that impact the safety of the system include the following:

- An introduced LVC super-mode for weapons control during LVC training would need to ensure that pilots could not accidentally fire a live weapon, thinking that it is simulated. Potential errors would need to be immediately detectable and recoverable.
- Cockpit controls used to initiate and clear non-live entities shown on sensor system displays need to be distinct from existing controls. If cockpit controls will be used by pilots to clear non-live entities when they become overwhelmed, the controls must be intuitive and easy to find and operate under stress.
- Pilots must be given continuous feedback on the status of non-live aircraft presentation on their sensor system displays. This design consideration is important because if the source of live aircraft information were to malfunction, the Link-16 Multifunctional Information Distribution System (MIDS) in particular, pilots need to immediately recognize that, although their displays are still populated with aircraft, those aircraft are all non-live and they are not seeing the live aircraft any longer.
- If non-live entities are to be represented using a new, unique symbol set, the symbols need to be designed such that they are not confusable with live aircraft symbols. Furthermore, the additions made to the non-live symbols must be designed so that if a live and a non-live track overlap, the pilot can see that there is real metal in that airspace.

Comparison with Hazards in Baseline Environment. In this study we did not examine the design of air combat aircraft cockpits. Human–machine interface designs in the cockpits of combat aircraft benefit from decades of work to improve safety in aviation. Nevertheless, the addition of software to automate the work of flying has resulted in pilot-automation miscommunications that have contributed to numerous mishaps and near-mishaps (Corrao 2009; Sarter et al. 1997; Williams 2006). System developers frequently count on humans in the system to adapt their work to the new technology (Hoffman et al. 2009). Frequently the humans do, but just as frequently, adaptation occurs through a process of lessons learned and trial and error. The resulting adaptations often interfere with getting work done efficiently and are known by names such as *kludge, work-around,* and *make-work* (Koopman and Hoffman 2003).

Change in Hazard Exposure. The specific interface changes to be introduced in support of LVC training will be novel, and if the "human-will-adapt" approach is taken for the design of LVC training system controls, feedback, symbology, and so forth, the resulting user interface will increase the hazard level of air combat training. Solutions must provide pilots with continuous knowledge and control of LVC training system status and artifacts, including non-live entities and simulated weapons, across all situations. To what extent the design solutions achieve that goal versus inadvertently adding to the risk of training depends partly on developers' methods.

The authors recommend an approach such as the Practitioner's Cycles Framework (Hoffman et al. 2010) whereby designs are evolved based on early and ongoing user interactions before they are implemented in a final form.

10.3.5 Exercise Management

LVC training technology will enable exercise management personnel to inject a greater number of aircraft into a given training exercise. The exercise management workload for these personnel will also increase, since they must monitor more participants while also keeping track of which are live and which are not. The increase in demands has the potential to undermine the effectiveness of exercise management and oversight. Thus, the effects will need to be monitored and mitigated by better exercise monitoring support tools, improved backup and support relationships among exercise management staff, and perhaps the definition of an additional exercise management role to ensure safety monitoring is not compromised.

The effect of LVC on the work of exercise management personnel will not be limited to a simple quantitative increase in attentional demands. Rather, LVC might also necessitate qualitative changes in their exercise management and oversight responsibilities. For example, new responsibilities will likely include decisions about whether to clear all non-live tracks or individual non-live tracks from participants' cockpit displays in order to decrease the complexity of a situation or to help a pilot regain SA. They might also include ensuring that interlopers, commercial and private aircraft that occasionally transit through exercise airspace, do not fly near or over the location of non-live aircraft. The non-live aircraft would not be visible to these interlopers, who are not linked into the exercise. Although judged unlikely by most pilot interviewees, it is possible for data correlation algorithms or an E-2 controller to correlate an interloper's radar return with a non-live track label as it passes over or near the track's location. The risk of this potential hazard will be mitigated if all tracks are assumed to be and are treated as live. As long as exercise participants treat all aircraft as though they are live, loss-of-separation incidents are unlikely to increase and hazard levels for live aircraft should not increase.

There are a number of technical challenges affecting LVC exercise management that will need to be resolved before LVC training is officially adopted. The ultimate form of the solutions to these technical challenges will determine the potential for hazards. A fundamental technical challenge is the selection or development of a technology for transmitting LVC exercise data. Issues to be considered include bandwidth requirements, security, and reliability. There is also a requirement to limit LVC exercise data transmission to exercise participants. In a nonexclusive LVC network (e.g. the current Link-16 system), all aircraft using that network for flight operations at the same time as an LVC training exercise is taking place will be able to see non-live entities even if they are not participating in the exercise. Nonparticipating pilots might not have been briefed on the presence of non-live entities in the area, and the disparity between their displays and their out-the-window view could result in substantial confusion.

Comparison with Hazards in Baseline Environment. Current exercise management practices have evolved over multiple decades to maintain a safe training environment with an acceptable level of risk. The training exercise monitor, a range training officer (RTO) or range safety officer (RSO), is supported in his or her safety objectives by adversary pilots who report witnessed training rule violations or suspected loss of SA by exercise participants. He or she is backed up by training rules and the responsibility of every exercise participant to declare "terminate" or "knock-it-off" if safety is in question. Complex, higher workload training exercises are staffed by both an RTO and RSO.

Although the transmission of exercise data is not always reliable in existing training exercises, work-arounds exist. For example, when pilots' data quality is reduced by Link-16 malfunctions, radar data are used to compensate. One pilot interviewee noted that an incomplete link is "not a reason to call [the exercise] off. . .it's basically a crutch for us. . .it just means that you're going to have to work even harder to make sure you know exactly where everybody is." As another example, for a period of time, certain adversary aircraft did not carry the Tactical Advanced Computer (TAC) pods that enable them to appear on RTO and RSO exercise monitoring displays. To compensate, those aircraft were paired with TAC pod-equipped aircraft during exercises. This work-around did not lead to any reported mishaps, but one interviewee did report experiencing confusion as an RTO.

Change in Hazard Exposure. This hazard category was assessed as posing no change in hazard level. Anticipated exercise management-related hazards were viewed as similar to existing hazards and manageable within the existing system, pending the effectiveness of LVC design and implementation solutions. Certainly, Link-16 malfunctions take on new significance in the LVC training environment. Loss of live track data in the midst of non-live tracks could be difficult to detect, and failed detection of the situation could easily lead to a mishap. LVC pilot interfaces will need to reliably and clearly inform pilots about the status of their live track data. The loss of non-live track data represents less of a threat to safety. However, there are no backup radar data to be used to compensate for its loss, and expert reviewers noted that the LVC training environment is qualitatively different from the live training environment. They recommended that the loss of non-live tracks be clearly and immediately communicated to exercise participants to reduce the risk of confusion or distraction.

Furthermore, additional research might be needed to assess the potential need for new exercise management roles and/or the possibility that both an RTO and RSO will be required, as is sometimes the case for very large live training exercises. Potentially, LVC exercises will include one or more simulation environment controller, or *puckster*, roles responsible for controlling the non-live entities and integrating them with live-force tactics and flow. An observation of high puckster workload during an LVC test exercise at Nellis Air Force Base suggests the use of more than one puckster (Observation, 16 August 2012). In addition, a report of delays between air intercept controller (AIC) communications and puckster inputs to non-live entities during a 2012 demonstration at the University of Iowa's Operator Performance Laboratory suggests a need to study LVC exercise management information flow.

10.3.6 Big Picture Awareness

In the context of air combat operations, the "big picture" typically refers to a visuospatial representation of all active aircraft and aircraft activity in a given air operation. In other words, it is a form of *global SA*. If a pilot has a good awareness of the big picture, he or she is keeping track of aircraft both within and outside his or her immediate surroundings – usually by monitoring the radar, other cockpit system indications, and communications – while considering their implications based on his or her experience and the preflight briefing. In large-force exercises, this awareness is essential to staying coordinated with the team, staying on the mission's timeline, providing timely assistance to other team elements, and anticipating what will happen next.

Communication is critical to the maintenance of global SA. Communications eavesdropping, or listening in to the radio chatter of other friendly forces, helps pilots to develop and maintain their big picture awareness (Neville et al. 2003). However, the natural language capabilities of constructive aircraft are limited relative to human language capabilities, and therefore it is likely that there will be significantly fewer communications and certainly less radio chatter generated by these aircraft during an exercise. While this limitation will not affect the initial implementation of LVC training, wherein only adversaries will be non-live, future evolutions of LVC training technology might introduce non-live friendly force aircraft. These non-live aircraft could be used, for example, to stand in for air platforms that are not available for a given exercise or to add additional sections to increase the complexity of tactical flow decision making. However, given pilots' reliance on communications eavesdropping for big picture awareness, the use of multiple constructive friendly force aircraft could negatively impact pilots' big picture awareness.

Comparison with Hazards in Baseline Environment. Air combat missions sometimes require communications silence or encounter communications jamming. This represents a precedent for operating under conditions of sparse communications. One interviewee discussed the difficulty of flying under those circumstances. This discussion suggested that hazard exposure is significantly heightened when communications are absent. This situation is not the same as reducing communications through the addition of non-live exercise participants. However, its difficulty, together with the role of eavesdropping in team SA (Neville et al. 2003; Vuckovic et al. 2004), indicates an increase in risk.

Change in Hazard Exposure. In the near term, this hazard category, which involves the maintenance of SA primarily through monitoring team verbal communications, does not pose a concern for the safety of training exercises since the US Navy does not plan to use constructive friendly force exercise participants in its initial implementation of non-live entities. However, as military aviators become more comfortable with LVC technology over time, the Navy might begin to experiment with using constructive aircraft in friendly force roles. If and when this possibility starts gaining momentum, its potential to produce a hazard situation will need to be investigated. In particular, research will be needed to determine whether certain friendly force roles need to be filled by live aircraft; whether a certain percentage of aircraft will need to be live during large-scale exercises to improve communication, and thereby, big picture awareness; and whether a loss in big picture awareness actually occurs. If a loss in big picture awareness is detected, it will be important to determine whether that loss poses a safety risk and, if it does, whether the risk is acceptable in light of other training benefits. It is possible that awareness will only be lowered for the location and activities of non-live aircraft, which may not be as serious a concern as lowered awareness of the overall big picture.

10.3.7 Negative Transfer of Training to the Operational Environment

Although this paper is concerned with the safety ramifications of introducing change into live naval air combat training, there are also training concerns that ultimately could degrade safety – that is, the likelihood of success and survival – in combat operations. LVC technology will presumably introduce artificialities into training. The nature of these artificialities will depend on the specific LVC engineering solutions produced but could include non-live tags added to track symbols of non-live aircraft, radar tracking that is better than what is possible with live aircraft systems, display graphics that designate live aircraft as *kill removal* as they try to exit the exercise area after being hit by a simulated missile, non-live tracks that disappear when they reach a certain distance from live friendly force aircraft, and a lower-complexity communications environment than would exist if all players were live and verbally communicating.

Notably, pilot interviewees expressed concern about losing the habits and mindfulness that they develop while training with live aircraft that present a real collision risk. This concern is of particular importance if non-live aircraft track symbols are distinguishable from live aircraft track symbols.

Another concern raised by participants was that artificially perfect radar tracking of non-live aircraft or other perfect sensor indications in the LVC training environment could reduce the quality of practice with translating real-world cockpit displays into an understanding of the big picture. The development and maintenance of big picture awareness, or global SA, is reportedly very challenging, and the associated skills are difficult to acquire. Brobst and Brown (1996) describe these F/A-18 pilot skills as "very time intensive to re-build once lost" (p. 36). Their acquisition should be monitored to assess potential impacts of LVC artificialities. Potentially, non-live aircraft might be used to benefit their acquisition by increasing the complexity of the big picture and providing more advanced big picture awareness training opportunities.

Developing skill in team communication in high-complexity environments may also be impacted if part of the team is not verbally participating. Although non-live aircraft initially would serve only as adversaries, if they eventually are used to populate the blue force in large numbers and with regularity, this could have a negative effect on *comm brevity* mindfulness and skill. Comm brevity skills reduce incidents of air wing team members "stepping" on one another's communications and help the teams coordinate despite very limited communication resources. Practice that encourages improvement of these skills increases safety (chances of survival and success) in real-world combat.

Comparison with Hazards in Baseline Environment. Artificialities in live air combat training currently exist to keep that training as safe as possible. For example, pilots know they can declare "knock-it-off" or "terminate" at any time. In addition, altitude blocks are used to keep aircraft separated, pilots are told to treat 5000 ft above ground level (AGL) as the ground, head-on missile shots are not allowed within a mile and a half, and aircraft are required to maintain 500 ft of separation during engagements (Naval Strike and Air Warfare Center n.d.), none of which would be true for combat. Artificialities also exist as a result of unavoidable constraints, such as the need to allow "killed" aircraft to exit the exercise airspace and the need to compensate for a limited number of aircraft by using one aircraft to represent many.

Change in Hazard Exposure. The effects of LVC artificialities on hazard exposure will depend on the roles and numbers of non-live aircraft participating in training exercises. Because of the criticality of skills for maintaining separation from and coordinating with a wingman, the Navy proposed that non-live aircraft would not be used in the role of wingman. They additionally proposed to use non-live aircraft only as adversaries, which avoids effects on team coordination and communication skills. They further proposed to only use non-live aircraft outside the merge to minimize bad habits such as focusing on sensor systems versus looking out the canopy.

On the positive side, LVC technology may negate the need for certain existing training exercise artificialities. For example, "killed" non-live aircraft could disappear from aircraft sensor systems in much the way they would in real-world combat, and, as an especially valuable increase in realism, non-live adversary aircraft could fly at more realistic speeds and execute more advanced tactics.

10.4 Conclusion

A fundamental premise of our evaluation work is that Navy air combat training is a complex system that has achieved, over time, a high level of resilience and stability. New technology may introduce novel disruptions that differ from what the system has evolved to handle. We found that most potential disruptions, or hazards, posed by the introduction of non-live aircraft into the training system could be mitigated by sources of resilience that already exist in the system. We also found that three categories of hazard will likely require adaptations to the system once non-live aircraft begin participating. Adaptations may take multiple forms in this sociotechnical system. They can include preliminary cockpit and exercise management design requirements for the future LVC training system acquisition program. They can also include extensions and adaptations to training rules to address virtual and constructive aircraft participation. The design of these mitigations was accomplished in follow-on work and is described in Neville et al. (2016).

This work demonstrates a human systems engineering approach that uses multidisciplinary methods to contribute to the preliminary design of an LVC-based air combat training system. The approach considers humans and other non-technology elements to be as much a part of the system as the technology elements. It uses a combination of qualitative ethnography, typically not associated with the engineering discipline, and a risk management method that is common in systems engineering. Both play important and complementary roles in understanding critical human–technology relationships within the system and their impact on system function.

The ethnographic portion of the evaluation can be classified as an example of action research (Susman and Evered 1978) in that it involved collaboration with representatives of the training systems' main stakeholder, the air combat community. Representatives included two naval air combat experts who participated in data collection and analysis. In the follow-on work cited above (Neville et al. 2016), domain experts led discussions with air combat interviewees to generate community-driven mitigations. The involvement of air combat community members contributed to the identification of system characteristics and mechanisms that provide the training system with its resilience and stability. An example is the local and highly redundant control over safety in the form of engagement and exercise termination codes that anyone in or involved with the exercise can transmit. Community member participants also revealed competing but balanced sources of tension within the training system, such as the tension between exercise fidelity and safety.

By engaging individual aviators, cross-scale interactions between downward and upward resilience are better foreseen and managed (Woods 2006). The manner by which organizational context either creates or facilitates the resolution of tensions or goal conflicts in the system (i.e. downward resilience) can interact with adaptations made by agents on the micro level, such as work-arounds or new tactics (i.e. upward resilience) that could undermine the organization's attempts to command compliance with broad standards. Take, for example, the hypothetical scenario in which the introduction of non-live entities into training is improperly handled at the organizational level and pilot workload is increased to unmanageable levels. In response, pilots might find work-arounds to reduce their workload, such as excessive use of the non-live "on–off" switch (currently under consideration as a safety measure) to reveal which aircraft are real and which are not and then treat non-live aircraft as less important.

This sort of work-around by pilots could render the Navy's attempts to command compliance with proposed training rules – for example, that all tracks shall be treated as live – as futile. It is, however, an example of the self-organization that takes place in complex systems to manage and protect the competing subsystem objectives and overall system goals. By involving pilots and other system personnel in the technology-introduction process, we were able to identify and assess ways the system might react to and self-organize in response to perceived threats and disruptions.

Hoffman and Woods (2011) propose five sources of tension that are common to complex systems and which achieve a form of system-specific balance. Balance is achieved through trade-offs and negotiations via self-organizing that is responsive to internal tensions, environmental pressures, and the need to stay within bounds of effective functioning. Non-live entities are a disruption that could affect the balance attained among sources of tension. For instance, adding a non-live aircraft "on–off" switch to the cockpit affects the *bounded effectiveness* trade space, where a given system attains a balance between the delegation of authority to the lowest level and high-level control over system elements (including pilot choices in this case). New LVC technologies and capabilities introduce new sources of control that reduce aircrew control. In response, pilots might feel compelled to seek more individual authority through use of the on-off switch.

An additional example involves the kill removal procedures governing the exit of aircraft from the training arena after being hit by a simulated missile. These procedures may not be required for non-live aircraft, yet they are part of the doctrine of the current training system. If changes are made to training rules to allow a killed non-live aircraft to just disappear, said changes may be worked out in what Hoffman and Woods call the *bounded ecology* trade space, where trade-offs are made between system specialization and generalizability/resilience. Training rules that are specialized to each type of aircraft, live and non-live – or even live, virtual, and constructive – are one side of that trade space. On the other side is a single set of training rules that does not distinguish between live and non-live aircraft. If either extreme is chosen, work-arounds and informal practices are unlikely to emerge because adherence to training rules is a fundamental requirement of air combat training. Flexibility in the form of informal adaptation is not an option. However, the training system does have built-in flexibility, i.e. a resilience mechanism, in the form of a process for community-driven training rule changes. Using this process, the system can adjust to a middle ground on this dimension that is a better fit.

The competition of tensions within trade spaces is complicated by interplay among organizational levels. Using the research approach described in this paper, cross-scale relationships between macro- and microlevel influences on system balance and resilience can be foreseen and managed. The manner by which organizational context either creates or facilitates the resolution of pressures and goal conflicts in the system can provide flexibility that supports rapid adaptation by agents on the micro level, such as the ability to try work-arounds or new tactics in response to sudden changes or events. When organizational flexibility is lacking, these work-arounds are more likely to involve unattractive choices and risk taking.

Our work demonstrates an approach to integrating new technology into a complex sociotechnical system, a challenging process involving numerous variables and many possible missteps leading to potential calamities but for which little guidance exists (Madni and Sievers 2014). By describing the types of disruptions and associated hazards that need to be monitored and evaluated in the air combat training system, we additionally contribute to the successful use of the LVC air combat training paradigm and to the advancement of high-fidelity military training (Brooks et al. 2002; Carroll 1999; George et al. 1999; Smith et al. 2007).

This evaluation has set the stage for a phased introduction of non-live entities that allows the LVC paradigm and air combat training system to coevolve, an integration strategy based on principles of resilience in complex systems (Benbya and McKelvey 2006; Kitano 2004). Future work, as noted previously, has involved the identification of hazard mitigations and their translation into recommended procedural changes and requirements for a future LVC air combat training system.

Beyond this particular effort, we hope to extend lessons learned about the effect of new technologies on established complex systems. The dominant paradigm for introducing new technologies and other changes into complex systems is one that places most of the burden of coping and adapting on the host system. The change is introduced, and the host system must weather negative consequences of misalignment, cope with new tensions, and adapt to a new status quo. As we describe above, complex systems self-organize in response to disruptions that destabilize balanced tensions or interfere with core system goals. However, systems do not always recover well, depending on the extent of the disruption and its management. For example, Trist and Bamforth (1951) describe the destabilizing impacts of new longwall mining technology on not just the mining operation, but the entire mining community. On the other hand, agents within a system often simply bar change from occurring. Examples abound of technology transition failures (Standish Group 2015), and it is no wonder: high costs are at stake; in some cases, national security is at stake; in many cases, human lives and well-being are at stake.

We need effective tools and methods to guide change in high-stakes complex systems. This evaluation provides a method for managing a change and its interactions with the system as a whole. It is grounded in complex systems theory, as described above, and provides a constructive option to the unproven and suspect assumption that technology users are simply resistant to change. This method and its use to explore sociotechnical system dynamics and responses to change represent a valuable contribution and progress along the path to improved transition of technology and change into our increasingly complex systems.

Acknowledgments

This research was sponsored by the Office of Naval Research (ONR). The authors wish to thank our extended research team members at Collins Aerospace and the Operator Performance Lab (OPL) at the University of Iowa, especially Dr. Tom Schnell (OPL). We also wish to thank John "Bam-Bam" Mooney and Derek "Baffle" Ashlock for their valuable input. Finally, we thank all of our interviewees for taking time out of their busy schedules to talk to us in-depth about the air combat training domain.

The views expressed in this paper are those of the authors and do not represent the official views of the organizations with which they are affiliated. An author's affiliation with The MITRE Corporation is provided for identification purposes only and is not intended to convey or imply MITRE's concurrence with, or support for, the positions, opinions, or viewpoints expressed by the author. This paper has been approved for Public Release; Distribution Unlimited; Case Number 43-341-14. ©2020 The MITRE Corporation. ALL RIGHTS RESERVED.

References

Ausink, J.A., Taylor, W.W., Bigelow, J.H., and Brancato, K. (2011). *Investment Strategies for Improving Fifth-Generation Fighter Training* (Document No. TR-871-AF). Santa Monica, CA: RAND Corporation http://www.dtic.mil/dtic/tr/fulltext/u2/a537970.pdf (accessed 21 August 2020).

Benbya, H. and McKelvey, B. (2006). Using coevolutionary and complexity theories to improve IS alignment: a multi-level approach. *Journal of Information Technology* 21 (4): 284–298.

Braun, V. and Clarke, V. (2006). Using thematic analysis in psychology. *Qualitative Research in Psychology* 3 (2): 77–101.

Brobst, W.D. and Brown, A.C. (1996). *Developing Measures of Performance for F/A-18 Aircrew Skills* (CRM 96-129.00). Alexandria, VA: Center for Naval Analyses.

Brooks, R.B., Breitbach, R.A., Pegg, S. et al. (2002). *Innovative C2 Training Solutions for Air Force Modular Control Systems* (AFRL-RH-AZ-PR-2002-0003). Mesa, AZ: Air Force Research Laboratory Warfighter Readiness Research Division.

COMNAVAIRFOR (2016). *NATOPS General Flight and Operating Instructions Manual* (CNAF M-3710.7). https://www.public.navy.mil/airfor/vaw120/Documents/CNAF%20M-3710.7_WEB.PDF (accessed 23 August 2020).

Carroll, L.A. (1999). Multimodal integrated team training. *Communications of the ACM* 42 (9): 68–71.

Corrao, P.A. (2009). An examination of the MH-60s common cockpit from a design methodology and acquisitions standpoint. Master's thesis. Naval Postgraduate School, Monterey, CA. Retrieved from the Defense Technical Information Center (ADA501110).

Department of the Navy, Office of the Chief of Naval Operations (2010). OPNAV instruction 3500.39C (OPNAVINST 3500.39C). https://www.public.navy.mil/airfor/vaw120/Documents/OPNAVINST%20 3500.39C%20(ORM%20Inst).pdf (accessed 23 August 2020).

Flanagan, J.C. (1954). The critical incident technique. *Psychological Bulletin* 5: 327–358.

George, G.R., Breitbach, R.A., Brooks, R.B. et al. (1999). Synthetic forces in the Air Force command and control distributed mission training environment, current and future. In: *Proceedings of the 8th Conference on Computer Generated Forces and Behavioral Representations*, 319–331. Orlando, FL: Simulation Interoperability Standards Organization.

Heinrich, H.W. (1931). *Industrial Accident Prevention*. New York: McGraw-Hill.

Hoffman, R.R. and Woods, D.D. (2011). Beyond Simon's slice: five fundamental trade-offs that bound the performance of macrocognitive work systems. *Intelligent Systems, IEEE* 26 (6): 67–71.

Hoffman, R.R., Neville, K., and Fowlkes, J. (2009). Using cognitive task analysis to explore issues in the procurement of intelligent decision support systems. *Cognition, Technology & Work* 11 (1): 57–70.

Hoffman, R.R., Deal, S.V., Potter, S., and Roth, E.M. (2010). The practitioner's cycles, part 2: solving envisioned world problems. *IEEE Intelligent Systems* 25 (3): 6–11.

Hollnagel, E., Woods, D.D., and Leveson, N. (2006). *Resilience Engineering: Concepts and Precepts*, 43–54. Burlington, VT: Ashgate.

Kitano, H. (2004). Biological robustness. *Nature Reviews Genetics* 5 (11): 826–837.

Klein, G.A., Calderwood, R., and MacGregor, D. (1989). Critical decision method for eliciting knowledge. *IEEE Transactions on Systems, Man and Cybernetics* 19 (3): 462–472.

Koopman, P. and Hoffman, R.R. (2003). Work-arounds, make-work, and kludges. *IEEE Intelligent Systems* 18 (6): 70–75.

Madni, A.M. and Sievers, M. (2014). System of systems integration: key considerations and challenges. *Systems Engineering* 17 (3): 330–347.

Naval Air Systems Command (2018). E-2 Hawkeye early warning and control aircraft. https://www. navy.mil/DesktopModules/ArticleCS/Print.aspx?PortalId=1&ModuleId=724&Article=2160986 (accessed 23 August 2020).

Naval Strike and Air Warfare Center (n.d.). *TOPGUN Training Rules*. Fallon Naval Air Station, NV: NSAWC.

Neville, K., Fowlkes, J.E., Nelson, M.M.W., and Bergondy-Wilhelm, M.L. (2003). A cognitive task analysis of coordination in a distributed tactical team: implications for expertise acquisition. In: *Proceedings of the Human Factors and Ergonomics Society Annual Meeting*, 359–363. Los Angeles, CA: Sage.

Neville, K., Sherwood, S., Mooney, J. et al. (2016). An assessment of a complex military training system's resilience to change. In: *Proceedings of the 2016 Annual Meeting of the Human Factors and Ergonomics Society*. Thousand Oaks, CA: Sage.

Pariès, J. (2006). Complexity, emergence, resilience. In: *Resilience Engineering: Concepts and Precepts* (eds. E. Hollnagel, D.D. Woods and N. Leveson), 43–54. Burlington, VT: Ashgate.

Rickard, R.N., Bennett, W.R., and Colgrove, C. (2013). Development of fighter aircraft live, virtual, and constructive (LVC) interface standards. In: *Proceedings of the Fall Simulation Interoperability Workshop*, 250–261. Red Hook, NY: Curran Publishing.

Sarter, N.B., Woods, D.D., and Billings, C.E. (1997). Automation surprises. In: *Handbook of Human Factors and Ergonomics*, 2e (ed. G. Salvendy), 1926–1943. Hoboken, NJ: Wiley.

Smith, E., McIntyre, H., Gehr, S.E. et al. (2007). *Evaluating the Impacts of Mission Training via Distributed Simulation on Live Exercise Performance: Results from the US/UK "Red Skies" Study* (AFRL-HE-AZ-TR-2006-0004). Mesa, AZ: Air Force Research Laboratory Warfighter Readiness Research Division.

Standish Group (2015). *The Chaos Report 2015.* https://www.standishgroup.com/sample_research_files/CHAOSReport2015-Final.pdf (accessed 23 August 2020).

Susman, G. and Evered, R. (1978). An assessment of the scientific merits of action-research. *Administrative Science Quarterly* 23: 582–603.

Trist, E.L. and Bamforth, K. (1951). Defences *(sic)* of a work group in relation to the social structure and of coal-getting: an examination of the psychological situation and some social and psychological consequences of the longwall method and technological content of the work system. *Human Relations* 4: 3–38.

Vuckovic, N.H., Lavelle, M., and Gorman, P. (2004). Eavesdropping as normative behavior in a cardiac intensive care unit. *JHQ Online* 5: 1–6.

Williams, K.W. (2006). *Human Factors Implications of Unmanned Aircraft Accidents: Flight Control Problems* (TR DOT/FAA/AM-06/8). Washington, DC: FAA OAM.

Woods, D.D. (2006). Essential characteristics of resilience. In: *Resilience Engineering: Concepts and Precepts* (eds. E. Hollnagel, D.D. Woods and N. Leveson), 21–34. Burlington, VT: Ashgate Publishing Company.

Section 4

Considering Human Characteristics

11

Engineering a Trustworthy Private Blockchain for Operational Risk Management

A Rapid Human Data Engineering Approach Based on Human Systems Engineering

Marius Becherer, Michael Zipperle, Stuart Green, Florian Gottwalt, Thien Bui-Nguyen, and Elizabeth Chang

University of New South Wales at Australian Defence Force Academy, Canberra, ACT, Australia

11.1   Introduction

This chapter presents a rapid human data engineering (HDE) approach based on human systems engineering (HSE) principles for the development of a big data management tool – known as trustworthy private blockchain. It not only places humans in the center of the blockchain-enabled human data integrated (HDI) system but also human-centered system design by engaging the subject matter expert and systems engineers for a human-centered trustworthy enterprise private blockchain development. It supports the human effort by automated data or transaction recording, processing, reporting, tracking, and analytics, from a trusted *single source of truth* data repository, to improve human performance by mitigating the operational risks in the context of compliance, assurance, financial accountability, and capability readiness, particularly for large complex enterprises.

11.2   Human Systems Engineering and Human Data Engineering

According to work by Handley (2019a, 2019b), Tolk (2012), and Mittal and Tolk (2020), the HSE approach applies to human principles, models, and techniques to optimize system design and performance by continuously taking human capabilities and limitations into consideration. We use this as the key principle for any large complex system development. One typical system is big data management, which leads us to adopt HSE for HDE and user data interaction design (UDID):

- HDE applies human principles, models, and techniques to optimize big data repository design and interaction including trust and security of the big data sources with continuous focus on human capabilities and limitations. This includes (i) trustworthiness of the big data repositories, (ii) understanding and interpretation of big data, and (iii) knowledge discovery and decision support from the big data.
- UDID addresses how easily humans can access all forms of big data including structured and unstructured, such as sound, video, transactions, individual profiles such as roles and responsibilities, and accountabilities, not only timeliness but also trustworthiness.

A Framework of Human Systems Engineering: Applications and Case Studies, First Edition.
Edited by Holly A. H. Handley and Andreas Tolk.
© 2021 The Institute of Electrical and Electronics Engineers, Inc. Published 2021 by John Wiley & Sons, Inc.

Today, no databases can provide such confidence, except the blockchain. Blockchain provides the trust of the data sources that were always needed but never existed, particularly in the last 30 years of Internet development and distributed networked economy. HSE principles open new and exciting opportunities for the designing of big data management systems, HDE, HDI, and UDID, particularly in the Fourth Industrial Revolution (4IR) era.

We present a rapid HDE through the use of the HSE approach to trustworthy private blockchain design, development, and deployment. There is a vast amount of work that needs to be completed on UDID, as presented in Mittal and Tolk (2020) along with metrics to evaluate the trust between humans and systems, for example, those presented by Freedy et al. (2007). Here, we will focus on the building of the trustworthy private blockchain or a trustworthy data repository that is secure, immutable, distributed, and efficient and effective for human-centered big data management. A specific application area is operational risk management.

11.3 Human-Centered System Design

The trustworthy blockchain is a human-centered big database system that primarily considers how human operators can use and trust the databases or the blockchain; how human operators can experience the *single source of truth*, data immutability, data transparency, and data auditability (Casino et al. 2019; Christidis and Devetsikiotis 2016); how humans are accountable for their transactions; and how humans can trust the ledgers and so forth. The human-centered system design focuses on how the blockchain system can be built so that the trust between the human operators, the hierarchy of management, the peer networks, and the big data can lead to efficiency and productivity in transaction processing, reporting, tracking, and analysis in this dynamic complex environment.

The human-centered system design in the HDE focuses on reducing human manual-based tasks such as the mental load on analyzing data, timely reporting, and decision making. The system automatically identifies human roles, responsibilities, and accountabilities and improves human operators' efficiency and transactions overload. It also addresses constraints imposed by a dynamic changing environment, human capabilities, and limitations as well as the slow progress imposed by a highly regulated hierarchy workforce.

In the project "Blockchain for Operational Risk Management" funded by the Department of Defence, we were guided by the human-centered system engineering (HSE) methodology. We worked together with frontline officers, subject matter experts, civil–military transactors, to trustworthy private blockchain system design, development, and deployment. In addition, we utilize rapid HSE for rapid development and deliveries, an approach against the traditional overlord IT procurement approach, where human operators are left out in terms of requirement definitions, technical specifications, and user experience, which results in complex contracts under management and incurring huge costs, delayed schedules, and merely useful products or tools at the end of a development life cycle, which has been detrimental to all stakeholders.

11.4 Practical Issues Leading to Large Complex Blockchain System Development

11.4.1 Human-Centered Operational Risk Management

One important operational risk in large complex enterprises is the compliance and assurance with regard to the accurate finance or asset or data/information management, which impacts the organization's reputation and capability preparedness. However, transaction recording, reporting, tracking, and analysis are the job of big data, either in physical or digital format. An example is the supply chain network in defense logistics and sustainment including multidimensional data management, which needs to address the whole capability life cycle, among military services and joint forces, as well as restricting workforces with a hierarchy. Managing and reporting such data is intensive, mandraulic, time consuming, and error prone. There have been no satisfactory tools proposed to support big data management particularly in defense logistics over the last 20–30 years. Not only does it hinder human performance but also wastes officers' valuable time away from undertaking higher-order tasks such as strategic planning or force design.

11.4.2 Issues Leading to Risk Management Innovation Through Blockchain

In the case of defense departments with billions of dollars of annual government-funded defense enterprise and assets held in its IT systems, they are distinctly exposed to risks, which exist in the form of findings being placed against defense financial statements. This can invariably result in an audit qualification being administered. In 2004, the Australian Department of Defense's auditing agency expressed a lack of confidence in auditing processes and placed findings against defense with qualified statements. Poor data management issues led to discrepancies among data and transactions for which no one was accountable. This led to millions of dollars expenditure over 10 years to rectify, and, to date, very little has been learned from this experience. The Australian Department of Defense had been operating under a relationship verified by the clear demonstration of fast tracking back to the potential financial qualification of statements once the governance regime was lifted. There were limited tools to help root cause analysis. The misalignment in the data across databases continues to haunt the Australia Department of Defense's reputation in financial accountability and poor data management, leading to poor compliance assurance and risk management. Defense's response was to re-instigate the overlord approach to data and IT management in 2014. A solution is still unavailable to all stakeholders.

Private blockchains are enterprise blockchains that address the needs for trustworthy data and transactions as well as accountability of human operators with a *single source of truth*. It is the solution for the next generation of enterprise data management and modernized workforce and human operations within the organization and between the organizations, and it is suitable for both public and private sectors. Enabling this characteristic requires trusted data sources and access between human operators or users of the private blockchain and among several groups of users at intra- and inter-organization settings.

In this case, a rapid HSE and HDE framework of compliance and assurance risk management tool was established as an extremely reactive response to the critical position that was identified pertaining to the Australian Department of Defense's financial statements.

11.4.3 Issues in Engineering Trustworthy Private Blockchain

The trust of private blockchain, such as Hyperledger Fabric, has encountered many challenges due to its centralized key management. Centralized data management creates the opportunity for data manipulation, data tempering, and multiple truths, leading to a discrepancy in reports and distrust among the stakeholders. Besides, it has a complex human hierarchy, employee type, and user identifications as part of private blockchain engineering for the workforce and the enterprise. Furthermore, there are very limited public resources available to support such human-centered systems engineering to private blockchain design and development.

To address this issue, the frontline operators must be closely involved with subject matter experts and domain knowledge workers in every stage of the private blockchain database system development.

We use Hyperledger Fabric, which is one available open-source approach for private blockchain (Mazumdar and Ruj 2019). Many researchers have tried to develop blockchain (Tama et al. 2017; Zheng et al. 2018), such as blockchain applications in stream data management with IoT (Conoscenti et al. 2016), big data management (Karafiloski and Mishev 2017), cybersecurity and trust management (Hebert and Di Cerbo 2019; Khan and Salah 2018; Seebacher and Schüritz 2017), and other data quality aspects (Koteska et al. 2017; Yli-Huumo et al. 2016). The key issues of trust and use of the private blockchain have barely been discussed or successfully implemented in the existing literature (Hebert and Di Cerbo 2019). In this chapter, we present the use of the HSE approach to private blockchain system development to achieve trust and accelerate government and large complex enterprise adoptions of private blockchain.

11.5 Framework for Rapid Human Systems–Human Data Engineering

The aims of the HSE framework mitigate human performance shortfalls and maximize system effectiveness by integrating well-defined HSE and HSI processes and activities (Handley 2019a). The framework presents from human-centered design (HCD) to HSE to HDE and HDI as presented in Figure 11.1. It includes important elements and relationships to the utilization of HSE and allows subject matter experts, domain specialists, system designers, and software engineers, as well as the hierarchy of program managers to provide design ideas and feedback from the prototype systems to determine appropriate requirements for the system and tool development.

The dimensions and descriptions of the framework are as follows:

A) Human-centered system design – This dimension represents the different levels and types of users that interact with the system designer and engineers to build the system. The system design has three phases with a problem area such as risk management, understanding enterprise policies and processes, and defining user and technical requirements of the system.
B) HSE – An engineering life cycle is presented from planning, design, implementation, testing, and maintenance to continuing system improvement.
C) HDE and HDI – These are the result of HSE that produced trustworthy blockchain databases that have the capability to manage operational risks. Humans are an integral part of all transactions and are recorded by blockchain as one entry of a ledger. This reinforces roles, responsibility, and accountability at any point of time and can be accessed at point of need.

The framework also presented the rapid development timelines including human engagement in the design and engineering processes along with prototype field validation of human data interaction as seen in Figure 11.2.

11.6 Human Systems Engineering for Trustworthy Blockchain

11.6.1 Engineering Trustworthy Blockchain

Trustworthy blockchain lies on the public key infrastructure (PKI), where key pairs are a central trust component. With this, PKI is used to satisfy several key concepts of trusted communications such as authenticity, message integrity, and confidentiality. The key and certification management depend on a central authority within this ecosystem to issue digital certificates and manage public

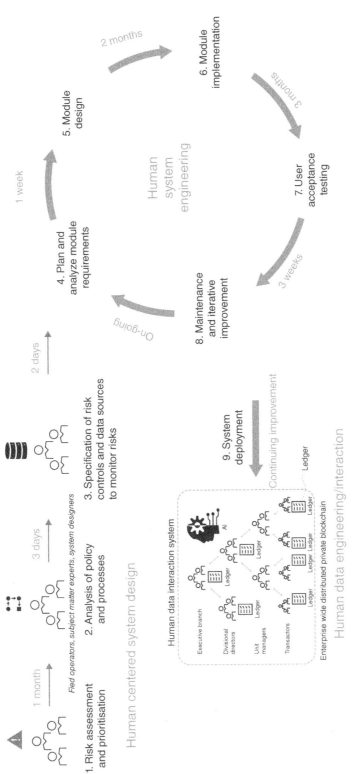

Figure 11.1 The rapid human-centered systems engineering and human data engineering framework.

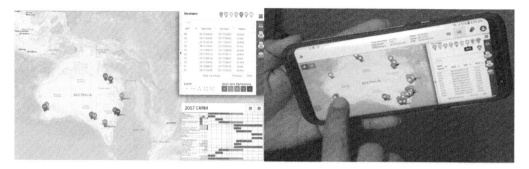

Figure 11.2 User data interaction design (UDID) of trustworthy blockchain illustrated in the location-based risk map.

and private key pairs. There is little information about the parameters used for validating trustworthy private blockchain like Hyperledger Fabric. Existing solutions of Private Hyperledger Fabric Blockchain deal with keys and certificates in a centralized orderer. Such solutions, however, violate one key significance of blockchain technology – that being decentralization through decentralized key management. In contrast, the private blockchains rely on centralized key management and the centralized authority. In order to develop a trustworthy private blockchain, we designed and engineered a unique PKI in private blockchains. Among other trust issues, we present the challenges faced in the implementation and solutions to keys and certificate management to ensure trustworthy communications with the blockchain including hosts, nodes, and participants of organizations.

11.6.2 Issues and Challenges in Trustworthy Private Blockchain

A public blockchain has its own reputation of trust in the public domain due to its distributed trust mechanism as well as high levels of security in terms of data immutability, decentralization, and no-central authority. An upsurge in adoption of blockchain in all types of government, commerce, and industry is aimed at achieving the trust between the organizations and individuals (Tapscott and Tapscott 2017).

A trustworthy blockchain network consists of numerous distributed nodes across different regions or continents. Different organizations that are part of the blockchain network can run their own blockchain nodes within their own enterprise IT infrastructure. This will allow users within their organizations to access various forms of data based on permission policies defined within the private blockchain network to which the nodes are connected. Policies reflected as smart contracts describe how the blockchain transactions are verified.

Unlike public blockchain, the private blockchain network is intended to be used indirectly by various users within the enterprise or consortium enterprises. Users that are interacting directly with the blockchain will be able to configure elements of the blockchain network such as the number of peers or leader peers (orderers) within their organizations.

There is little information about the parameters used for validating trustworthy private blockchain like Hyperledger Fabric such as:

- How to establish trust between the enterprise existing physical security infrastructure and the security infrastructure enforced by the blockchain and how to integrate both infrastructures.
- Users or peers can dynamically join and exit the public blockchain network anywhere and at any time. However, in the private blockchain, how is trust established in the newly joined users, peers, and organizations in such a dynamic environment?

- How does one ensure the trust of the dynamic policies in the permission-based private blockchain? Unlike a public blockchain, which is a permission-less blockchain, the policy for transactions is not permitted to change. However, a private blockchain allows the users to change the policies of its blockchain network.
- What endorsement should be set as the trustworthy policies and how to decide on existing, updated, and configured policies?
- How is trust determined in the private blockchain databases and where is the backup of the blockchain hosted?
- What restrictions should be set for creating a user account and is the user account holder trustworthy, and how can trust be established to handle deletion of accounts?
- What happens when nodes are compromised?

11.6.3 Concepts Used in Trustworthy Private Blockchain

The concepts used for engineering trustworthy private blockchain systems are:

- *User* – An individual member that uses a blockchain system or applications within the blockchain network.
- *Entity* – An individual user, an actor, or a node in the blockchain.
- *Identity* – Information associated with a certificate about the permission of a user in a blockchain network.
- *Peer* – A Hyperledger Fabric network node in the private blockchain that has a copy of the ledger.
- *Orderer* – A higher-level peer or a super node that is responsible for ordering transactions into blocks (or to a ledger/blockchain).
- *Certificate authority (CA)* – A specific entity that issues digital certificates to other entities.
- *Certificate* – A document that binds a public key to a particular entity of the private blockchain network. It includes information such as the name of the certificate holder and the CA that issued the certificate.
- *Organization* – An entity that can join a blockchain network by providing its service to the network. It has its own organizational structure and manages its own entities (users, admins, peers, etc.).
- *Consortium* – The set of all organizations within a blockchain network that can form alliances or partnerships, also known as channels.
- *Channel* – A private communication between different organizations, also known as subnet. Each channel contains its own distributed, blockchain ledger, set of entities involved, and policies governing interactions between the entities. Multiple channels can exist within a consortium.
- *Membership Service Provider (MSP)* – The set of certificates that are used to define an organization and its entities (users, admins, peers, etc.).
- *Consensus* – The mechanism for reaching agreement between different entities about transactions and/or blocks in the private blockchain such as Hyperledger Fabric. The private blockchain consensus mechanism involves many more steps than a public blockchain, an example of which is Bitcoin.
- *Smart contract* – A specific business logic that is in line with the data models that are stored/retrieved in the blockchain ledger.
- *Endorsement* – Peer executing some smart contracts and determines whether later steps should be followed to commit the transactions to the blockchain ledger.

- *Policy* – An endorsement policy that defines which entities must endorse a transaction within the transactional flow of a private blockchain such as Hyperledger Fabric.
- *Hardware security module (HSM)* – A physical computing device that can be used to store sensitive information (e.g. digital keys). HSM are tamper resistant and can be configured to destroy all keys if a security breach occurs.
- *Software development kit (SDK)* – A collection of software tools used for software development.
- *Gateway* – Network servers that expose a REST API SDK server to allow users to interact with the blockchain applications (e.g. create transactions, query the databases).
- *Cluster* – A group of services that are connected and communicate with each other to fulfill common objectives.
- *Root certificate authority (RCA)* – The RCA issues certificates for the trustworthiness of the subordinate entity.
- *Intermediate certificate authority (ICA)* – Subordinate certificate authority that issues certificates instead of RCA to increase security through a void.
- *Docker* – An open-source software tool involving containerization technology.
- *Docker Secrets* – A concept to store sensitive data within Docker Swarm.
- *Reverse proxy* – Serves as a proxy to hide the accessible services from external actors.
- *Public key infrastructure* – A system for issuing, distributing, and validating certificates and managing key pairs to secure communication.

11.6.4 Prototype Scenario for Trusted Blockchain Network

This section provides our design of a trustworthy private blockchain with Hyperledger Fabric blockchain. The blockchain is composed of peers, users, certificate authorities, gateways, numerous databases, and reverse proxies all running inside Docker containers. These components can communicate with each other to form a single Hyperledger Fabric channel for testing, which forms the blockchain and its ledger. Three virtual machines are used for hosting the blockchain network, and the network topology is static in this deployment plan for simplicity. The deployment plan outlines the key points of three topics including prototype scenario, PKI, and network topology.

The prototype scenario involves two different organizations, Org1 and Org2, that wish to form a blockchain network. A third trusted organization, referred to as OrdererOrg, is used to facilitate trust between these two organizations in terms of setting up the blockchain networks and providing the ordering service. Overall, the purpose of the blockchain network is to provide transparency with asset management data utilizing the key properties of a blockchain such as an immutable and distributed ledger to record the transactions from those projects. Each virtual machine will hold network components of at least two of the organizations to showcase the distributive nature of the blockchain (Brinkman 2017), management of security features, the underlying PKI, and fault tolerance capabilities. The blockchain network will be resilient to a single virtual machine failure with minimal or no consequences. The different network entities will be Docker containers that can contain each across the three different virtual machines via Docker Swarm.

11.6.5 Systems Engineering of the Chain of Trust

In this section, we provide our proposal of trustworthy distributed key management for private blockchain through a unique design of private and public key infrastructure that contains several steps to create the chain of trust between users or peers, between organizations, and within consortium partnerships.

A specific sequence of steps must be followed to configure and set up the blockchain network in production mode. Most of the complexity in deployment arises from PKI, with different network nodes requiring their certificates at certain times. The entire deployment procedure is, therefore, divided into manageable subsections with supporting diagrams to help illustrate the process.

Hereby, only the conceptual blockchain network is described as the nodes and can be distributed arbitrarily (but evenly to avoid single points of failure) on the virtual machines provided by enterprise. Furthermore, most of these steps are automated using scripts and are only shown for future reference for more complex deployment settings.

11.6.6 Design Public Key Infrastructure (PKI) for Trust

The PKI of a Hyperledger Fabric network involves all entities of the blockchain network. All network node credentials (private/public key pairs) will be generated inside an HSM or software vault. These keys will use the Elliptic Curve Digital Signature Algorithm (ECDSA) with curve "prime256v1" and size 256 bits to meet current state-of-the-art requirements (Barker and Roginsky 2019). All private keys of peers, orderers, and CAs will be stored in an HSM. However, the RCA credentials will be stored in a separate HSM that can be taken offline to increase security. In contrast to this, less sensitive certificates will be centrally managed using Docker Secrets within a Swarm for simplified deployment. A single administrator will preferably manage these Swarms having control of a subset of their organization's network nodes. Swarms from different locations can communicate with each other by exposing a reverse proxy that redirects requests to different services inside its network.

Besides the Fabric nodes, all user credentials will also be generated and stored dynamically in an HSM or software vault. This process will involve users first registering via the application, which will call an endpoint on one of the gateways. This endpoint can be used to indirectly communicate with either an HSM or a software vault to generate the user's public/private key pair. Hence, all user credentials will be stored server-side, and only a permissioned administrator will have access to perform these user operations. The servers must have temporary read access to their user's private key to sign blockchain transactions.

11.6.6.1 Design of Certificate Authority (CA)

In practice, the entire deployment process will likely be a rare event; manual execution of these steps is acceptable. However, for testing purposes, scripts should be used to automate all key generation and key signing.

The RCA is added first to the network to create the CA credentials within the HSM, referred to as RCA-HSM (Schuh et al. 2012). The certificate can be self-signed as well as signed from existing PKI. Afterwards, the setup of credentials for ICAs allows the respective organization to join the network. This involves generating the public/private key pairs for each organization within their appropriated HSM (i.e. Org1-HSM for Org1). Overall, each organization requires two different CA certificates for further steps: endorsement CA and TLS CA.

Each generated key pair inside the ICA-HSM has to be signed from the RCA as shown in Figure 11.3. After the signing of the RCA, certificates will be produced for each of the ICAs (Hyperledger 2019). With this, all ICA credentials have been set up, and their Docker containers should be initialized alongside their databases used to track issued certificates.

At this point, the RCA should then be taken "offline" through disabling the HSM to avoid external access to RCA. Now, the design of the certification authorities is completed as shown in

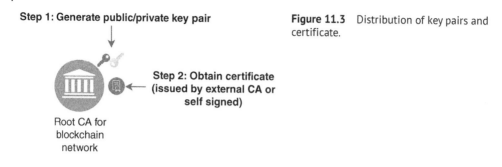

Step 1: Generate public/private key pair

Step 2: Obtain certificate (issued by external CA or self signed)

Root CA for blockchain network

Figure 11.3 Distribution of key pairs and certificate.

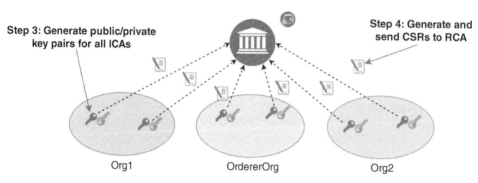

Step 3: Generate public/private key pairs for all ICAs

Step 4: Generate and send CSRs to RCA

Org1   OrdererOrg   Org2

Figure 11.4 Key and CSR generation for ICAs.

Figure 11.4. Authentication between the ICAs and their databases is secured via Transport Layer Security (TLS) and an additional authentication method such as username/password, whereas the secrets are stored in the organization's HSM. A reverse proxy will be used to provide load balancing for each CA cluster.

11.6.6.2 Design the Trusted Gateways

The gateways are considered next in the deployment plan. Hereby, the Docker containers for these gateways getting authenticated through appropriate credentials in Docker Secrets enroll their preconfigured organization administration. It is essential that each organization enroll the preconfigured admin for both the endorsement CA and the TLS CA. The current implementation for these servers will automatically enroll the admin on start-up. The gateways should also be deployed in clusters but with different administrators that use the HSM to store/retrieve user credentials.

Specifically, public/private key pair and a signing request for the certificate are generated for each administrator when a gateway is initialized. The requests are sent to the appropriate CAs, outputting certificates for the administrators. Any CA node within a cluster can issue a certificate. This abstraction hides the internal network configuration from the gateways alongside allowing future additions of users, peers, and orderers to follow the same process. The CAs respond with their certificate chains and the new admin certificates to the gateways, which comprise the MSP information for their organization. These certificates should be published (e.g. via a central database) to allow other organizations to know about their organization MSP information, which is required to create channels. All communication between the gateways and a CA cluster will go through the associated reverse proxy cluster as shown in Figure 11.5.

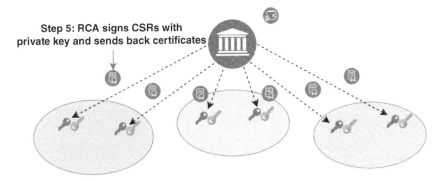

Figure 11.5 Design of trusted gateways.

11.6.6.3 Involving Trusted Peers and Orderers

After the admin credentials have been set up, peers and orderers can then be added to the respective organizations. Different steps are required to initialize the peers and orderers though some elements can be configured concurrently (Raj 2018a, 2018b). In this section, peer configuration is shown first for simplicity.

The key pairs for endorsement and TLS have to be generated first for each participating peer and orderer inside their organization's HSM. Associated certificates for each of these key pairs must also be requested through the gateways to the respective CAs (Barker and Roginsky 2019; Thakkar et al. 2018).

Each peer and user receive signed certificates alongside the MSP information of their organization. With all the cryptographic materials required (Barker and Roginsky 2019), the peers can then be initialized. This involves initializing the world state databases and configuration of each peer to receive their credentials from the appropriated HSM, in detail, credentials from the admin, the CA, endorsement certificate, and the TLS certificate. Afterward, the gateway from the orderer organization queries the MSP information of all organizations in the entire blockchain network.

An important observation is that the gateways could potentially be used to call endpoints of the HSMs to generate the keys and CSRs for new peers and orderers. These servers could handle all blockchain interactions and cryptography generation (indirectly), which leaves bringing up the peer and orderer containers as separate steps for a network administrator. However, the initialization of Docker containers with the correct configurations could also be automated (Tam 2019). This means that network administrators could also use the gateways in future to add/remove nodes without having to call HSM endpoints themselves.

Given the MSP information of all participating organizations, the orderers can then be initialized (Singh 2019). Hyperledger Fabric requires the orderers to be "bootstrapped" with a *Genesis Block* that will be used as the *Genesis Block* for new blockchains when initialized. This is only used to specify which organizations can create channels (i.e. blockchains). The certificates of each organization's MSP are used to generate a *Genesis Block*. Now, the peers and orderers in the respective organizations are setups as illustrated in Figure 11.6.

11.6.6.4 Facilitate Trust Through Channels

The channel creation involves sending the previously generated channel creation transaction through one of the gateways. Currently, the transaction is automatically accepted and sent to the other organizations' gateways, which will agree to sign the transaction with their administrator's credentials. Once the signed transactions are received, the gateway sends the result to the orderer Cluster. Consequently, the initial channel consists of the orderers defined in the prospective

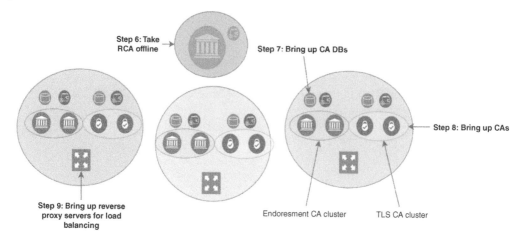

Figure 11.6 Design of trusted peers and orderers to form a preliminary blockchain network.

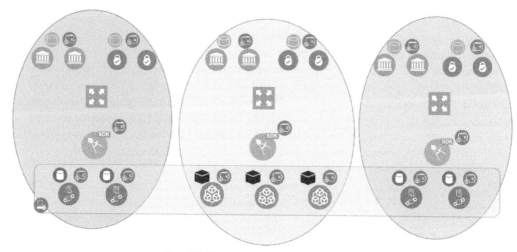

Step 27: Peers are joined to the channel

Figure 11.7 Trusted channel among the consortium.

channel transactions. By this, the blockchain ledger will contain the *Genesis Block* as well as additional configuration.

Peers of each organization can then be added to the channel as shown in Hyperledger (2019). Note that the correct gateway must be used to add peers, as only peers with certificates from the same organization as that gateway will be verified correctly. Furthermore, a minimal amount of default peers must be defined within connection profiles of the gateways. This information should define the peers that will be used as anchors for the organization. Organizations will also only need to define information regarding their own peers. If an organization obtained the peer information of another organization, it would still not be able to join that peer to the channel as there is a mismatch between the peer's organization listed in its certificate and that of the administrator attempting to perform that join. Once complete, the peers and users will be logically connected in a shared channel as shown in Figure 11.7.

11.7 From Human System Interaction to Human Data Interaction

The HSE-based human data interaction architecture of the trustworthy blockchain is shown in Figure 11.8. It is designed through the close collaboration between the subject matter expert and system design engineers and putting humans' activity and capability as the center of the big data system specification, design, implementation, and deployment.

11.8 Future Work for Trust in Human Systems Engineering

This section describes the notable trust issues in the engineered system such as large complex automated systems. According to Tolk et al. (2009), trust refers to the expectation of another in the context of human interaction with systems engineered to provide decision support. Human trust in engineered systems is the most important question and should be a foundation for human–machine integration and cooperation and a centerpiece of HSE. Work in this area has just started to scratch the surface (Tolk et al. 2009), despite the significant amount of work that has been done including social and technical systems (Getty et al. 1995; Handley 2019a), agents and decision support systems (Tolk 2012; Tolk et al. 2009), human–machine interaction (Madhavan and Wiegmann 2007), fault tolerance and reliability (Lee and Anderson 2012), and the like. The next section provides possible future work of trust in HSE.

11.8.1 Software Engineering of Trust for Large Engineered Complex Systems

Figure 11.9 presents where HSE is in terms of a trend based on work undertaken by Schuh et al. (2012) tracking development between 1970 and 2010 with Chang and Jensen (2017b) considering 2020–2030 onward. The intelligence of the complex system has been improved through artificial intelligence (AI) and big data management. Significant progress has been made through software engineering efforts particularly in the areas of public blockchain, machine learning, AI, cloud computing, and so forth. It is likely that software will play a critical role in building trust for future engineered systems particularly in the areas of privacy, security, and safety of large complex systems.

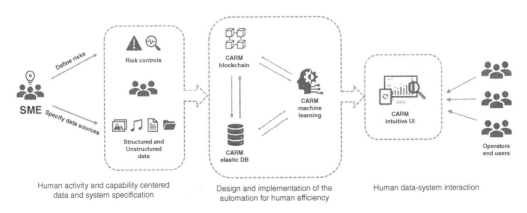

Figure 11.8 Systems engineering of human data interaction of the trustworthy blockchain.

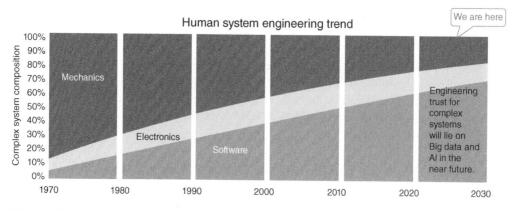

Figure 11.9 Future of human systems engineering trend based on VDMA prediction. *Source:* Chang and Jensen (2017b) and Schuh et al. (2012).

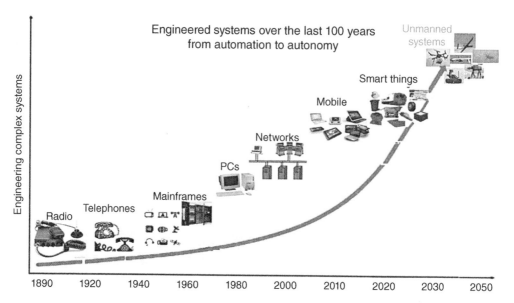

Figure 11.10 Future of human systems engineering trend. *Source:* Chang et al. (2008) and Chang and Jensen (2017a).

11.8.2 Human-Centered AI for the Future Engineering of Intelligent Systems

Human-centered AI will be a new area in the HSE, addressing ethical issues of humanity and trust of any large complex system development and deployment particularly in the area of embedded systems, IoT, cyber–physical systems, and future cyber warfare systems. Figure 11.10 presents 100 years of systems engineering and the need to address human-centered AI in future engineering of intelligent systems.

11.8.3 Trust in the Private Blockchain for Big Complex Data Systems in the Future

Public blockchain systems not only provide community-based distributed trust among peers and agents in the networks but also address critical issues of human trust to the large engineered massive networked database system, such as Ethereum. It is the achievement of the trust of a big data repository, a type of large complex system, after 40 years of ICT evolution because it creates trust from humanity. However, this does not apply to a private blockchain due to key management infrastructure.

The trust involved with the blockchain system focuses on key management. This refers to where and how the trust credentials of the network and users should be handled. The keys must be secured for all entities as they are the core mechanism behind the trust of the private blockchain. If a private key is compromised, an attacker could effectively impersonate the owner of that private key. For a CA, this implies the generation of arbitrary certificates. Hence, all certificates issued under that CA would need to be revoked. For peers and users, "fake" results could be created for transaction endorsements and block orderings. Furthermore, data transmitted via TLS for nodes would be insecure for them if their TLS private keys were compromised. Attackers that obtain the private keys of end users could perform queries against the ledger data and effectively create arbitrary transactions under the name of the key owner.

To ensure trust, HSMs should be used to store the private keys of all entities in the network. However, HSMs can be costly and should be considered based on different business needs. For example, since the host computer for a gateway requires read access to end user keys for signing, it suggests that a software key management such as Vault from HashiCorp could be used.

Bugs in the Current Hyperledger Fabric. Since Hyperledger Fabric is open-source software, it relies on contributors to implement its functionalities. In terms of future product release, the trust issues exist with backward and forward compatibility and data migration issues.

Smart contract implementations incorporate certain levels of risk, and the assurance in incorrect endorsements of transactions is one of the issues. Furthermore, overlooked chaincode could result in organizations accepting the smart contract to which they agree but do not fully understand.

The trust in the private blockchain server and issues with the gateway is related to software capability, admin power, and route exposure. Lack of software capability to authenticate end users is detrimental. Users that are incorrectly authenticated would have indirect access to another user's private key. They would then be able to perform blockchain operations as another user. To mitigate this, a strong input validation system should be used for user passwords. This means users must create passwords of acceptable length and entropy. Furthermore, an appropriate password hash algorithm should be applied. For instance, the "bcrypt" password hashing function can be used with a number of rounds. The number should be dependent on the host machine's specifications and large enough to deliberately slow down password verification/hashing to a set limit. More recent password hashing functions have been released, such as Argon2, that should be considered to maximize security.

11.9 Conclusion

In this chapter, we used our prototype systems as a demonstration to discuss identified trust issues to address private blockchain. We adopted an HSE approach to the development of the trustworthy private blockchain and its application for operational risk management in large enterprises. Through our engineering process, we have discovered the potential trust issues in the private blockchain. We engineered the trust infrastructure through unique design of public and private key management systems. We believe key management in the private blockchain is critical to human trust in the private blockchain, and this will impact the adoption of blockchain in large, complex extended enterprise environments such as the banking sector, defense sector, and any large public and private enterprises. Finally, we presented suggested future work in the building of trust in HSE.

Acknowledgment

This research is funded by the Department of Defense, Australian Government. We would particularly like to thank Deputy Secretary of CASG Defense, Tony Fraser, for the Innovation Award December 2019 and Director General Ed Louis of the Materiel Logistics at the DoD for the financial support of the project between 2018 and 2020.

References

Barker, E. and Roginsky, A. (2019). *Transitioning the Use of Cryptographic Algorithms and Key Lengths*. https://doi.org/10.6028/NIST.SP.800-131Ar2.

Brinkman, R. (2017). *Blockchain: How Should You Organize Your Peers?* The Green Grid. https://www.thegreengrid.org/en/newsroom/blog/blockchain-how-should-you-organize-your-peers (accessed 21 June 2020).

Casino, F., Dasaklis, T.K., and Patsakis, C. (2019). A systematic literature review of blockchain-based applications: current status, classification and open issues. *Telematics and Informatics* 36: 55–81.

Chang, E. and Jensen, A. (2017a). 25 years of e-Business Engineering, where we are now and where we are going, Keynote speech. *14th IEEE International. Conference on e-Business Engineering*, Shanghai, Chain (November 2017).

Chang, E. and Jensen, A. (2017b). Predictive analytics for warehousing and logistics operations. Invited Keynote Speech, Smart Warehouse Summit 2017, Sydney, Australia (24–25 October 2017).

Chang, E., Dillon, T., and Calder, D. (2008). Human system interaction with confident computing. The mega trend. *2008 Conference on Human System Interactions,* Krakow, Poland (25–27 May 2008).

Christidis, K. and Devetsikiotis, M. (2016). Blockchains and smart contracts for the internet of things. *IEEE Access* 4: 2292–2303.

Conoscenti, M., Vetro, A., and De Martin, J.C. (2016). Blockchain for the internet of things: a systematic literature review. In: *2016 IEEE/ACS 13th International Conference of Computer Systems and Applications (AICCSA), Agadir, Morocco (12 June 2017)*, 1–6. IEEE.

Freedy, A., DeVisser, E., Weltman, G., and Coeyman, N. (2007). Measurement of trust in human-robot collaboration. In: *2007 International Symposium on Collaborative Technologies and Systems, Orlando, USA (25 May 2007)*, 106–114. IEEE.

Getty, D.J., Swets, J.A., Pickett, R.M., and Gonthier, D. (1995). System operator response to warnings of danger: a laboratory investigation of the effects of the predictive value of a warning on human response time. *Journal of Experimental Psychology: Applied* 1 (1): 19.

Handley, H.A. (2019a). A socio-technical architecture. In: *The Human Viewpoint for System Architectures*, 27–38. Springer.

Handley, H.A. (2019b). *The Human Viewpoint for System Architectures*, vol. 35. Springer.

Hebert, C. and Di Cerbo, F. (2019). Secure blockchain in the enterprise: a methodology. *Pervasive and Mobile Computing* 59: 101038.

Hyperledger (2019). *Channels.* Hyperledger Fabric. https://hyperledger-fabric.readthedocs.io/en/release-1.4/channels.html (accessed 21 June 2020).

Karafiloski, E. and Mishev, A. (2017). Blockchain solutions for big data challenges: a literature review. In: *IEEE EUROCON 2017 – 17th International Conference on Smart Technologies, Ohrid, Macedonia (6–8 July 2017)*, 763–768. IEEE.

Khan, M.A. and Salah, K. (2018). IoT security: review, blockchain solutions, and open challenges. *Future Generation Computer Systems* 82: 395–411.

Koteska, B., Karafiloski, E., and Mishev, A. (2017). Blockchain implementation quality challenges: a literature. *SQAMIA 2017: 6th Workshop of Software Quality, Analysis, Monitoring, Improvement, and Applications*, Belgrade, Serbia (11–13 September 2017).

Lee, P.A. and Anderson, T. (2012). *Fault Tolerance: Principles and Practice*. Springer-Verlag.

Madhavan, P. and Wiegmann, D.A. (2007). Similarities and differences between human–human and human–automation trust: an integrative review. *Theoretical Issues in Ergonomics Science* 8 (4): 277–301.

Mazumdar, S. and Ruj, S. (2019). Design of anonymous endorsement system in hyperledger fabric. *IEEE Transactions on Emerging Topics in Computing* https://doi.org/10.1109/TETC.2019.2920719.

Mittal, S. and Tolk, A. (2020). *Complexity Challenges in Cyber Physical Systems: Using Modeling and Simulation (M&S) to Support Intelligence, Adaptation and Autonomy*. Wiley.

Raj, V. (2018a). *Hyperledger Fabric Peer Roles*. medium.com (24 November). https://medium.com/@raj_63596/hyperledger-fabric-peer-roles-b4477451f97c (accessed 21 June 2020).

Raj, V. (2018b). *Hyperledger Fabric Architecture: Explained in Detail*. Skript (29 January). https://dev.to/skcript/hyperledger-fabric-architecture-explained-in-detail-32bb (accessed 21 June 2020).

Schuh, G., Neugebauer, R., and Uhlmann, E. (2012). Future trends in production engineering. In: *Proceedings of the First Conference of the German Academic Society for Production Engineering (WGP), Berlin, Germany, 8th–9th June 2011*. Springer Science & Business Media.

Seebacher, S. and Schüritz, R. (2017). Blockchain technology as an enabler of service systems: a structured literature review. *International Conference on Exploring Services Science*, Rome, Italy (24–26 May 2017).

Singh, S.P. (2019). *Detail Analysis of Raft & Its Implementation in Hyperledger Fabric*. https://medium.com (May). https://medium.com/@spsingh559/detail-analysis-of-raft-its-implementation-in-hyperledger-fabric-d269367a79c0 (accessed 21 June 2020).

Tam, K. (2019). Hyperledger Fabric. https://hyperledger-fabric.readthedocs.io/en/release-1.4/identity/identity.html#what-are-pkis (accessed 21 June 2020).

Tama, B.A., Kweka, B.J., Park, Y., and Rhee, K.-H. (2017). A critical review of blockchain and its current applications. *2017 International Conference on Electrical Engineering and Computer Science (ICECOS)*, Palembang, Indonesia (22–23 August 2017).

Tapscott, D. and Tapscott, A. (2017). How blockchain will change organizations. *MIT Sloan Management Review* 58 (2): 10.

Thakkar, P., Nathan, S., and Viswanathan, B. (2018). Performance benchmarking and optimizing hyperledger fabric blockchain platform. *2018 IEEE 26th International Symposium on Modeling, Analysis, and Simulation of Computer and Telecommunication Systems (MASCOTS)*, Milwaukee, USA (25–28 September 2018).

Tolk, A. (2012). *Engineering Principles of Combat Modeling and Distributed Simulation*. Wiley.

Tolk, A., Madhavan, P., Jain, L., and Tweedale, J. (2009). *Agents and Decision Support Systems*, vol. 14. Wiley Online Library.

Yli-Huumo, J., Ko, D., Choi, S. et al. (2016). Where is current research on blockchain technology? – a systematic review. *PloS One* 11 (10): e0163477.

Zheng, Z., Xie, S., Dai, H.-N. et al. (2018). Blockchain challenges and opportunities: a survey. *International Journal of Web and Grid Services* 14 (4): 352–375.

12

Light's Properties and Power in Facilitating Organizational Change

Pravir Malik

First Order Technologies, LLC, Berkeley, CA, USA

12.1 Introduction

The practice of integrating the action of light with any challenging situation can be used as a technique to expedite holistic change. This is because light is not just the physically proven electromagnetic universal phenomenon, as we have come to scientifically understand it, but exists at different speeds as can more easily be perceived by the mind's eye. The result of the relationship of light at different speeds can also be seen as responsible for the existence of quanta and for an ongoing causal quantum-level computation that results in the generation of different types and categories of information as has been explored in some detail in the nine-book *Cosmology of Light* series (Malik 2017a, 2017b, 2017c; Malik and Pretorius 2017b; Malik 2018a, 2018b, 2018c, 2019a, 2019b) and in a series of related IEEE and other journal articles (Malik et al. 2017a; Malik and Pretorius 2018a, 2019; Malik 2020). Hence, light has a causal power that, if leveraged in the right way, can holistically affect a large portfolio of organizational challenges.

Further, it will be found that the information and possibilities embedded in light can be categorized by four organizational principles. One can easily get a glimpse of this by understanding that four fundamental aspects of experienced reality arise because of the finite speed of light as will be explored in Section 12.1. These are the notions of past, present, future, and matter that are surrogates for the deeper organizational principles of presence, power, knowledge, and harmony, respectively. There is concrete evidence that this group of four organizes matter and life as will be explored in Section 12.2, as manifest in the fourfold structuring of the electromagnetic spectrum, quantum particles, the periodic table, any living cell, and fundamental capacities of self, among other emergences. Hence there is an ubiquity of light, present also in all of matter and life, and light can be thought of as existing in every iota of existence. Being able to perceive light differently allows us to leverage it differently, and this is what gives power to begin to bring about a different category of inherently holistic organizational change, as will be explored in Section 12.3. Such change is holistic because of the fact that light connects everything and because everything, including changes in information about situations, arise due to light.

This chapter, hence, will examine implicit properties and a fundamental mathematical model of light (Section 12.1), the process by which properties of light organize all material and life existence (Section 12.2), and then, based on a series of workshops conducted primarily at Zappos.com

A Framework of Human Systems Engineering: Applications and Case Studies, First Edition.
Edited by Holly A. H. Handley and Andreas Tolk.

LLC, among other places through 2019, will suggest ways in which this technology can be integrated into day-to-day organizational functioning to facilitate a range of issues and challenges typically faced by an organization (Section 12.3). Section 12.4 will offer a brief summary and conclusion.

12.2 Implicit Properties and a Mathematical Model of Light

Practically, the finite speed of light implies that light will take a finite amount of time to travel from one point to another. This is significant even when viewed at the atomic scale. If there is a source of electromagnetic radiation in the nucleus or due to the electrons changing orbits around a nucleus, that radiation will be experienced only a finite time later. It can be inferred that this phenomenon is related to quanta. Energy of quanta is specified by (12.1), where E is energy of quanta, h is Planck's constant, and υ is frequency of the radiation – hence Equation 12.1:

$$E = h\upsilon$$

Eq. 12.1: Energy of quanta

Further, the speed of light c, frequency of radiation υ, and wavelength of radiation λ are related by Equation 12.2:

$$c = \upsilon\lambda$$

Eq. 12.2: Speed of light

Combining (12.1) and (12.2) the inverse relationship between c and h can be observed in Equation 12.3 for any fixed level of E:

$$E = \frac{hc}{\lambda}$$

Eq. 12.3: Inverse relationship between c and h

Assume now a thought experiment to bring home the creative nature of light and why all of life would possibly emerge from it.

Imagine light traveling at an infinite speed. If this were so, then the inverse relationship between c and h would necessitate that h approach 0. In other words, matter would be unable to form. Conversely if light were to slow down, to approach c, then h would progressively increase to some threshold where energy is able to sustain itself in quanta and matter would emerge. So finite speed of light creates quanta, which creates matter. Therefore, all of matter emerges from light.

Further, in this view, it can be seen that the Big Bang is nothing other than light slowing down and creating matter as a result of that (Malik 2018a). So, if all matter is an action of light, then any universe arising is only a result of it. There are then likely some overarching properties that would be true of light and therefore also true of everything that was to arise in any such universe. It may even be that these properties would be fundamental and would determine structure of matter and everything that emerges out of it.

So, what can be inferred about the properties of light?

The speed of light is known to have implications on the experienced nature of reality. The finiteness, c, at 186 000 miles per second in a vacuum creates an upper bound to the speed with which any object may travel. This also implies that objective reality will be experienced as a past, a present, and a future, from the point of view of that object (Einstein 1995). These characteristics – a past, a present, and a future – can therefore be thought of as implicit in the nature of light and become part of objective reality because of the speed of light.

Further, as just suggested, *c* also creates a lower bound when inverted ($1/c$) being proportional to Planck's constant, *h*. *h* as we know pegs the minimum amount of energy or quanta required for expression at the subatomic level (Isaacson 2008). Planck's constant, *h*, therefore, allows matter to form (Lorentz 1925) and for the reality of nature with a past, present, and future to also be progressively experienced as a phenomenon of connection between seemingly independent islands of matter. This characteristic of "connection" is therefore also proposed to be implicit in the nature of light and becomes part of objective reality because of the speed of light.

As suggested in *Connecting Inner Power with Global Change* (Malik 2009), a "future" implies the notion of a cause, or seed, or direction and suggests the "mentality" or meaning that perhaps drives the emergence of phenomena. A "present" implies the working out of the play of forces and suggests the "vitality" of nature where the most energetic or powerful force will express itself over others. A "past" can be viewed as established reality as defined by what the eye or other lenses of perception can see. Such lenses see what has already "physically" been formed in time.

These implicit characteristics of the nature of light as experienced at the layer of reality set up by a finite speed of light may hence be summarized by Equation 12.4, where C_U refers to the speed of light of 186 000 miles per second, which has created the perceived nature of reality, *U*:

$$C_U : \left[\text{Physical, Vital, Mental, Connection} \right]$$

Eq. 12.4: Nature of reality due to light at *C*

It is known however that at quantum levels the nature of reality is at least characterized by wave–particle duality. Light itself (Feynman 1985) and matter (De Broglie 1929) may be experienced as both particles and waves (Ekspong 2014). Such duality, as will be explored shortly, is related to the notion of quantum. But for matter to be experienced as waves implies that *h* has become a fraction of itself, h_{fraction}. This further implies that *c* must have become greater than itself, c_N, such that the inequality specified by Equation 12.5 holds:

$$c_N > c_U$$

Eq. 12.5: Speed of light inequality

Note that what is implied here is that just as there is a nature of reality specified by *U* that is the result of the speed of light being 186 000 miles per second, so too there is another nature of reality specified by *N* that is the result of a speed of light greater than 186 000 miles per second. The existence of other realities is perhaps consistent with recent developments in physics with the notion of property spaces being separate from but influencing physical space as explored by Nobel physicist Frank Wilczek (2016), among others. But further in *Slow Light* Perkowitz's recent treatment of today's breakthroughs in the science of light (Perkowitz 2011), he states: "Although relativity implies that it's impossible to accelerate an object to the speed of light, the theory may not disallow particles already moving at speed c or greater."

It stands to reason that current instrumentation, experience, and normal modes of thinking having developed as a by-product of the characteristics so created in the layer of reality *U* may be inadequate to access *N* without appropriate modification.

The notion of wave–particle duality already challenges the notion of normal thinking perhaps because wave-like phenomena is a function of faster than c motion and particle-like phenomena is a function of less than or equal to c motion. That these may be happening simultaneously is reinforced by principles such as complementarity in which experimental observation may allow measurement of one or another but not of both (Whitaker 2006).

But then taking this trend of a possible increase in the speed of light to its limit, this will result in a speed of light of infinite miles per second. The question is, what is the nature of reality when light is traveling at infinite miles per second? In any space-time continuum, be it an area or volume, regardless of scale, light originating at any point will instantaneously have arrived at every other point. Hence light will have a full and immediate *presence* in that space-time continuum. Further, that light will *know* everything that is happening in that space-time completely instantaneously – that is know what is emerging, what is changing, what is diminishing, what may be connected to what, and so on – or have a quality of *knowledge*. It will connect every object in that space-time completely and therefore have a quality of connection or *harmony*. Finally, nothing will be able to resist it or set up a separate reality that excludes it and hence it will have a quality of *power*.

These implicit characteristics of the nature of light as experienced at the layer of reality set up by light traveling infinitely fast may hence be summarized by Equation 12.6, where c_∞ refers to the speed of light of ∞ miles per second, which has created the perceived nature of reality, ∞:

$$c_\infty : \left[\text{Presence, Power, Knowledge, Harmony} \right]$$

Eq. 12.6: Nature of reality due to light at ∞

But it can also be noticed from (12.4) that "physical" is related to presence, "vital" is related to power, "mental" is related to knowledge, and "connection" is related to harmony.

The question then, is how do these apparent qualities at ∞ precipitate, translate into, or become the physical-vital-mental-connection-based diversity experienced at U? This may be achieved through the intervention or action of a couple of mathematical transformations. First, the essential characteristics of presence, power, knowledge, and harmony that it is posited exist at every point instant by virtue of the ubiquity of light at ∞ will need to be expressed as sets with up to infinite elements. Such a precipitation is none other than an act of quantization since something implicit in the layer where light travels faster is collecting in "quanta" to be expressed more materially in the layer where light travels slower. Second, elements in these sets will need to combine together in potentially infinite ways to create a myriad of seeds or signatures that then become the source of the immense diversity experienced at U. This, similarly, is also an act of quantization or the action of a "quantization function." This suggests that all that is seen and experienced at U may be nothing other than "information" or "content" of light and as such that there are fundamental mathematical symmetries at play where everything at U is essentially the same thing that exists at ∞.

It may also be inferred that wherever wave–particle duality exists, it does so because of a more observable quantum translation from one speed of light to another through the device of quanta.

Assuming that the first transformation occurs at a layer of reality K where the speed of light is c_K, such that $c_U < c_K < c_\infty$, this may be expressed by Equation 12.7:

$$c_K : \left[S_{\text{Pr}}, S_{\text{Po}}, S_K, S_H \right]$$

Eq. 12.7: Nature of reality due to light at K

S_{Pr} signifies "set of presence" and may have elements associated with the qualities of being present everywhere or of creating a physical basis in or on which other functions or characteristics can manifest. Hence S_{Pr} may have elements as expressed by Equation 12.8:

$$S_{Pr} \ni \left[\text{Service, Diligence, Perseverance, Stability,} \ldots \right]$$

Eq. 12.8: Set of presence

S_{Po} signifies "set of power" and may have elements associated with the qualities of being powerful or of the play of vitality and experimentation that creates all possibility. Hence S_{Po} may have elements as expressed by Equation 12.9:

$$S_{Po} \ni \left[\text{Power, Energy, Adventure, Experimentation,} \ldots \right]$$

Eq. 12.9: Set of power

S_K signifies "set of knowledge" and may have elements associated with the qualities of knowledge or the search and codification of knowledge. Hence S_K may have elements as expressed by Equation 12.10:

$$S_K \ni \left[\text{Knowledge, Making of Laws, Spread of Knowl.,} \ldots \right]$$

Eq. 12.10: Set of knowledge

S_H signifies "set of harmony" and may have elements associated with the qualities of harmony or creating relationship and love. Hence S_H may have elements as expressed by Equation 12.11:

$$S_H \ni \left[\text{Harmony, Relationship, Love, Specialization,} \ldots \right]$$

Eq. 12.11: Set of harmony

Assuming that the second transformation occurs at a layer of reality N where the speed of light is c_N, such that $c_U < c_N < c_K < c_\infty$, this may be expressed by Equation 12.12:

$$c_{N:} f \left(S_{Pr} \times S_{Po} \times S_K \times S_H \right)$$

Eq. 12.12: Nature of reality due to light at N

The unique seeds are therefore a function, f, of some unique combination of the elements in the four sets S_{Pr}, S_{Po}, S_K, S_H.

The relationship between the layers of light may be hypothesized by the following matrix (Equation 12.13):

$$\begin{bmatrix} c_\infty : \left[\text{Pr, Po}, K, H \right] \\ \left(\downarrow R_{C_K} = f \left(R_{C_\infty} \right) \right) \\ c_K : \left[S_{Pr}, S_{Po}, S_K, S_H \right] \\ \left(\downarrow R_{C_N} = f \left(R_{C_K} \right) \right) \\ c_{N:} f \left(S_{Pr} \times S_{Po} \times S_K \times S_H \right) \\ \left(\downarrow R_{C_U} = f \left(R_{C_N} \right) \right) \\ c_U : \left[P, V, M, C \right] \end{bmatrix}_{\text{Light}}$$

Eq. 12.13: Relationship between layers of light

The matrix suggests a series of transformations leading from the ubiquitous nature of light implicit in a point – presence, power, knowledge, and harmony – to the seeming diversity of matter observed at the layer of reality U, which is fundamentally the same presence, power, knowledge, and harmony projected into another form of itself.

The first transformation is summarized by Equation 12.14:

$$R_{C_K} = f\left(R_{C_\infty}\right)$$

Eq. 12.14: First transformation

This is suggesting that the reality at the layer specified by the speed of light c_K, R_{C_K} is a function of the reality at the layer specified by the speed of light c_∞. This transformation translates the essential nature of a point into the sets described by (12.8–12.11). Note that (12.14) is essentially a quantization function, in that something of the reality of light existing at R_{C_∞} is translated into reality experienced at R_{C_K}.

The second transformation is summarized by Equation 12.15:

$$R_{C_N} = f\left(R_{C_K}\right)$$

Eq. 12.15: Second transformation

This is suggesting that the reality at the layer specified by the speed of light c_N, R_{C_N} is a function of the reality at the layer specified by the speed of light c_K. This transformation combines elements of the sets into unique seeds as suggested by (12.12). This transformation can also be thought of as the result of a quantization function such that something of R_{C_K} is collected as unique seeds at R_{C_N}.

The third transformation is summarized by Equation 12.16:

$$R_{C_U} = f\left(R_{C_N}\right)$$

Eq. 12.16: Third transformation

This is suggesting that the reality at the layer specified by the speed of light c_U, R_{C_U} is a function of the reality at the layer specified by the speed of light c_N. This transformation builds on the unique seeds suggested by (12.12) to create the diversity of U as specified by (12.4). This transformation is therefore also the result of a quantization function such that the seed aspect of R_{C_N} is translated into the immense diversity experienced at R_{C_U}.

In this framework the notion of wave–particle duality hence may become complementary block–field–wave–particle quadrality where block refers to phenomenon resident to ∞, field to phenomenon resident to N, wave to phenomenon resident to K, and particle to phenomenon resident to U. The essential translation from one level to the next is due to a series of quantization functions so that (12.13) summarizes an algorithm for life (Malik and Pretorius 2019), where an implicit quaternary basis of presence, power, knowledge, and harmony sets up potentially infinite number of elements derived from sets of presence, power, knowledge, and harmony.

12.3 Materialization of Light

The preceding section elaborated a fourfold architecture to light. This architecture can be thought of as organized by four implicit properties of light. These properties can in turn be thought of as organizational principles, which it can be extrapolated, drive all the complex materialization of matter, life, and beyond. This section will suggest some concrete examples of such fourfold materialization starting from the subtle to the overt.

12.3.1 The Electromagnetic Spectrum

As relates to the electromagnetic spectrum, the four characteristics of light are expressed by the range of wavelengths implicit in the electromagnetic spectrum (knowledge), its implicit energy gradient (power), the speed c with which it is propagated (harmony), and the potential for mass (presence) also implicit in it, respectively.

Specifically, from the range of different types of waves from radio waves to gamma rays, determined as a function of frequency and wavelength of light, the electromagnetic spectrum itself is a repository of knowledge of a vast range of phenomena experienced in the universe as indicated by Equation 12.17:

$$\text{Knowledge} \propto \left[f(\lambda) \right]$$

Eq. 12.17: Knowledge aspect of electromagnetic spectrum

But further, there is a large variance in the frequency of light in the electromagnetic spectrum, and this provides for a large variation in energy or power implicit in the electromagnetic spectrum as suggested by Equation 12.18, where v is the frequency of the electromagnetic wave:

$$\text{Power} \propto hv$$

Eq. 12.18: Power aspect of electromagnetic spectrum

As already suggested the principle of harmony is related to the speed of light at U, C_U, due to possibilities of connection between mass-based entities existing in the reality so set up by light, and as highlighted by Equation 12.19:

$$\text{Harmony} \propto C_U$$

Eq. 12.19: Harmony aspect of electromagnetic spectrum

Finally, mass potential can be expressed in the following way since $E = mc^2$ or $m = \dfrac{E}{c^2}$ and substituting hv for E as in Equation 12.20:

$$\text{Presence} \propto hv\big/c^2$$

Eq. 12.20: Presence aspect of electromagnetic spectrum

Further, (12.12) has suggested how unique seeds can be created. Elaborating an equation for such seeds that fulfills (12.12) can be adapted from Malik et al. (2015) and is summarized in Equation 12.21:

$$\text{Seed} = Xa + \overline{Yb_{0-n}} \text{ where } \begin{bmatrix} X \in \left[S_{Pr}, S_{Po}, S_K, S_H \right] \\ Y \in \left[S_{Pr}, S_{Po}, S_K, S_H \right] \\ a, b \text{ are integers; } a > b \end{bmatrix}$$

Eq. 12.21: Unique seeds

Reproducing (12.12), C_N: $f(S_{Pr} \times S_{Po} \times S_K \times S_H)$, it can be seen that (12.21) combines elements from each of the sets of the essential characteristics of light to conceivably spawn an infinite set of seeds. $\overline{Yb_{0-n}}$ implies that the Y or secondary element may be repeated multiple times.

Now as an example, an application of (12.17) suggests seeds for the knowledge element or structure of the electromagnetic spectrum in Equation 12.22:

$$\text{Seed}_{\text{EM Spectrum Structure}} = Xa + \overline{Yb_{0-n}} \text{ where } \begin{bmatrix} X \in [S_K] \\ Y \in [S_{\text{Pr}}, S_{\text{Po}}, S_K, S_H] \\ a, b \text{ are integers; } a > b \end{bmatrix}$$

Eq. 12.22: Structure of electromagnetic spectrum

12.3.2 Quantum Particles

At the quantum level it is suggested that quarks are a precipitation of light's property of knowledge. Quarks are the only fundamental particle contributing to creating the nucleus of an atom. Protons, which determine atomic number, are composed of two "up" quarks and one "down" quark. Atomic number in turn uniquely identifies the element from the periodic table. Hence, an atomic number of 47, for example, specifies that the element is silver. In other words, it can be suggested that the unique properties of an element – the knowledge of what it is and how it will behave in the universe – are related to the quark. This is summarized by Equation 12.23:

$$\text{Knowledge} \propto f(\text{quarks})$$

Eq. 12.23: Knowledge aspect of quantum particles

Electrons can be considered as a precipitation of light's property of power. If the apparent characteristics of leptons are considered, they appear to be point-like particles without internal structure (Olive et al. 2014). While quarks only exist in composite particles with other quarks, leptons are solitary particles. The best-known lepton is the electron. In his book *Representing Electrons: A Biographical Approach to Theoretical Entities*, Arabatzis (2006) details the characteristics of electrons. The electron may be considered as a surrogate for the lepton class. The electron appears to be the associated with the flow of energy and power. Further they appear to be the adventurers, easily leaving the atom they are a part of. They also lock or form bonds with other atoms through the force of attraction and repulsion. This is summarized by Equation 12.24:

$$\text{Power|Energy} \propto f(\text{leptons})$$

Eq. 12.24: Power aspect of quantum particles

Bosons are thought of as force carriers. They allow all known quantum particles to interact. The three fundamental bosons in this category are the photon, the W and Z bosons, and the gluon. The carrier particle of the electromagnetic force is the photon. The carrier particle of the strong nuclear force that holds quarks together is the gluon. The carrier particle for the weak interactions, responsible for the decay of massive quarks and leptons into lighter quarks and leptons, is the W and Z bosons. Bosons can be thought of as the precipitation of what creates relationship and harmony at the quantum level. Hence, they can be thought of as the precipitation of light's property of harmony as summarized by Equation 12.25:

$$\text{Harmony} \propto f(\text{bosons})$$

Eq. 12.25: Harmony aspect of quantum particles

This leaves the other discovered fundamental particle the Higgs boson. In ordinary matter, most of the mass is contained in atoms, and the majority of the mass of an atom resides in the nucleus, made of protons and neutrons. Protons and neutrons are each made of three quarks. It is the quarks that get their mass by interacting with the Higgs field (Olive et al. 2014). Hence the Higgs boson can be thought of as the mass giver. In other words, it is what gives presence to the quarks, and it can be thought of as the precipitation of light's property of presence as summarized by Equation 12.26:

$$\text{Presence} \propto f\left(\text{Higgs\_boson}\right)$$

Eq. 12.26: Presence aspect of quantum particles

Just as there are multiple particles in each of the other "families," it is likely that there will be multiple particles in the Higgs boson family. Recent research at CERN indicates that the Higgs boson may have a cousin (Overbye 2015).

Further, as an example from the layer of quantum particles, the seed for the family of leptons with the primary *x*-element deriving from the set of power is suggested by Equation 12.27:

$$\text{Seed}_{\text{Leptons}} = Xa + \overline{Yb_{0-n}} \text{ where } \begin{bmatrix} X \in \left[S_{\text{Po}}\right] \\ Y \in \left[S_{\text{Pr}}, S_{\text{Po}}, S_{K}, S_{H}\right] \\ a, b \text{ are integers; } a > b \end{bmatrix}$$

Eq. 12.27: Signature of leptons

12.3.3 The Periodic Table and Atoms

All atoms in the periodic table can be classified by the *p*-group, *d*-group, *d*-group, and *f*-group.

The *p*-group of elements are those with a valence shell specified by the *p*-orbital, indicating that the probability of the existence of an electron is equally likely on either side of the nucleus. There are some very significant elements in this group that are part of the metal, metalloid, nonmetal, halogen, and noble gas subgroupings. Carbon, nitrogen, oxygen, and silicon are some of the sample elements. In some sense this grouping summarizes all the element possibilities within it. It is perhaps that the possibility of ideas behind all elements has precipitated in this group, and one can hypothesize that this group may be a reflection of the set of knowledge, forming archetypes from which all other elements are created.

Philosophically, the one probability cloud (*s*), to be discussed shortly, becoming two (*p*) signifies an essential polarity created within a unit space. Considering the hypothesis that the form is a "switching" function that attracts function into form, then this dual manifestation may be viewed as the prerequisite condition by which a larger number of such "switches" also come into being. These "essential two" created along three dimensions of space may allow a threshold meta-function experimentation to come into being. Being the first instance of this variability in space, it could be that it therefore becomes an attractor for all the essential element archetypes to precipitate.

But further, the essential elements that allow both thinking and virtual thinking machines to come into being are also contained within this group. Carbon is the basis of DNA and of all life. The fact that silicon (Si), directly below it in the periodic table and therefore sharing essential qualities, is considered the basis of all virtual thinking machines is perhaps significant and may reinforce the notion that the *p*-group is a precipitation of light's property of knowledge, as summarized by Equation 12.28:

$$\text{Knowledge} \propto f\left(p\_\text{orbital}\right)$$

Eq. 12.28: Knowledge aspect of periodic table

The *d*-group comprises the transition metals. These metals are generally hard and strong, exhibit corrosive resistance, and can be thought of as workhorse elements. Many industrial and well-known elements sit in this group: titanium, chromium, manganese, iron, cobalt, nickel, copper, zinc, silver, platinum, and gold, among others.

The *d*-orbital itself is a probability space characterized by four lobes around the nucleus. Four lobes occurring in five possible planes around the nucleus will likely create a space of stability, since there is a possibility of four lobes creating the four vertices of a tetrahedron that has been implicitly positioned as one of the most stable shapes (Fuller 1982). Work done in crystal field theory (UCDAVIS-CFT 2015) reinforces this concept. The general stability of the transition metals is reinforced by the d-orbital arrangement.

Much of the constructed world around us is created from these elements. Further, most of the series in the group easily lose one or more electrons to form a vast array of compounds. It can be seen that these metals exist for service, to help bring about perfection in the constructed world, to help much of the machinery in which they are used, and to assist the processes dependent on them to be completed with diligence. Hence, these transition metals appear to be a precipitation of light's property of presence, as suggested by Equation 12.29:

$$\text{Presence} \propto f\left(d\_\text{orbital}\right)$$

Eq. 12.29: Presence aspect of periodic table

Exploring the *s*-group, one sees that it consists primarily of alkali metals and alkali earth metals. These groups are extremely electropositive, easily losing electrons and forming positive ions and releasing a lot of energy while doing so. In his book *Essential Elements*, Tweed (2003) refers to these groups as the "violent world of the s-block." Gray (2009), in *The Elements*, points out that stars shine because they are transmuting vast amounts of hydrogen into helium, both of which are *s*-block elements. This characteristic of easily released energy that the elements of this group share suggests that the *s*-group may parallel or be a precipitation of light's property of power as suggested by Equation 12.30:

$$\text{Power} \propto f\left(s\_\text{orbital}\right)$$

Eq. 12.30: Power aspect of periodic table

Philosophically the *s*-orbital as a probability cloud indicates the equal likelihood that an electron can be anywhere in a symmetrical sphere around a nucleus. Since all other orbitals can be thought of as occurring within the cloud specified by the *s*-orbital, in some sense, this is like an imprint or precipitation from meta-levels that allows future and more varied meta-functions to more easily precipitate at the level of *U*. The elements that are part of the *s*-group may be thought of as the adventurers with courage who venture into a brave new world to create some foundation by which all other element creations can follow. The fact that H and He constitute 98% of the universe (Heiserman 1991) relative to other elements therefore makes sense in this view, especially since H and He provide the fuel with which the star furnaces manufacture all other elements.

The *f*-group comprises the lanthanides and actinides. Philosophically, the *f*-orbital consists of six probability lobes around the nucleus in seven different planes, implicitly suggesting the notion of extended relationship and collectivity: the attempt to build larger and larger bonds within a small space. Continuing to draw the link with the quaternary architecture, it is likely that this group is a precipitation of light's property of harmony.

Thinking about lanthanides, some interesting facts may reinforce this notion:

- First, the spin of electrons in the valence shell is aligned, creating a very strong magnetic field. The notion of creating a strong magnetic field seems to be consistent with the notion of engendering a collectivity through the ordered attraction and repulsion of elements.
- Second, these elements curiously occur together in nature often in the same ores and are chemically interchangeable (Gray 2009), also suggesting the notion of forming a tight intragroup collectivity.

The following observations as it relates to actinides can be made:

1) Actinides are inherently radioactive. This implies that these elements have inherently crossed a threshold of stability and have the urge, over their own half-lives, to decompose into other elements. This natural urge may suggest some boundary conditions on the notion of collectivity and nurturing, giving insight into these conditions.
2) The entire actinide group, as opposed to the lanthanide group that is inherently stable, is unstable. It is curious that both these should be part of the f-group and they must provide insight into boundary conditions into the notion of collectivity in elements.

Equation 12.31 summarizes the correlation with light's property of harmony:

$$\text{Harmony} \propto f\left(f\_\text{orbital}\right)$$

Eq. 12.31: Harmony aspect of periodic table

The seed for the family of lanthanides with an f-orbital valence shell with the primary x-element deriving from the set of harmony, as an example from this category of atoms, is suggested by Equation 12.32:

$$\text{Seed}_{\text{Lanthanides}} = Xa + \overline{Yb_{0-n}} \text{ where } \begin{bmatrix} X \in [S_H] \\ Y \in [S_{Pr}, S_{Po}, S_K, S_H] \\ a,b \text{ are integers; } a > b \end{bmatrix}$$

Eq. 12.32: Signature of lanthanides

12.3.4 A Living Cell

In *The Machinery of Life*, Goodsell, an associate professor of molecular biology at the Scripps Research Institute (Goodsell 2010), suggests that every living thing on Earth uses a similar set of molecules to eat, to breathe, to move, and to reproduce. There are molecular machines that do the myriad things that distinguish living organisms that are identical in all living cells. This nanoscale machinery of cells uses four basic molecular plans with unique chemical personalities: nucleic acids, proteins, lipids, and polysaccharides.

Nucleic acids basically encode information. They store and transmit the genome, the hereditary information needed to keep the cell alive. They function as the cell's librarians and contain information on how to make proteins and when to make them. They are, hence, the keepers of a cell's knowledge, its wisdom, and its ability to make laws and the vehicle to spread knowledge within cells and to the next generation of cells. Nucleic acids can therefore be thought of as a precipitation of light's property of knowledge at the cellular level as specified by Equation 12.33:

$$S_{K(\text{cell})} \ni \left[\text{Knowledge, Wisdom, Law Making, Spread of Knowl...}\right]$$

Eq. 12.33: Nucleic acids as set of knowledge

This relationship may also be summarized more simply as Equation 12.34:

$$\text{Knowledge} \propto f\left(\text{nucleic acids}\right)$$

Eq. 12.34: Knowledge aspect of a living cell

Proteins are the cells' workhorses. Proteins are found working in any part of the cell. Proteins are built in thousands of shapes and sizes, each performing a different function. As Goodsell describes, "some are built simply to adopt a defined shape, assembling into rods, nets, hollow spheres, and tubes. Some are molecular motors, using energy to rotate, or flex, or crawl. Many are chemical catalysts that perform chemical reactions atom-by-atom, transferring and transforming chemical groups exactly as needed." With their wide potential for diversity, proteins are constructed to perform most of the everyday tasks of the cells. In fact, human cells build around 30 000 different kinds of proteins to execute on the diverse array of cellular level tasks.

Proteins, hence, exist for service, to bring about perfection at the level of the cell, are characterized by extreme diligence and perseverance and so on. Proteins can therefore be thought of as a precipitation of light's property of presence at the cellular level as specified by Equation 12.35:

$$S_{Pr(cell)} \ni \left[\text{Service, Perfection, Diligence, Perseverance,}\ldots\right]$$

Eq. 12.35: Proteins as set of presence

This relationship may also be summarized more simply as Equation 12.36:

$$\text{Presence} \propto f\left(\text{proteins}\right)$$

Eq. 12.36: Presence aspect of a living cell

Lipids by themselves are tiny molecules, but when grouped together form the largest structures of the cell. When placed in water lipid molecules aggregate to form huge waterproof sheets. These sheets easily form boundaries at multiple levels and allow concentrated interactions and work to be performed within a cell. Hence, the nucleus and the mitochondria are contained within lipid-defined compartments. Similarly, each cell itself is contained within a lipid-defined boundary.

Lipids are therefore promoters of relationship, of harmony in the cell, of nurturing the cell-level division of labor, of allowing specialization and uniqueness to emerge, and hence perhaps even of early forms of compassion and love. This function of harmonization suggests that lipids can therefore be thought of as a precipitation of light's property of harmony at the cellular level as specified by Equation 12.37:

$$S_{H(cell)} \ni \left[\text{Love, Compassion, Harmony, Relationship,}\ldots\right]$$

Eq. 12.37: Lipids as set of harmony

This relationship may also be summarized more simply as Equation 12.38:

$$\text{Harmony} \propto f\left(\text{lipids}\right)$$

Eq. 12.38: Harmony aspect of a living cell

Polysaccharides are long, often branched chains of sugar molecules. Sugars are covered with hydroxyl groups, which associate to form storage containers. As a result, polysaccharides function as the storehouse of cell's energy. In addition, polysaccharides are also used to build some of the most durable biological structures. The stiff shell of insects, for example, is made of long polysaccharides.

Polysaccharides function to create energy, power, courage, and strength, thereby readying the cell for adventure, among related functions. Providing energy and strength, polysaccharides can be thought of as a precipitation of light's property of power at the cellular level as specified by Equation 12.39:

$$S_{Po(cell)} \ni \left[\text{Power, Courage, Adventure, Justice,} \ldots \right]$$

Eq. 12.39: Polysaccharides as set of power

This relationship may also be summarized more simply as Equation 12.40:

$$\text{Power} \propto f\left(\text{polysaccharides}\right)$$

Eq. 12.40: Power aspect of a living cell

The seed for the family of proteins with the primary x-element deriving from the set of presence, as an example from the layer of cells, is suggested by Equation 12.41:

$$\text{Seed}_{protein} = Xa + \overline{Yb_{0-n}} \text{ where } \left[\begin{array}{c} X \in \left[S_{Pr} \right] \\ Y \in \left[S_{Pr}, S_{Po}, S_K, S_H \right] \\ a, b \text{ are integers; } a > b \end{array} \right]$$

Eq. 12.41: Signature of a protein

As an illustration, (12.41) could yield a vast number of functional proteins: in fact, it may be possible that the 30 000 or so known proteins created by the human cell could each be specified by a signature equation of this nature.

12.3.5 Fundamental Capacities of Self

Fundamental capacities of self, such as sensations, wills, desires, feelings, urges, and thought, can also be seen to be emergences of the essential organizing principles implicit in light.

Sensations are those things we experience with our senses. We see things, hear things, smell things, taste things, and can touch things. This ability to enter into relationship with objects through sensation is nothing other than a result of the emergence of light's property of presence. We become present to presence through the device of sensation. Sensation can be thought of as the means by which this property of light – presence – molds or ingrains itself in us as human beings. Its potentiality, all of which is contained in this aspect of light, becomes available to us through the power of sensation. Equation 12.42 generally represents the family of sensations. Some of the elements in the sets that it would comprise of may be "tangible" and "take notice of," among others:

$$\text{Sig}_{sensation} = Xa + \overline{Yb_{0-n}} \text{ where } \left[\begin{array}{c} X \in \left[S_{System_{Pr}} \right] \\ Y \in \left[S_{System_{Pr}}, S_{System_{P}}, S_{System_{K}}, S_{System_{N}} \right] \\ a, b \text{ are integers; } a > b \end{array} \right]$$

Eq. 12.42: Sensation

But there is also a deeper experience of sensation that is possible. When we see things, for instance, what are we seeing? Is it just the surface rendering of the play of matter, or do we see that the fullness of light is still there, with all its potentiality and possibility, in the smallest thing we look at? Do we see that the whole universe and more is present in all its fullness in the least thing that we easily ignore, or belittle, or loathe? When we touch things, is it the seeming concreteness of the play of the particles or atoms or chains of molecules that we touch? Or is it the love and light and the vastness of all that IS that allows itself to be as a small corner that we touch so as to make infinity be felt by something so finite?

Such a deeper contact offered through sensation suggests a subset of (12.42) with secondary elements perhaps described as "fullness of light" and "contacting infinity," among others, thus also yielding an equation:

$$\text{Sig}_{\text{deeper-sensation}} = Xa + \overline{Yb_{0-n}} \quad \text{where} \quad \begin{bmatrix} X \in \left[S_{\text{System}_{P_r}} \right] \\ Y \in \left[S_{\text{System}_{P_r}}, S_{\text{System}_P}, S_{\text{System}_K}, S_{\text{System}_N} \right] \\ a, b \text{ are integers; } a > b \end{bmatrix}$$

Eq. 12.43: Deeper sensation

This relationship may also be summarized more simply as Equation 12.44:

$$\text{Presence} \propto f\left(\text{sensations}\right)$$

Eq. 12.44: Presence aspect of fundamental capacities of self

Urges and desires and wills are similarly a play of the emergence of light's property of power. In the mystery of focus, the vastness of light has projected itself in us into an apparent smallness that is in reality everything that is. And this smallness is trying through urge and desire and will to connect viscerally or even intentionally to other smallnesses that similarly are nothing other than the fullness of light projected into a small smorgasbord of selected function. So the urge or desire for food, or companionship, or of possession, or of climbing a peak, is nothing other than light's compressed property of power, trying to reach more of the fullness that it is through a fulfillment of the urge or desire or will that it masquerades as. Hence urges can be represented as Equation 12.45:

$$\text{Sig}_{\text{urges}} = Xa + \overline{Yb_{0-n}} \quad \text{where} \quad \begin{bmatrix} X \in \left[S_{\text{System}_P} \right] \\ Y \in \left[S_{\text{System}_{P_r}}, S_{\text{System}_P}, S_{\text{System}_K}, S_{\text{System}_N} \right] \\ a, b \text{ are integers; } a > b \end{bmatrix}$$

Eq. 12.45: Urges

Elements may be of the type of "grasp," "possess," and "deeply connect," among others. This relationship may also be summarized more simply as Equation 12.46:

$$\text{Power} \propto f\left(\text{urges}\right)$$

Eq. 12.46: Power aspect of fundamental capacities of self

Feelings and emotions are a play of the emergence of light's property of harmony. Its instrument is the heart, and it generates an array of emotions that are an indication or active radar of whether we are moving toward or away from a reality of harmony, whether based on our small self or some larger self of light. Gradually, by navigating with these emotions and feelings, we can get to a state where we always feel positive emotions, which basically means we have more truly entered into relationship with some larger continent of light. An equation for feelings is as represented by Equation 12.47:

$$\text{Sig}_{\text{feelings}} = Xa + \overline{Yb_{0-n}} \quad \text{where} \quad \begin{bmatrix} X \in \left[S_{\text{System}_N} \right] \\ Y \in \left[S_{\text{System}_{P_r}}, S_{\text{System}_P}, S_{\text{System}_K}, S_{\text{System}_N} \right] \\ a, b \text{ are integers; } a > b \end{bmatrix}$$

Eq. 12.47: Feelings

This relationship may also be summarized more simply as Equation 12.48:

$$\text{Harmony} \propto f\left(\text{feelings}\right)$$

Eq. 12.48: Harmony aspect of fundamental capacities of self

Thoughts are a play of the emergence of light's property of knowledge. Through the thought we can become greater or conceptualize things greater or begun to enter into relationship with some things other than our small self. Thought allows us to connect to more "othernesses" or even the oneness of the reality of light. An equation for thoughts, Equation 12.49, is the following:

$$\text{Sig}_{\text{thoughts}} = Xa + \overline{Yb_{0-n}} \quad \text{where} \quad \begin{bmatrix} X \in \left[S_{\text{System}_K}\right] \\ Y \in \left[S_{\text{System}_P}, S_{\text{System}_P}, S_{\text{System}_K}, S_{\text{System}_N}\right] \\ a, b \text{ are integers}; \ a > b \end{bmatrix}$$

Eq. 12.49: Thoughts

This relationship may also be summarized more simply as Equation 12.50:

$$\text{Knowledge} \propto f\left(\text{thoughts}\right)$$

Eq. 12.50: Knowledge aspect of fundamental capacities of self

12.4 Leveraging Light to Bring About Organizational Change

The previous section illustrates how the four properties implicit in light of knowledge, power, presence, and harmony act as organizational principles to structure progressively more complex layers of matter, starting from the electromagnetic spectrum and ending in molecular plans at the level of living cells. By extrapolation all tissues, composed of living cells, and therefore all organs, and therefore the very body of all living things, are nothing other than a materialization of light. It was also suggested that fundamental capacities of self are ordered by the four principles implicit in light. Such ordering at the level of matter and life also implies that all collectivities comprising groups of human beings, and therefore also, larger and larger collections of collectivities bordering on and including civilizations, can also be reseen as emerging from light.

Further, as human beings and through force of habit, it is easy for us to always approach an issue or a problem in the same way. We each have a set of beliefs, assumptions, and behaviors, with which we process circumstance. As a result, in the arbitration of possibility, we tend to only reinforce dynamics at U, at the bottom layer in (12.13). In other words, we rarely invoke the other layers of light present in (12.13) because usually a conception or activation of these is not part of our usual processing instrumentation. If, however, we consciously allow light to enter into our conception of circumstances, then the arbitration of possibility taps into super-sets of functionality and possibility resident at meta-layers of light, as a result of which the potential outcomes may be superior than in the absence of the availability of such super-sets. This has implications for HSE in that systems can be engineered to prod individuals to process circumstance differently by helping them to overcome habitual patterns and perceptions faster. Experimentations using an emotional intelligence-based system, aligned with the math introduced in this chapter, have been used in a complex medical environment to help individuals shift emotional patterns (Malik and Pretorius 2018b).

Imagining light or reseeing that everything is in fact emergent from light hence viscerally connects us with light, and this initiates an automatic and more complete real-time quantum computation in which the information field accompanying or defining a circumstance or event is changed.

Recall that quanta are a bridge between a faster-moving layer of light and a slower-moving one. Quanta in fact collect possibility from a faster-moving layer of light and make it more materially available in a slower-moving layer of light. Reseeing any situation as constituted in light more easily connects us with the full complexity imagined by (12.13) and allows possibility present in meta-layers to become more relevant in arbitrating possibility that occurs at *U*. Such an arbitration involving quanta connecting multiple layers of light is none other than an act of quantum computation. Contemporary schemes involving quantum computation are more based on probability, as proposed by quantum scientists such as MIT's Seth Lloyd (2007), in contrast to the more deterministic approach (Malik et al. 2017a) framed by possibilities existing in each layer of (12.13), as proposed here.

The output of any computation is always information of some sort. Further, all circumstance, intimately composed of emergent layers of light and always influenced by meta-layers of light, is encapsulating nothing but a sophisticated array of information. Reseeing a situation from the bottom up, starting at the level of the electromagnetic spectrum, through quantum particles, through atoms, through molecules, and so on, is in fact a master act that allows the greater logic, possibilities, and information content of light-based emergent and meta-layers to automatically become active in the reseen situation. This further allows the reseen situation to admit of a larger set of dynamics than just the beliefs, perceptions, and assumptions with which it is seen by an observer, thereby allowing the information field of that reseen situation to change. The changing of the information field allows different potential outcomes to emerge than would have been possible in the absence of such an act of reseeing.

Hence, no matter what the situation and no matter how complex it is, it is possible to alter it through a conscious act of reseeing or reenvisioning it. Note though that meta-layers of light, involving super-sets, unique seeds, and implicit properties, have a different logic than the layer of reality created by light traveling at *c* and when invoked will likely result in an outcome aligned with a more global perspective than a perspective engendered by a more myopic individual view. In other words, while the outcome of a reseen situation will be enhanced, a person or group of people involved in such reseeing should not expect that they can mold outcomes based on their perceptions, beliefs, and assumptions alone. While people can act to change a circumstance through an act of reseeing it, they will need to give up expectations on likely outcomes and trust instead that the best thing from a more global perspective is what will happen. Light, a priori, is envisioned as being ubiquitous, connecting things in a single embrace, and with a power and knowledge that is radically different than what any individual or group of individuals is capable of.

Over the last one year, the author has conducted a number of workshops both within and outside Zappos.com LLC. These have involved people with diverse backgrounds representative of all functions within a complex Internet technology organization. Typically, up to 15 people have participated in a single 2.5-hour workshop. Most participants have no background in any of the materials they are exposed to in a workshop and essentially come in cold to a workshop. The first half of the workshop summarizes the theoretical basis of light, as laid out in Sections 12.1 and 12.2. The materials though are presented pictorially and in a nonmathematical way. The first half of the workshop aims at addressing two questions through interactive inquiry. First, what is light made up of? Second, what is made up of light? At the end of the first half of the workshop, usually participants will experience some "aha" moment and conceptually acknowledge that in fact light can be in everything and that everything can emerge from light. Once this is experienced, the participants are ready for the second half of the workshop.

The second half, by contrast, is practical and involves reenvisioning objects and situations as light. While the guided sessions of the workshop start off with simple reenvisioning, seeing, for example, a rose as emerging from light, these reenvisionings progressively become more complex involving work projects or even unresolved conflict situations. In the self-practice part of the workshop, which usually happens closer to the end of the second half, situations envisioned by participants often expand to encompass the full range of issues typically experienced by an organization.

The following summarize light maxims that structure the workshops and can be thought of as individual takeaways aimed at helping participants facilitate ongoing change:

1) *Fluency in light dynamics is like a muscle that needs to be built*: The biggest obstacle to change is one's own continued reinforcement of individual perceptions, beliefs, and assumptions. In reality each individual is highly creative. The only problem is that the reality being created is always created using the same materials derived from one's own limited perceptions, beliefs, and assumptions. Therefore, the outcome is always the same and it appears as though nothing is really changing. The very possibility of a continuous and persistent quantum computation though reinforces that change is possible at every instant. The only thing stopping it is one's own programming. Actively using light dynamics, to the point where it is an automatic feature in individual processing of circumstances, can therefore be a very powerful thing. But such an active use of light dynamics will not happen just by itself or through one or two guided sessions. It has to be taken on seriously and practiced consistently for it to become second nature and for it to begin to influence change in a meaningful way.

2) *Learn to perceive light differently – light is in everything, and everything arises from light*: With eyes closed or open, it is possible to see that light is in everything and that everything arises from light. Even though light is everywhere and even though in its known manifestation of the electromagnetic spectrum, it is foundational to most technologies that drive the world, and even though without it life would not exist, we just take it for grated without even contemplating what it really is. Yes, artists, scientists, and mystics have focused on it to try to penetrate something of its mystery. But such penetration does not require us to be either artist, or scientist, or mystic. Since we are ourselves intimately built from it, it is each individual's natural right and privilege to know it once one begins to see it and resee all in it. The more one resees in light, the more one connects with its global and freeing nature, and progressively even, becomes something or someone entirely different.

3) *Creating one's own light sphere will help facilitate any kind of change:* Every person has a natural center within them, and often this is within the heart. In reference to (12.13) the center of a person can be thought of as that place where the unique seed formed through action of metalayers of light resides within a person. This unique seed can be thought of as the anchor or center of the personality unique to a person. To create one's light sphere, one begins by focusing on this center and revisualizing it as a small sphere of light. Slowly and gradually one can envision this light sphere growing so that more and more of one's own body and being is encompassed by it. Gradually one can further expand one's light sphere to become larger than oneself. At each step of the expansion, it is important to try to experience the nature of light – its warmth, its glow, its dynamism, and its love. From the bottom up, given that every single part of us is emergent from light, as discussed in detail in Sections 12.2 and 12.3, envisioning such a light sphere connects us with our own ever-present source of light, which is of the nature of light in its fullness and as embodied, at least mathematically, by (12.13). Connecting with and nurturing this light sphere allow us to participate more consciously and actively in the persistent and ongoing quantum computation that influences all things. It is like turning on and keeping on our own personal computing device, just as we turn on, keep on, and use computing devices such as smartphones and computers. Except the computational machinery that animates the

light sphere is intimately connected with all other life and all other objects in a way in which we cannot imagine, and the dynamics that it has access to and can influence are essentially unbounded and become bounded only by the paucity of our own imagination.

4) *Any person or situation or condition can be invited into one's light sphere and be reseen in light*: One's light sphere is essentially dynamic, with living light flows and a body of light. Experiencing light within one's light sphere connects one, even if one does not quite feel this, with the essential oneness and nature of light. And being so connected it is easier to resee anything or anyone or any circumstance within one's light sphere. In fact, the process of beginning to bring about holistic change to any circumstance is simply to begin to invite that circumstance into one's personal light sphere. It is then necessary, without judgment, without any expectation of outcome or result to see that the light in one's light sphere is completely filling and flowing through the object or person or circumstance that has been invited into the light sphere. That invited entity has to be reseen in light. If one sees that there are thoughts or emotions or feelings or sensations that arise from the invited entity, or even from the reseeing of oneself within one's light sphere, one needs to see that all these too are nothing but light itself. This needs to be reseen again and again. The invited entity or situation within one's light sphere needs to continually be flooded with light. This action needs to continue until one feels that it cannot be continued anymore.

5) *Light has its own logic and will change the information field of anything in one's light sphere*: The nature of light, regardless of the speed it may be traveling at, is one. When one connects with some image of light in oneself, one is essentially connecting with the full nature of light, whether one feels or sees or realizes this or not. That is the power of reseeing in light. It immediately brings one into full contact with the multidimensional power of light. Light is multidimensional along the separate dimensions of knowledge, of power, of harmony, and of presence or service. Each of these dimensions generates a vast amount of information, as suggested by Equations (12.7)–(12.12). When one resees something or someone or some situation in light, then the essential observer–observee dynamic as suggested by numerous experiments in quantum physics (Lanza and Berman 2009) can basically alter the active information fields, as a result of which the active dynamics related with that something or someone or situation can change. The change is not something that can be architected as per one's own narrow wants. But light itself, having been invoked, will begin to alter the information field in subtle ways and as per the vaster view and logic contained in its natural ubiquity.

6) *The power to positively influence change is limited only by the relationship with one's own light sphere*: It is easy to resign to circumstance and to continue to see everything with the embedded lenses that are habitually used to see and to think and to feel. But the very repetition of the act of seeing or feeling or thinking in the same way is nothing other than a highly creative act in which past models of personal reality are reinforced in every new moment to recreate the range of effective dynamics in that moment. Hence, the more one invests in one's own light sphere, being able to dynamically resee any situation or person or groups of persons or objects as light and flooded with light, and the more one is able to see that light has its own logic and therefore that personal expectation and wanting of a particular outcome or result can be quite ineffective, the more powerful light dynamics within one's own light sphere can become, and the more effectively can circumstance, regardless of scale and complexity, be directly influenced to change into something better from a more global and impersonal perspective. Being able to change things, then, comes down to developing a relationship with one's own light sphere.

These six light maxims also surface additional design considerations to be incorporated in any HSE framework.

12.5   Summary and Conclusion

Even though light is present everywhere, drives most of our technologies, and is responsible for all of life on earth, we hardly know it, understand it, nor cultivate it. Yet it is among the most easily accessible of all phenomena, and as this chapter suggests, has intimately architected the emergence of all of matter, life, and even what may lie beyond these.

The infinite information codified in light is suggested to be due to the all possibility of its nature as perceived when light travels infinitely fast and is organized along four axes or dimensions that we call knowledge, power, presence, and harmony. The infinite diversity that exists when light travels at *c* is suggested to be due to a process of materialization involving quanta. Such materialization ordered along the four implicit properties or dimensions within light becomes apparent as one perceives the natural fourfold ordering that exists at increasing levels of complexity through matter and life, as made evident in the electromagnetic spectrum, quantum particles, atoms, molecular plans, and fundamental capacities of the self.

The arbitration of possibility is suggested to be due to a constant quantum computation that happens at every moment and points to the inherently creative aspect of existence, whether in life, or in matter, or in individuals. Habitual thinking and perceiving in individuals easily reinforce the reality that an individual tends to be anchored to. But this can easily change through a process of reseeing or reenvisioning everything in light.

The power to change things starts here and starts now. The most important protagonist in that change is not leadership, but the individual, who by virtue of an intimate connection with light can make change happen without permission and as the visceral urge to do so arises. HSE efforts need to design for this possibility proactively. Instead of being victimized, an individual can through practice, more effectively change the information field associated with any circumstance, as a result of which a different outcome is likely to result. This is the ultimate act of power and has to become part and parcel of the circumstance-processing capability of each individual. This is what the light technology presented here aims to inculcate, and this is how real change can be effectively facilitated by each person, all the time. This insight also suggests additional factors to be considered in any HSE effort. It is likely that such additional factors will have a positive effect by not only chipping away at human limitations but also allowing existing human capabilities to be heightened.

References

Arabatzis, T. (2006). *Representing Electrons: A Biological Approach to Theoretical Entities*. Chicago: University of Chicago.

De Broglie, L. (1929). *The Wave Nature of the Electron*. Nobel Lecture.

Einstein, A. (1995). *Relativity: The Special and General Theory*. New York: Broadway Books.

Ekspong, G. (2014). *The Dual Nature of Light as Reflected in the Nobel Archives*. Nobel Media AB 2014 (15 October). Nobelprize.org.

Feynman, R.P. (1985). *The Strange Theory of Light and Matter*. Princeton, NJ: Princeton University Press.

Fuller, B. (1982). *Synergetics: Explorations in the Geometry of Thinking*. New York: MacMillan Publishing Co.

Goodsell, D. (2010). *The Machinery of Life*. New York: Springer.

Gray, T. (2009). *The Elements: A Visual Exploration of Every Known Atom in the Universe*. New York: Black Dog & Leventhal Publishers.

Heiserman, D. (1991). *Exploring Chemical Elements and Their Compounds*. New York: McGraw-Hill.

Isaacson, W. (2008). *Einstein: His Life and Universe*. New York: Simon and Schuster.

Lanza, R. and Berman, B. (2009). *Biocentrism: How Life and Consciousness Are the Keys to Understanding the True Nature of the Universe*. Dallas: Benbella Books.

Lloyd, S. (2007). *Programming the Universe: A Quantum Computer Scientist Takes on the Cosmos*. New York: Vintage.

Lorentz, H.A. (1925). *The Science of Nature*, vol. 25, 1008. Springer.

Malik, P. (2009). *Connecting Inner Power with Global Change: The Fractal Ladder*. New Delhi: Sage Publications.

Malik, P. (2017a). *A Story of Light*. Amazon Kindle & Google Books.

Malik, P. (2017b). *Emergence*. Amazon Kindle & Google Books.

Malik, P. (2017c). *Quantum Certainty*. Amazon Kindle & Google Books.

Malik, P. (2018a). *Super Matter*. Amazon Kindle & Google Books.

Malik, P. (2018b). *Cosmology of Light*. Amazon Kindle & Google Books.

Malik, P. (2018c). *The Emperor's Quantum Computer*. Amazon Kindle & Google Books.

Malik, P. (2019a). *The Origin and Possibilities of Genetics*. Amazon Kindle & Google Books.

Malik, P. (2019b). *The Second Singularity*. Amazon Kindle & Google Books.

Malik, P. (2020). Light-based interpretation of quanta and implications for quantum computing. *2020 IEEE 9th Annual Computing and Communication Workshop and Conference (CCWC)*, Las Vegas, NV (7 January).

Malik, P. and Pretorius, L. (2017b). *Oceans of Innovation*. Amazon Kindle & Google Books.

Malik, P. and Pretorius, L. (2018a). Symmetries of light and emergence of matter. *Indian Journal of Science and Technology* 11 (14): 1–11. https://doi.org/10.17485/ijst/2018/v11i14/110789.

Malik, P. and Pretorius, L. (2018b). A case study validation of the application of a generalised equation of innovation in complex adaptive systems. *South African Journal of Industrial Engineering* 29 (1): 1–20. https://doi.org/10.7166/29-1-1780.

Malik, P. and Pretorius, L. (2019). An algorithm for the emergence of life based on a multi-layered symmetry-based model of light. *2019 IEEE 9th Annual Computing and Communication Workshop and Conference (CCWC)*. doi:https://doi.org/10.1109/CCWC.2019.8666554.

Malik, P., Pretorius, L., and Winzker, D. (2015). A mathematical basis of innovation. *Conference Proceedings IAMOT 2015*.

Malik, P., Pretorius, L., and Winzker, D. (2017a). Qualified determinism in emergent-technology complex adaptive systems. In: *2017 IEEE Technology & Engineering Management Conference (TEMSCON), Santa Clara, CA*, 113–118. IEEE https://doi.org/10.1109/TEMSCON.2017.7998363.

Olive, K.A. et al. (2014). Particle data group. *Chinese Physics C* 38: 090001.

Overbye, D. (2015). Physicists in Europe find tantalizing hints of a mysterious new particle. *New York Times* (15 December).

Perkowitz, S. (2011). *Slow Light*. London: Imperial College Press.

Tweed, M. (2003). *Essential Elements: Atoms, Quarks, and the Periodic Table*. New York: Walker & Company.

UCDAVIS-CFT (2015). *Description of Orbitals*.

Whitaker, A. (2006). *Einstein, Bohr and the Quantum Dilemma: From Quantum Theory to Quantum Information*. Cambridge: Cambridge University Press.

Wilczek, F. (2016). A Beautiful Question: Finding Nature's Deep Design. New York: Penguin Books.

Section 5

From the Field

13

Observations of Real-Time Control Room Simulation

Hugh David with an editor introduction by Holly A. H. Handley

Chartered Institute of Ergonomics and Human Factors, Birmingham, UK

13.1 Introduction

13.1.1 What Is a "Real-Time Control Room Simulator"?

Real-time control room simulation is a method of investigating the design and operation of control rooms. It is sometimes called "man-in-the-loop" simulation, a phrase that should be avoided since not only is it sexist, but it implies an exclusively mechanical view of human activity. A real-time control room simulator consists principally of a copy of the control room, with inputs provided by computer data processors, with "simulator pilots" simulating the actions of vehicles under control (usually under computer prompting) and skilled participants simulating operators of adjacent control centers, where needed.

Traditionally, simulators have mostly been dedicated to a specific system, although the simulator at the EUROCONTROL Experimental Centre, originally designed in the early 1970s, was the first digital general-purpose simulator. The EUROCONTROL real-time simulator can be configured to provide up to 40 working positions with 20 simulation staff, showing 200 aircraft simultaneously in up to 20 sectors. A simulation study may take up to five weeks, with up to four 60- to 90-minute runs on five days per week. It is still simulating many different air traffic control rooms, although all parts of it have been renewed several times over the years (like Jack Hobbs's cricket bat – three new blades and two new handles but still the same bat.)

13.1.2 What Is It Used For?

Real-time control room simulators are used for elementary and advanced training and to test proposed modifications to the system controlled. They will, in the future, be used increasingly for skill maintenance, including practicing for emergencies. (Although emergencies are normally

Editor's Note: Hugh David, an expert in control room simulations, provided this chapter based on his observations of users while writing his book *Control Room Simulation – The Real Life of Real-Time*, David (2017). It appears here in its entirety in Hugh's colorful writing style as a set of practitioner's field notes recording his "real-life" experiences. We are happy to provide you this unedited chapter both for your information and enjoyment as one of our final chapters. As such the views expressed in this chapter are the author's own.

A Framework of Human Systems Engineering: Applications and Case Studies, First Edition.
Edited by Holly A. H. Handley and Andreas Tolk.
© 2021 The Institute of Electrical and Electronics Engineers, Inc. Published 2021 by John Wiley & Sons, Inc.

unexpected, their probable nature can be predicted with a reasonable level of accuracy, if not absolutely.) Real-time simulators can also be used where a system is being planned to avoid costly mistakes in constructing the new system or forensically (if proper records have been kept) to investigate disasters where the actual control room has been destroyed or is otherwise unavailable.

13.1.3 What Does It Look Like?

The physical design of control rooms is now sufficiently standardized to form the subject of an international standard (ISO 11064-1 2000, ISO 11064-2 2000, ISO 11064-3 1999, ISO 11064-4 2004, ISO 11064-5 2008, ISO 11064-6 2005, ISO 11064-7 2006). Workstations have one, two, or occasionally three large screen displays in a single row. Most control rooms now consist of several rows of workstations arranged in pairs, facing a large display at the end of the room. This display may show an overall view of the plant, a schematic of the process, a number of CCTV displays, or anything of common concern to the operators. A supervisor's position is placed behind the work positions, usually slightly raised to provide a view of all working positions. The visitors' gallery (if any) is placed behind the supervisor's position and ideally should be separated from the control room by a transparent panel. Figure 13.1 shows a representative modern control room. This could be either a real or simulated control room.

Figure 13.1 Contemporary control room. *Source:* David (2017).

13.1.4 How Will They Develop?

Modern control systems tend to centralize control, often at locations remote from the area of activity. This is partly due to a desire to reduce costs and partly due to security and safety considerations. Modern systems are digital – controlled by digital computers. The software of these systems is increasingly modular. Data input, checking, and cyber defense modules are essentially similar in many fields. Control room display, control input, and feasibility checking are equally modular and determined largely by human abilities, modified to a decreasing extent by tradition, wishful thinking, and willful blindness. The future development of real-time control room simulation will probably diverge into general-purpose simulators, designed to provide services to all users of standardized control rooms, and special-purpose simulators, incorporated into operational control facilities.

13.2 Future General-Purpose Simulators

Figure 13.2 provides a schematic of a future general-purpose simulator. Although this is necessarily a simplified version, it shows the main features of such a simulator. A number of different sets of "static data" describing the real world would be available. In air traffic control, for instance, these data would include definitions of the performances of different types of aircraft; digitized files of the positions of airfields, positions of beacons, and routes if these are still in use; flight plans; and potential weather patterns. It would also include the algorithms necessary to simulate traffic. In a

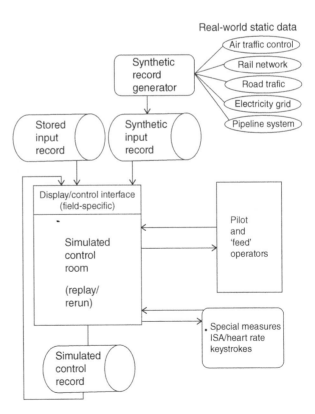

Figure 13.2 Future general-purpose simulator. *Source:* David (2017). CreateSpace Independent Publishing Platform © Amazon.com.

pipeline system, it would contain digitized maps of the system, details of pumping stations, communications links, and remote sensors. It would include schedules of planned pumping operations and, for training purposes, algorithms to simulate the effects of system failures. Similar sets of data would be available for other fields of interest, such as rail networks, road networks, and electricity grids.

A synthetic record generator would draw on the field-specific information to generate preplanned traffic, including particular situations of interest. In air traffic control, these might include potential conflicts between aircraft. (The generator will ensure that the problems posed were reasonable, for example, that aircraft were not already in conflict when they entered the area simulated or that three aircraft did not converge on one entry point at the same time). This generator would produce a synthetic input record, a "script" in the same form as that produced in the real control room.

The synthetic input record, or a stored real input record, is played to the simulated control room via a display/control interface. This will usually be the same in most control rooms in a particular field of activity. In addition to the operators in the simulated control room, there may be additional "simulator pilots" who simulate the onboard crew of vehicles in the system using prompts generated by the simulator and inserting the verbal orders they receive in coded form. These "simulator pilots" are trained but need not be fully qualified to perform the "piloting" task in the real world. Equally, there may be a requirement for "feed" operators, who simulate adjacent areas of responsibility. These are usually experienced qualified operators. As such, they do not appreciate the names of "ghost" or even "dummy" operators, sometimes employed.

In a general-purpose simulator, some additional standard measurement technique will be applied. The measurement of simulations – or of real operations – is a difficult problem and will be dealt with in detail later.

A simulation run will produce a simulated control record in the same format as used in the real world. It will contain additional measures not required in the real world, such as records of incomplete or canceled input orders. The same analysis and data reduction programs will be used as are applied to real operational records, with additional programs for additional measures.

13.2.1 Future On-Site Simulators

Figure 13.3 provides a schematic of a future on-site special-purpose simulator. Again, this is a simplified version. On the right-hand side, the actual working control room is shown, taking information from the real world, concentrating it, and processing it. (This processing includes various data checking and anti-hacking measures, which should not be discussed here.) An input record is created at this time, as the information is fed into the field-specific display/control interface, which refers to the real-world static data to provide meaningful information to the operators. The operators act on the real world, forming a closed loop. Their actions are recorded in a control record, which is correlated with the input record.

On the left-hand side, the on-site simulator facility uses an input record, either the current or an archived version, to present the situation via the same display/control interface to operators for training using producing a simulated control record, which can be analyzed by the same suite of analysis programs. Additionally, it will be possible to rerun a recorded sequence of events, substituting control records of inputs for the operator inputs, to allow "post-mortems" of accidents or near accidents, for forensic, safety, and instructional purposes. The working positions used will be those provided for use where operational working positions become unserviceable or require routine maintenance or updating. Very little additional hardware or software will be required to add a simulation facility to a modular digital system.

Figure 13.3 Future on-site simulator
Source: David (2017). CreateSpace
Independent Publishing Platform
© Amazon.com.

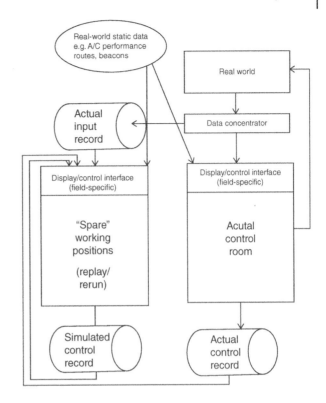

13.3 Operators

Having described in schematic form the probable future development of real-time simulation hardware and software, with the proviso that real systems will always face unexpected problems and may need specific solutions, we can now consider the most important part of the system – the operators.

It is vital to remember that operators are not the "naïve subjects" of experimental psychology. They are usually highly trained, responsible, experienced experts in their particular field. Their opinions may be decisive for the future of the system. They will often have reached what is sometimes called the "samurai" state of operation, where they are so familiar with their task that they are not consciously aware of how they do it. They usually respect "scientists" and assume they know what they are talking about. However, they are usually good judges of people and quick to detect bluff or bullshit. They often have a "tribal" attitude, with more or less conscious admission rituals. Investigators should be prepared to ingest alcohol in larger or smaller quantities and tolerate with apparent enjoyment attitudes they would regret in normal life. Gender differences may be an advantage or handicap depending on circumstances best left unspecified.

Like all human beings, they are not equipped with a "pause" or a "reset" function, but they learn consciously or unconsciously and cannot disregard what they have learned. This presents severe methodological problems. Most simulations are set up using the familiar models of experimental design, developed in agriculture and experimental psychology. These models suppose that the experimenter can vary some parameters of the situation while holding all others constant. In practice, there are many uncontrolled and uncontrollable factors in any large-scale simulation so that the "residual variation" is always large.

Most simulations require familiarization runs at the start, while operators adapt to the unavoidable differences from their normal routine. (They will often be away from home, working during the day only, instead of doing shift work, with equipment that differs in detail or radically from their norm and with colleagues whom they do not know.) In most organizations, operators work in teams, working the same shifts for months and years. Informal social patterns develop, for example, "Bill wakes up slowly – Jill cannot stay late – Momo needs to be free at prayer times – Fred always has a hangover on Mondays." It is rarely practical for a complete team to be detached for a simulation, so a "shakedown" is always necessary.

13.4   Data

The raw material for the simulation (the data on which it operates) is always subject to problems. A random sample of traffic will always contain some anomalous events. It may turn out that the real situation is significantly different from what management supposes. Many organizations have rule books that have been amended and modified in detail over the years so that it is, in fact, impossible to follow them in practice. Recourse to the advice of participants may help. But Dubey (2000) points out that participants often feel that they "own" the data, since it is their daily concern, and may raise objections to data as a means of expressing disquiet with other aspects of the simulation that they feel are determined by management.

Because traffic is inherently variable, different traffic samples vary in difficulty. The degree of difficulty is generally difficult to define. It is tempting to use the same traffic sample on several occasions – perhaps with different possible organizations. It has been found that operators do not usually recognize a repeated sample, particularly if it is disguised (for example, in air traffic control, by changing aircraft call signs). However, neurological studies suggest that learning occurs on a subconscious level.

In air traffic control, traffic patterns tend to be repetitive. A particular flight leaves a given airport and flies a standard route to the same destination each day. Controllers spend considerable time learning these regularities, and it can be argued that purely random traffic is therefore unrepresentative. Not all traffic is repetitive, however. A possible solution is to develop a sample of about twice the expected traffic level and make random selections of different percentages of traffic. This also solves the problem that the task must not be so easy that any organization can cope, nor so hard that no organization can cope. Similar considerations probably apply in other potential fields of application.

13.5   Measurement

Once a simulation has been established, we need to measure it. To measure it, we need to know what we are looking for. Generally, simulations are used to establish the capacity of a system, to check that as the proposed system is workable, or to assess its effect on the operators. It may be useful to bear in mind the difference between "stress" (the demands of the tasks required) and "strain" (the effects of "stress" on the operators). There are almost as many definitions of "stress" and "strain" as there are authors. This text follows the engineering definition rather than the journalistic one.

The classic experimental paradigm is to measure the speed and accuracy of subjects' performance. In simulation, the speed is normally determined by the process being simulated, and the degree of accuracy required is such that errors are too few for statistically (or practically) significant differences to be established. The experimental psychology approach is therefore not useful.

We usually attempt to make objective and subjective measurements. Many methods have been tried, and a few have stood the test of time. There are no absolute standards and no mechanical assessment procedure. Knowledge, judgment, and sensitivity are required to reach a useful conclusion.

13.5.1 Objective Measures

It is clearly preferable to have objective measurements, where they may be relevant. Such measures tend to measure the stress on the operators rather than the strain experienced by the operators. They also tend to depend on the nature of the process being controlled. Some general points can however be made.

13.5.1.1 Recommended

Among objective measures may be the amount of time spent in communication, the number of communications, and any delays or potential conflicts observed. Even these obvious measures need careful supervision. For example, one simulation run showed unusually high use of the frequency on which controllers communicate with the aircraft. Investigation showed that the (Italian) controllers, after a mostly liquid farewell lunch, were singing "O Sole Mio" in chorus on the frequency. These are measures of stress.

Another measure used in air traffic control studies is the frequency of "conflicts" (dangerously close approaches of aircraft). In normal operation, serious conflicts are very rare, although the formal limits are sometimes infringed where no danger exists. As a measure, conflict frequency is of little use, since if it is high enough for statistical analysis, it is far too high to be realistic. Similar considerations may apply in other fields of study.

One measure that may sometimes be useful is heart rate. It is now possible to measure heart rate with nonintrusive methods using standard sports or self-training applications to perform remote monitoring and recording. Although we know that there are many factors that will affect heart rate, both physical and psychological, it is possible to define a threshold heart rate for operators, above which the situation merits investigation. It is tempting, but ultimately unrewarding, to try to filter out interfering variables.

13.5.1.2 Not Recommended

There are (at least) three lines of investigation that have been developed but should, in the light of experience, be avoided. Secondary tasks have been used in experimental psychology to measure the extent to which an experimental subject is occupied by a task. The subject is asked to carry out a secondary task, such as mental arithmetic, as much as he can while doing his primary task. Although this may work empirically when studying simple tasks, it assumes that the human operator has the characteristics of a machine, with a fixed maximum capacity. In the real (or simulated) world, operators are carrying out complex tasks, planning their future workload, monitoring displays, recognizing situations, worrying about their mortgage or their marriage, and carrying out a variety of conscious or unconscious activities.

They have no rigid upper limit to their activity.

Embedded secondary tasks can however sometimes be identified. These are tasks that are not strictly necessary but normally required. Often the system can carry them out if the operator omits them. Dropping these tasks may be a valid indication of overload. Identifying such tasks requires familiarity with the system, and such tasks do not always exist.

A considerable effort has been devoted to the identification of psychophysiological measures of strain. These involve the measurement of body activity such as heart rate, electroencephalographic activity, or the measurement of hormones associated with stress in blood or saliva. Heart rate in itself may be a useful measure, as mentioned above. Other measures, such as respiration rate, sinus arrhythmia (the minor irregularities observed in normal beat-to-beat heart rate, rather than the major irregularities indicative of imminent heart failure), and muscle tremor, are all intimately linked and require elaborate analysis for very little added information.

Considering electroencephalography (EEG), there are some basic methodological problems. Although we know more about the activity of the brain than we did a decade ago, we still do not know enough to understand what is actually going on in practical terms. We know that the brain is not a computer but has a complex and specialized structure. Much research has been carried out using Fourier analysis of EEG signals. These are essentially correlational studies, which may show that certain frequencies are associated with specific activities but shed no light on the functional relationship. We also find that individuals vary greatly in their EEG activity.

Hormonal studies have similar problems. Although we know a lot about the complex relations of hormones within the human body, we find that they are affected by many intervening factors, such as the time of day, diet, coffee drinking, menstrual cycle, and psychological type. Some of these parameters are intensely personal and subject to considerations of personal privacy, which operators should not be asked to sacrifice without more justification than simulation studies provide.

An additional ethical question arises where it becomes clear from such measures that an operator is at or over the boundaries of normal physiological responses and may be at risk of physical injury. Equally, it may become clear that the operator is using drugs or other substances that may affect their performance. Drawing attention to the problem may cost the operator his job and be a breach of confidentiality. Not drawing attention may risk the life of the operator and potentially the lives of many more. If you are not a qualified medical practitioner acting in your professional capacity, you may be legally as well as morally liable for these consequences.

Eye movement studies have become considerably easier in recent years, requiring less calibration and less intrusive equipment. They remain very expensive. It is now practically possible to determine where the operator is looking with within about 10 cm on a desk display. It is not, so far, possible to identify which aircraft the operator is looking at on a radar display or to correlate that information to other simulation data. There are also interpretation problems. It is not necessarily true that an operator spends less time looking at a better organized display. As an operator pointed out to me, he may simply find it somewhere to rest his eyes while thinking of something else.

There are sometimes simplifications that can be made. For example, when an operator is, as air traffic controllers say, "losing the picture," their gaze tends to jump around the display, making more and shorter fixations. Another simple observation may be that operators never look at some display. Practically, an eye movement study may suggest which displays should be at the center of the operators' field of vision, because they are used most, and which displays are looked in sequence so that they should be adjacent.

Although the attraction of modern "cutting-edge" equipment is considerable, it is sometimes possible to achieve the same results by systematic observation, either online or by analyzing a video recording (David 1986). It is also considerably cheaper, quicker, and less intrusive.

13.5.2 Subjective Measures

The obvious way to determine how a simulation affects the operators is to ask them. Although it is generally true, operators' opinions may need to be treated with some reserve. Oppenheim (1992) is a useful guide to the design of questionnaires and the conduct of interviews. Gawron (2000) is a very comprehensive handbook of human performance measures.

In general, it seems preferable to use a simple, rapid "online" method to assess the overall difficulty of the task, particularly when this varies with the arrival of traffic, combined with a more elaborate form immediately following the exercise. Most simulations start with an empty task load, building up to a level, and would lead to a similar decline of traffic at the end. It is usual to discard records for the period of traffic buildup and to halt the simulation when no further traffic load is expected.

13.5.2.1 Recommended

The instantaneous self-assessment (ISA) method requires a five-key keyboard or a touch-sensitive or point-and-click display incorporated into the working display. The operator is required to assess his own workload at regular intervals (usually two or three minutes) when cued by the flashing of the keys. The five keys are marked as in Figure 13.4. The display flashes until the operator responds or for 20 seconds. The operators' responses and the time taken to respond are recorded and form a practical record of the operators' subjective responses. The operators' responses may be shown to the simulation supervisor in graphic form to provide warnings of difficulties during the simulation. When analyzing the data, "VERY HIGH" or "HIGH" readings are more significant than "LOW" or "VERY LOW" readings so that simple averaging can be misleading. Although the mean of two "VERY HIGH" and eight "VERY LOW" readings is less than that of 10 "FAIR" readings, the first simulation is clearly more problematic than the second.

The NASA Task Load Index (NASA-TLX) is applied immediately after each exercise and requires the operators to rank the simulation on six scales, as in Figure 13.5. As originally devised, the scores on the six scales are weighted by a prior ranking derived by paired comparisons of scales for each operator. However, Hart (2001 ["The NASA-Tlx scales may be analysed independently"]) in a personal communication confirmed practical experience that the scales could be analyzed separately. The "Physical Demand" scale, for example, is rarely significant in real-time simulations. The Performance scale runs in the opposite direction to the other scales and is sometimes misunderstood by operators.

13.5.2.2 Not Recommended

The Subjective Workload Assessment Technique (SWAT) is a self-reporting technique, similar to ISA above. The operator is asked to report how stressed (s)he is on three scales – time load, mental effort load, and psychological stress load. Each scale has three levels. SWAT requires a ranking of

Value	Level	Symbol	Colour	Definition
5	VERY HIGH	+ +	Red	Completely occupied, some tasks missed
4	HIGH	+	Amber	Almost completely occupied, but task can be completed
3	FAIR	=	Green	Steady, reasonable workload. Some breaks
2	LOW	–	Cyan	Little work, much spare time
1	VERY LOW	– –	Magenta	Practically no work, boredom, lack of stimulus

Figure 13.4 ISA key values.

Mental demand	How much mental and perceptual activity was required (e.g. thinking, deciding, calculating, remembering, looking, searching, etc.)? Was the task easy or demanding, simple or complex, exacting or forgiving?
	LOW |__|__|__|__|__|__|__|__|__|__|__|__|__|__|__|__|__|__|__|__| HIGH
Physical demand	How much physical activity was required (e.g. pushing, pulling, turning, controlling, activating, etc.)? Was the task easy or demanding, slow or brisk, slack or strenuous, restful or laborious?
	LOW |__|__|__|__|__|__|__|__|__|__|__|__|__|__|__|__|__|__|_ _|__|__| HIGH
Temporal demand	How much time pressure did you feel due to the rate or pace at which the tasks or task elements occurred? Was the pace slow and leisurely or rapid and frantic?
	LOW |__|__|__|__|__|__|__|__|__|__|__|__|__|__|__|__|__|__|__|__| HIGH
Performance **Careful!**	How successful do you think you were in accomplishing the goals of the task set by the experimenter (or yourself)? How satisfied were you with your performance in accomplishing these goals?
	GOOD |__|__|__|__|__|__|__|__|__|__|__|__|__|__|__|__|__|__|__|__| **POOR**
Effort	How hard did you have to work (mentally and physically) to accomplish your level of performance?
	LOW |__|__|__|__|__|__|__|__|__|__|__|__|__|__|__|__|__|__|__|__| HIGH
Frustration level	How insecure, discouraged, irritated, stressed and annoyed versus secure, content, relaxed and complacent did you feel during the task?
	LOW |__|__|__|__|__|__|__|__|__|__|__|__|__|__|__|__|__|__|__|__| HIGH

Figure 13.5   NASA-TLX scales.

the different scales by a complex method, requiring a computer-based weighting program. Unfortunately, the designated supplier of this software appears to be defunct – rendering the method generally unavailable. (In any case, a comparative study suggests that it is significantly more intrusive than ISA and does not produce independent scales in use [David and Pledger 1995; Pledger 1994].)

Considerable effort has been put into attempts to measure "mental fatigue." The Konzentrations-Leistungs-Test and the Grammatical Reasoning Test (David 1985) are examples of such tests. Unfortunately, operators recover from any such fatigue at the levels acceptable in simulation, sufficiently fast that any test carried out even immediately after a simulation shows no significant effect.

Structured interviews require a trained interviewer to ask an operator for his opinions and suggestions on a series of specific aspects of the exercise, according to a prepared script of questions. Interviewers are trained to draw out the operator and may follow up unplanned aspects if they seem important. It is a major skill of the interviewer to extract an impression of what the operators really feel about the organizations being examined. (Structured interviews can only be held after enough exercises have been completed for the operators to have formed opinions.)

Operators, like any other interviewees, are extremely sensitive to minor, nearly subliminal signals from the interviewer, which may lead to operator effects. There is a moderate risk that the interviewer may not be able to obtain definite responses to some questions. Inexperienced interviewers or those unfamiliar with the culture, language, or attitudes of the operators may lose their confidence and will obtain no useful data. Operators generally do their best to give truthful, realistic answers to the problems they see. There is, however, a moderate risk that they may give responses that are either what they think the group of which they are part will think or what they think the observer wants to hear (Langford and Deana 2003; (Oppenheim 1992).

Peer assessment, where equally competent operators observe the operators and judge the extent to which they are under strain, is very expensive in personnel and is subject to bias because the assessors tend to judge by what they would have done, rather than what the operators in fact did.

13.6 Conclusion

Real-time simulation is essentially an empirical process. Although considerable efforts have been made to place it on a "scientific" basis, they have not generally been successful.

Simulations are subject to many controlled, uncontrolled, or unknown disturbances. It is rarely possible to carry out sufficient trials to produce statistically significant results in the classic sense. Bayesian statistics may be applied, although they employ an unfamiliar concept of probability that is not always acceptable. The final conclusions of a real-time simulation are not always based on formal experimental methods. The judgment of the simulation leader is usually the determining factor. Fortunately, in most simulations, the results are so clearly unambiguous that the deficiencies in scientific terms do not matter.

Disclaimer

The views expressed in this chapter are the author's own and are not necessarily endorsed by EUROCONTROL or its management.

References

David, H. (1985). *Measurement of Controllers' Mental State in a Real Time Simulation Environment.* EEC Report No. 183. Bretigny sur Orge, France: EUROCONTROL Experimental Centre.

David, H. (1986). Comparison of three methods of assessing eye movements of air traffic controllers. *Ergonomics Society Annual Conference 1986*, Durham, England.

David, H. (2017). *Control Room Simulation: The Real Life of the Real Time.* R+D Hastings: Amazon Books.

David, H. and Pledger, S. (1995). *Speech Recognition and Keyboard Input for Control and Stress Reporting in an Air Traffic Control Simulation.* Contemporary Ergonomics 1995. London: Taylor and Francis.

Dubey, G. (2000). *Social Factors in Air Traffic Control Simulation.* EEC Report No. 348. Bretigny sur Orge, France: EUROCONTROL Experimental Centre.

Gawron, V.J. (2000). *Human Performance Measures Handbook.* Mahwah, NJ: Lawrence Earlbaum Associates.

ISO 11064-1 (2000). *Principles for the Design of Control Centres.* Geneva, Switzerland: International Standards Organisation.

ISO 11064-2 (2000). *Principles for the Arrangement of Control Suites*. Geneva, Switzerland: International Standards Organisation.

ISO 11064-3 (1999). *Control Room Layout ISO 11064-3:1999/Cor 1:2002*. Geneva, Switzerland: International Standards Organisation.

ISO 11064-4 (2004). *Layout and Dimensions of Workstations*. Geneva, Switzerland: International Standards Organisation.

ISO 11064-5 (2008). *Displays and Controls*. Geneva, Switzerland: International Standards Organisation.

ISO 11064-6 (2005). *Environmental Requirements for Control Centres*. Geneva, Switzerland: International Standards Organisation.

ISO 11064-7 (2006). *Principles for the Evaluation of Control Centres*. Geneva, Switzerland: International Standards Organisation.

Langford, J. and Deana, M. (2003). *Focus Groups*. London: Taylor and Francis.

Oppenheim, A.N. (1992). *Questionnaire Design, Interviewing and Attitude Measurement*. London: Cassel.

Pledger, S. (1994). Effects of self-assessment on performing air traffic control. M.Sc. thesis. Loughborough University, Loughborough, Leicestershire.

14

A Research Agenda for Human Systems Engineering

Andreas Tolk

The MITRE Corporation, Charlottesville, VA, USA

14.1 The State of Human Systems Engineering

The case studies and examples provided in this volume clearly demonstrate a wide range of contributions by human-centered disciplines to systems engineering disciplines, from traditional systems engineering (Sage 1992) to system of systems engineering approaches (Keating et al. 2003) and enterprise architecture engineering efforts (Bernard 2012) to the new paradigm of mission engineering (Sousa-Poza 2015). All these systems engineering disciplines are recognizing the role of the human when using the system or when being part of the system, as systems are becoming increasingly sociotechnical and complex.

Engineers often struggle with the fact that the human being, which defies precise specification and prescription, is a core system component, even if not standardizable. While some degree of emergence and complexity can be observed in most technical systems, the moment humans are part of the system, this is guaranteed to increase. Humans have often been the source of "messy details" that do not follow prescribed system behavior. The complexity they introduce is contrary to the orderly reductionism sought by traditional engineering approaches. Traditional engineering approaches therefore tend to be inadequate for the engineering of systems that encompass technology and human organizations, i.e. sociotechnical systems such as found in healthcare, city planning, hospital utilization, policy assessment, etc. (Nemeth et al. 2004).

Another aspect that increases complexity and adds unexpected trajectories through the system state space is that humans do not always follow the specifications of the intended use, but affordance is often more important: how can I use the technical parts of the system and their functionality best to fulfill my current needs. As a result, humans often use systems in a different way than the systems engineer originally intended. Particularly in systems of systems environments, where systems are used in a new and often highly dynamic and agile context, systems will not only be used in well-engineered, preconceived ensembles, but they will also self-organize in collectives, possibly exposing emergent attributes and behavior (Mittal et al. 2018).

The general topic of the examples given above is that humans introduce complexity. They do not always follow standards and specifications. They differ not only in physical traits but also in their behaviors, expectations, choices, emotions, and level of skill. Yet, they are a vital and undeniable part of almost all systems. Human-centered disciplines have knowledge, methods, and tools that

A Framework of Human Systems Engineering: Applications and Case Studies, First Edition.
Edited by Holly A. H. Handley and Andreas Tolk.
© 2021 The Institute of Electrical and Electronics Engineers, Inc. Published 2021 by John Wiley & Sons, Inc.

can help systems engineering adapt to and meet these challenges. For example, there is the work that has accumulated in the field of cognitive systems engineering. Hollnagel and Woods (1983) proposed the discipline of cognitive systems engineering to address the human gap in systems engineering, as covered in more detail in Woods and Roth (1988). Cognitive systems engineering focuses on principles and methods for engineering systems that perform cognitive work, such as information processing and decision making. This discipline considers a system to consist of all interacting elements – not just the technical. The recent rise of the Internet of Things (Ashton 2009) and of cyber–physical systems (Baheti and Gill 2011) introduced new challenges for systems engineering methods. The parallel development of more and more decentralized computing in the form of cloud computing, sensor clouds, fog computing, and related technology, as discussed by Sehgal et al. (2014), added another dimension of challenge. Once again, human-centered disciplines offer tools that can help systems engineering rise to these challenges.

We understand today that a much broader approach than relying purely on traditional engineering methods is needed to address the engineering of modern sociotechnical systems. We must address the various skills and capabilities, both physical and cognitive, to ensure that the technical parts of the system collaborate effectively and efficiently with the human parts. Human–machine interfaces and interactions are only a small facet of the challenge of cooperation in human–machine teams, as envisioned by Hoc (2000), among others, and exemplified for cyber–physical systems in Mosterman and Zander (2016). We need to include the anthropological perspectives as well, as humans have the potential to act differently across instances of the same set of local information, as a result of context sensitivity that extends to organizational, cultural, and historical contexts (Avison and Myers 1995). These observations can and need to be extended to organizations, adding another layer of complexity, ranging soon from human–machine challenges to organization system of systems challenges.

Within the leading chapter to this volume, we introduced a framework for human systems engineering. We want to use this framework and the various contributions of the experts who contributed chapters to this volume to identify gaps and shortcoming and, more importantly, a research agenda to close these gaps and overcome the shortcomings, following the examples given by Tolk and Rainey (2015). With this agenda, we hope to contribute to a discussion on the next steps, encourage doctoral students to look for relevant research topics, and provide inputs for professional organizations to identify challenges.

14.2 Recommendations from the Chapter Contributions

The authors of the various chapters of this volume implicitly and explicitly identified research topics that need to be better addressed. These research points are grouped by the topic of research, trying to provide a more general call for research action than the individual chapters do.

14.2.1 Data and Visualization Challenges

The use of big data methods to collect, analyze, and provide information to the human is addressed for traditional visual displays as well as for novel displays, such as wearable solutions used in the public safety and other domains. Generally, the data needs to be evaluated – sometimes in real time – and noteworthy events and aggregated valuable information need to be displayed.

A key challenge in the realm of big data is context-sensitive information filtering, integration, and management. This is especially difficult for systems operating in complex environments in which context is continuously shifting. Not all data are always equally important. Their relevance depends on the context of the situation and the goals of the user and may vary with the age of the data available. Interactive and immersive visualization tools and techniques, as discussed in Bach et al. (2017) and Cordeil et al. (2019), need to be further explored regarding their ability to provide information to the human when it is needed and in the form needed for the given context. Understanding the way information provided is perceived by people will require close collaboration with experts of the human sensory systems, such as perceptual psychologists. The way information is perceived and used is also highly influenced by the autonomic nervous system; experts in this domain will also be needed. This is highly important when humans are part of the system: how do the short- and long-term stress responses influence individuals and what does this mean for the system. Experts in perception and stress response will need to translate their research results into applicable solutions and best practices for systems engineers.

The second topic deals with the availability of data, technically and legally. While the literature on big data extracted from social media or other huge and heterogeneous data sources is growing, human systems engineering may encounter a new challenge: how to understand if the correct data is available. What if the available data is heavily biased or otherwise misleading? Can we develop tests that can be applied to avoid the use such data?

Finally, real-world data tends to be incomplete, contradictive, vague, and rarely collected with the required accuracy and resolution. How can trust in the data be increased? How can we ensure that by cleaning the data, we are not losing data that could easily hint toward an unexpected event, as often observed in complex environments? Data scientists often start with the assumption that the answer lies in the data, but the way we collect data already limits the questions we can ask, as we use a data model to collect data that specifies the abstraction level and scope of the data to be collected. If an important piece of evidence is exposed by data not perceived to be important when the data model is created, how can we collect the related important information? Philosophy of science can help to ask the right questions to come closer to a possible solution or at least derive heuristics (Gelfert 2016).

14.2.2 Next-Generation Computing

Many systems of interest to the human systems engineering community will be software-intensive systems. Within the information age, our ability to store, communicate, and compute information grew exponentially over the last years. Today, digital communication surpasses every other means by far. Some examples are given in Hilbert and López (2011), estimating that 99.9% of telecommunication has been digital since 2007 and 94% of the technological memory uses digital formats. Human systems engineering must take advantage of this data richness under the constraints discussed in the section on data and visualization.

What is going to support better access to digital information for the human are developments in sensing, imaging, and communication. As observed in Passian and Imaan (2019), accessibility through cloud, fog, and mist computing technologies and the advancement in nanotechnology allow not only for continuous integration into the data world, but it also allows for new, potentially more human-centered access of data, e.g. using augmented reality, context-sensitive displays, etc. Some recent examples are given in Fraga-Lamas et al. (2018).

The danger is sensory overload, but the technological opportunities seem to be endless. However, what is technically feasible and what is useful for the human is the subject of ongoing research. Again, tight collaboration between experts in human perception and sensorics with technology experts is pivotal to maximize the benefit for human systems engineering.

14.2.3 Advanced Methods and Tools

Four interrelated but individual methods and supporting tools of additional interest for future research are utilized in the chapters, namely, modeling and simulation, artificial intelligence, digital twins, and autonomy. These technologies are used individually or in concert to drive a control room of the future, evaluate and compare alternative cockpits of aircraft and spacecraft, and help to support latest breakthrough solutions in engineering.

Modeling and simulation is often discussed as its own discipline but shows exceptional potential when combined with systems engineering methods. Tolk et al. (2017) show the potential when model-based systems engineering is extended to enable the execution of the models as simulation not only individually but also as simulated entities with a portfolio of enterprise-wide operations. Recent years have seen some research on the possible alignment of artifacts that will allow us to standardize the use of simulation within systems engineering beyond the scope of proprietary tool solutions, such as Mittal (2006), Shuman (2010), or recently Traoré (2019).

What may be more important than technological advances could be the application of modeling and simulation in the era of complexity and human systems engineering. In flight training, simulators are used to immerse pilots early into a high-fidelity representation of their environment, as close as possible to the real thing. The philosophy is to learn the theory, e.g. by studying the manuals, build a mental model, and then practice first in the virtual, and later in the real environment. However, we may miss a chance to use simulation to develop the best mental model, particularly when the system the human will use, or will be part of, is complex. Introducing the trainee gradually to these new systems, using different increasing degrees of complexity and difficulty to build up an experience, is often underutilized. Again, experts in human learning can be of significant help to avoid such shortfalls in human systems engineering approaches.

Artificial intelligence is needed whenever technical systems expose intelligent behavior. This can be the behavior of a robot, the behavior of an entity within the system – or the simulation thereof, or any of its components. In recent years, machine learning and deep learning experienced a revival, as the computational power of today allows us to utilize huge amounts of data to train neural networks and evaluate the data using multivariable correlation analysis tools. Although initial approaches are established to make artificial intelligence interpretable, such as summarized in Došilović et al. (2018), this topic remains a rich research field. How can we build trust into machine decisions?

Digital twins were originally developed in the manufacturing domain. They are digital replicas that allow us to couple the system with its virtual counterpart in support of continuous collection and evaluation of data to not only monitor the system but also to continuously optimize it within the desired process. They allow "what-if" analysis and evaluations of alternatives in real time without having to interrupt the supported process. They are often understood as a bridge between the physical world and its cyber representation. It is possible to build the digital twin before the original system, which loops us back to the executable versions of systems engineering artifacts. The number of digital twins representing other systems or components is growing daily. How can they be used to better support human-centric processes in the future? Can digital twins replace standardized interfaces with individual interfaces for each user of the system? Can the technology be used to better adapt the system behavior to the individual need of humans within or using the system? How can they support better training?

Finally, autonomy has been addressed as well. The US Armed Forces used these categories to identify the following definition for classes of autonomous systems (Williams 2008):

- Human-operated systems are fully controlled by humans. All activities result from human initialization, eventually based on provided sensor information.

- Human-assisted systems perform activities in parallel with human inputs, augmenting the human's ability.
- Human-delegated systems perform limited control activities. Humans can overrule the system at any time.
- Human-supervised systems conduct all activities needed to perform a given mission, but they inform the human consistently, including providing explanations for decisions.
- Mixed initiative systems are capable of human–machine teams and can take over given tasks independently.
- Fully autonomous systems require no human intervention or presence. They conduct all activities, across all ranges of conditions.

Besides ensuring the level of required autonomy, research is needed regarding what level of autonomy will be needed and accepted by humans in the various application domains of human systems engineering. Does it make sense to strive for full autonomy, or does it make more sense to put the emphasis on human–machine teams, in which case the machines will have a lower degree of autonomy?

14.2.4   Increased Integration of Social Components into System Artifacts

Human-centered disciplines are interested in social components as much, if not more, as technical components. The systems engineering community has an abundance of experience with frameworks and languages that can be applied to do so, such as the System Markup Language (SysML), the Universal Markup Language (UML), and business process modeling (BPM). The rigor and precision of these languages have been intriguing for some time, as documented by books like Eriksson and Penker (1999). Only recently, after working with members from the humanities, the experience was captured in Tolk et al. (2018). Contributions for a recent Defense Advanced Research Projects Agency on computational representation of human and social behavior are captured in Davis et al. (2019).

Despite these success stories, questions remain: Are these engineering artifacts the best way for addressing humans in human systems engineering? Are there better alternatives to elicit knowledge and capture it unambiguously for better integration? Is the critical decision method, designed to gather data on the bases for proficient performance of naturalistic tasks, applicable (Klein et al. 1989)? What are better alternatives?

14.3   Uniting the Human Systems Engineering Stakeholders

A common topic of challenges is the need to bring the various stakeholders together. They are coming from different backgrounds and various disciplines and often do not share common methods and tools. Systems engineering and system of systems engineering focus more on the technical aspects; enterprise architectures emphasize the organizational aspects; human–machine interfaces and human–machine teams focus on the human views. The community came a long way from seeing people as a disturbing factor in an otherwise well-specified and well-behaved system to the need of systems to take physical and cognitive diversity of humans into account and adapt to them. Nonetheless, to better support all aspects of affordance, an even tighter connection to the humanities – including anthropology, psychology, and sociology – will be needed, adding to the variety and diversity of terms and tools, as these disciplines unquestionably differ, often significantly, from traditional engineering approaches.

The following three subsections address fundamental aspects that must be understood and supported to allow human systems engineering to overcome some of the current shortfalls. First, a truly transdisciplinary approach is needed. Second, this approach is utilizing common formalisms, and these common formalisms allow thirdly for common metrics to be used to evaluate and validate approaches.

14.3.1 Transdisciplinary Approach

Traditional disciplines are highly specialized in their topics and successfully help to grow knowledge over decades. However, from the second half of the twentieth century, the challenges resulting from relying too much on reductionism led to a desire for more collaboration, creating new knowledge connecting the contributing traditional disciplines (Apostel et al. 1972). A recent National Academy of Sciences report on interdisciplinary research identified four primary drivers, namely, (i) the recognition of the inherent complexity of nature and society and the inability of reductionism to cope with these challenges, (ii) exploring problems and questions that are not confined to a single discipline, (iii) growing societal problems that require a broader approach on a shorter timescale, and (iv) emergence of new technologies that are applicable in more than one discipline (National Academy of Sciences 2004).

A recent overview of the various interdisciplinary ideas and applications in academia and industry has been compiled in Frodeman et al. (2017). Within this compilation, examples for various degrees of collaboration between the participating disciplines are observed and mapped to the taxonomy proposed by Klein (2017), who distinguishes between multidisciplinarity, interdisciplinarity, and transdisciplinarity as the main categories.

In multidisciplinary teams a loose and temporary coupling of disciplines to solve a common problem is observed. This juxtaposes disciplines and their methods. The multidisciplinary team coordinates and sequences their contributions, which often are conducted within the hosting discipline. The contributions complement each other, often only connected via the exchange of results. The definition of terms is aligned with a focus on the collaborative effort; the information exchange is established ad hoc by the team. Once the common problem is coped with, the team usually disengages.

Interdisciplinarity collaboration creates a closer linkage between disciplines. Data are aligned, and processes are orchestrated, supporting a much higher degree of integration. The interdisciplinary team focuses on identifying overlapping domains of knowledge and uses integrated solutions in its participating disciplines. By doing so, they are building permanent bridges that link domains together. In addition to the research results, common concepts and common semantics of terms are established that become part of the originating disciplines. This allows them to collaborate on a whole set challenges, and teams often continue to work on the mutual extension of the knowledge of their disciplines, resulting in common methods and tools.

The strongest coupling of disciplines is captured by the transdisciplinary approach, which transcends, transgresses, and transforms the disciplines and specialties into something new. Concepts, terms, and activities are not only described in common terms, but also they are systematically integrated, and new interactions across the sectors of original disciplines are defined. While knowledge components are used to complement each other in multidisciplinary teams, transdisciplinarity hybridizes the knowledge. This teams are often permanent, leading to new disciplines. Complexity science is an example.

In order to succeed, human systems engineering must become transdisciplinary. While a coordinated use of methods and tools of the human-centered systems engineering disciplines is a necessary and valuable start, they will not be sufficient to address the challenges in a sustainable fashion. The first step is a better alignment of our terminology. The use of common formalisms will provide a foundation for transdisciplinarity, but the concerned effort for this closest form of collaboration needs to come from the community.

14.3.2 Common Formalisms

A powerful example of what will happen in the absence of a common formalism, or even a common terminology, is described in Chen and Crilly (2016). The authors describe their analysis of two teams, both working in the complexity domain. The first team worked on synthetic biology challenges, while the second team was engineering swarm robotics solutions. The research showed that the problem space of both teams overlapped significantly, even more than the problem domains of the parent disciplines of biology and engineering. However, as the terminology of these disciplines is different, the complexity teams were not aware of the possibility to share and reuse results from the other group. We know similar observations from our daily experiences as well. For example, the term virtual reality has many interpretations in the various communities. While technology-oriented groups mainly think about head-mounted displays or virtual glasses of the latest generation, the cognitive scientist is more interested in how the brain processes the new workload. That does not mean that we all have to use the same terms, but we must be aware of their meaning in the different contexts they are used in.

While changing the terminology of established disciplines is unlikely, capturing the research in a mathematically rigorous formalism is a possibility. Such a formalism ensures high-quality research, dissemination of the results, and the general contribution to the archived body of knowledge that provides a comprehensive and concise representation of concepts, terms, and activities needed to make up a professional scientific domain. If this formalism can be read and processed by machines, a significant amount of work when it comes to components matching for compositions, or selection of alternatives, becomes possible. The "digital engineering initiative" within the US Department of Defense (DoD) is an example to strive for a common description of contributing elements to the DoD Enterprise, which contributes to "deliberate planning, analyzing, organizing, and integrating of current and emerging operational and system capabilities to achieve desired operational mission effects" as envisioned for mission engineering (Defense Aquisition University 2017) to allow for operational agility of the military commander.

A common formalism allows one to comprehend, share, and reproduce research results to increase knowledge over time. Without such precise communication means, the human systems engineering community is doomed to run into the same problems as discussed by Chen and Crilly, particularly when working with the humanities to improve the understanding of humans in all their roles. Furthermore, such a formalism allows us to transfer solutions more easily, such as improvements in the use of artificial intelligence, data science, machine learning, and other common methods of interest.

In this context, the use of ontologies and taxonomies is a valuable first contribution, as these formal presentations of knowledge facilitate communication. The foundations for such representations have been compiled in Sowa (1999). The framework used to organize the various contributions to this volume is an example of a possible taxonomy to address the challenges. Other examples that may also contribute are the contributions to the classification of human–robot interaction, such as provided by Yanco and Dury (2002) and updated by Yanco and Dury (2004). These later taxonomies are starting to look at teams of humans and robots, and we can generalize this to humans and systems, and how the task to be accomplished will influence the team structure and the distribution of subtasks.

14.3.3 Common Metrics

Traditional systems engineering starts from a set of requirements. A physical architecture provides the system components and their interconnections to host the functions needed to fulfill the requirements. The functional architecture defines the orchestrated set of these functions. Finally, a set of metrics is defined that measures the performance in order to evaluate if the requirements indeed are met by the engineered system. We need such a set of metrics as well, not only for the engineered components but also for the humans interacting with or within the system, as well as for the human–machine collaboration parts. While it makes sense to start with the established sets of measures of performance (how well are the system's functions performed) and measures of effectiveness (how well does the system contribute to the operational objective of the portfolio it is part of), this is only a first step. The use of trusted private blockchain technology in the context of human systems engineering is one example.

Developing metrics to test and evaluate organizational resilience is another topic of interest. In the context of resilience for cyber–physical systems, some examples are given in Haque et al. (2019). They identify requirements for metrics, mainly focusing on resilience, that should be applicable in the context of human systems engineering as well, but additional research is needed. However, the metrics cannot be limited to the cyber–physical components but must include the behavior of humans as well as organizations. A long-term goal may be to replace the current risk assessment and mitigation methods with an organizational resilience framework, which will help to move from optimized point solutions toward robust and stable solutions that can remain useful even in complex and changing environments. Ultimately, we do not only want to achieve resilient systems, but we want to engineer antifragile systems as introduced by Taleb (2012). These systems do not only survive changes in their environment, they even get better under stress by adapting their components and rules accordingly. Such a framework must represent resilience at the system level, where system resilience is a function of technical and social interactions and dynamics. The research summarized in Kantur and İşeri-Say (2012) can help with some of the organizational aspects, but a comprehensive approach bringing together systems engineering, system of systems engineering, and human and organizational disciplines together is still missing.

14.4 Summary

In this final chapter, a selection of research topics challenges has been compiled into several categories. The enumeration is obviously neither complete nor exclusive, and the categorization is just a first approach of grouping these challenges. Nonetheless, we hope that members of the scholastic community will contribute to the improvement of this first topology of challenges as well as the framework for human systems engineering itself. One observation is that we have already seen additional ideas to extend the framework and to apply higher fidelity in the various dimensions in some of the contributing chapters to this volume. However, what has not yet been approached sufficiently is the question of which measures of merit to be applied, including which parameters to collect and which metrics to use.

Furthermore, engineering management research has shown that for new ideas to be successfully applied in real-world solutions, three components are needed:

- New methods must be supported by a set of mature tools with interfaces that make it easy to integrate them into the existing workflow, preferably with little additional work for the workforce. The more parameters that can be collected automatically, the better.

- The workforce needs to be educated and trained in new processes and tools. They may seem obvious to the process creator and the tool developer, but the diversity of the workforce requires a good education to maximize the use of the innovative ideas.
- The management needs to buy into the new ideas. If the management is not integrated into the new processes and does not actively promote the change, there is a high likelihood that everybody will fall back to the old solutions, as soon as first obstacles are encountered.

This should not imply that the introduction will happen in well-defined sequences. Cognitive systems engineering and complexity science show us that new ideas and the hosting work systems will coevolve in a very agile process with many interdependencies and feedback loops.

Finally, human systems engineering may learn some valuable lessons from the emerging field of socio-environment systems modeling, which addresses some interesting overlaps with our human systems engineering challenges. In their recent research paper, Elsawah et al. (2020) identify eight grand challenges that also look for a common formal approach to allow bridging epistemologies across disciplines, integrated treatment of modeling uncertainties, combining qualitative and quantitative methods and data sources, dealing with various scales and scaling, capturing systemic changes in the system of interest, integrating the human dimension, leveraging new data types and data sources, and elevating the resulting models and impacts on policy and decision making. All these topics should be addressed in a research agenda for human systems engineering as well.

As observed in the introductory remarks to this volume, the necessity to improve methods to integrate the human element into systems engineering will only increase. The integration of the human as the critical element of the system, as a component or user, remains one of the major challenges. With the cases of human systems engineering presented in this volume, we hope to contribute ideas that contribute to the formation of a transdisciplinary paradigm and to a more robust, varied, and comprehensive set of tools that allow us to tackle modern complex sociotechnical systems more effectively.

Disclaimer

The author's affiliation with The MITRE Corporation is provided for identification purposes only and is not intended to convey or imply MITRE's concurrence with, or support for, the positions, opinions, or viewpoints expressed by the author. This paper has been approved for Public Release; Distribution Unlimited; Case Number 19-01906-20.

References

Apostel, L., Berger, G., Briggs, A., and Michaud, G. (1972). *Interdisciplinarity: Problems of Teaching and Research in Universities*. Paris: ERIC.

Ashton, K. (2009). That "internet of things" thing. *RFID Journal* 22 (7): 97–114.

Avison, D.E. and Myers, M.D. (1995). Information systems and anthropology: and anthropological perspective on IT and organizational culture. *Information & People* 8 (3): 43–56.

Bach, B., Sicat, R., Beyer, J. et al. (2017). The hologram in my hand: how effective is interactive exploration of 3d visualizations in immersive tangible augmented reality? *IEEE Transactions on Visualization and Computer Graphics* 24 (1): 457–467.

Baheti, R. and Gill, H. (2011). Cyber-physical systems. *The impact of control technology* 12: 161–166.

Bernard, S.A. (2012). *An Introduction to Enterprise Architecture*. AuthorHouse.

Chen, C.-C. and Crilly, N. (2016). Describing complex design practices with a cross-domain framework: learning from synthetic biology and swarm robotics. *Research in Engineering Design* 27 (3): 291–305.

Cordeil, M., Cunningham, A., Bach, B. et al. (2019). IATK: an immersive analytics toolkit. In: *IEEE Conference on Virtual Reality and 3D User Interfaces*, 200–209. Osaka, Japan: IEEE Press.

Davis, P.K., O'Mahony, A., and Pfautz, J. (2019). *Social-Behavioral Modeling for Complex Systems*. Hoboken, NJ: Wiley.

Defense Acquisition University. (2017). *Defense Acquisition Guidebook*.

Došilović, F.K., Brčić, M., and Hlupić, N. (2018). Explainable artificial intelligence: a survey. In: *41st International Convention on Information and Communication Technology, Electronics and Microelectronics (MIPRO)*, 210–215. IEEE Press.

Elsawah, S., Filatova, T., Jakeman, A.J. et al. (2020). Eight grand challenges in socio-environmental systems modeling. *Socio-Environmental Systems Modeling* 2: 16226.

Eriksson, H.-E. and Penker, M. (1999). *Business Modeling with UML: Business Patterns at Work*. Hoboken, NJ: Wiley.

Fraga-Lamas, P., Fernández-Caramés, T.M., Blanco-Novoa, Ó., and Vilar-Montesinos, M.A. (2018). A review on industrial augmented reality systems for the industry 4.0 shipyard. *IEEE Access* 6: 13358–13375.

Frodeman, R., Klein, J.T., and Pacheco, R.C. (2017). *The Oxford Handbook of Interdisciplinarity*. Oxford University Press.

Gelfert, A. (2016). *How To Do Science with Models: A Philosophical Primer*. Cham, Switzerland: Springer International Publishing.

Haque, M.A., Shetty, S., and Krishnappa, B. (2019). Cyber-physical system resilience: frameworks, metrics, complexities, challenges, and future directions. In: *Complexity Challenges in Cyber Physical Systems* (eds. S. Mittal and A. Tolk), 301–337. Hoboken, NJ: Wiley.

Hilbert, M. and López, P. (2011). The World's technological capacity to store, communicate, and compute information. *Science* 332 (6025): 60–65.

Hoc, J.-M. (2000). From human-machine interaction to human-machine cooperation. *Ergonomics* 43 (7): 833–843.

Hollnagel, E. and Woods, D.D. (1983). Cognitive systems engineering: new wine in new bottles. *International Journla on Man-Machine Studies* 18 (6): 583–600.

Kantur, D. and İşeri-Say, A. (2012). Organizational resilience: a conceptual integrative framework. *Journal of Management & Organization* 18 (6): 762–773.

Keating, C., Ralph Rogers, R.U., David Dryer, A.S.-P., and Rabadi, G. (2003). System of systems engineering. *Engineering Management Journal* 15 (3): 36–45.

Klein, J.T. (2017). Typologies of interdisciplinarity: the boundary work of definition. In: *The Oxford Handbook of Interdisciplinarity*, 2e (eds. R. Frodeman, J.T. Klein and R.C. Pacheco), 21–34. Oxford University Press.

Klein, G.A., Calderwood, R., and Macgregor, D. (1989). Critical decision method for eliciting knowledge. *IEEE Transactions on Systems, Man, and Cybernetics* 19 (3): 462–472.

Mittal, S. (2006). Extending DoDAF to allow integrated DEVS based modeling and simulation. *Journal of Defense Modeling and Simulation* 3 (2): 95–123.

Mittal, S., Diallo, S.Y., and Tolk, A. (2018). *Emergent Behavior in Complex Systems Engineering: A Modelingand Simulation Approach*. Hoboken, NJ: Wiley.

Mosterman, P.J. and Zander, J. (2016). Cyber-physical systems challenges: a needs analysis for collaborating embedded software systems. *Software and Systems Modeling* 15 (1): 5–16.

National Academy of Sciences (2004). *Facilitating Interdisciplinary Research*. Washington, DC: National Academies Press.

Nemeth, C.P., Cook, R.I., and Woods, D.D. (2004). The messy details: insights from the study of technical work in healthcare. *IEEE Transactions on Systems, Man, and Cybernetics* 34 (6): 689–692.

Passian, A. and Imaan, N. (2019). Nanosystems, edge computing, and the next generation of computing systems. *Sensors (Basel)* 19 (18): 4048.

Sage, A.P. (1992). *Systems Engineering*. Hoboken, NJ: Wiley.

Sehgal, V., Patrick, A., and Rajpoot, L. (2014). A comparative study of cyber physical cloud, cloud of sensors and internet of things: their ideology, similarities and differences. In: *IEEE International Advance Computing Conference*, 708–716. Gurgaon, India: IEEE.

Shuman, E.A. (2010). Understanding executable architectures through an examination of language model elements. In: *Summer Computer Simulation Conference*, 483–497. San Diego, California: SCS Press.

Sousa-Poza, A. (2015). Mission engineering. *International Journal of System of Systems Engineering* 6 (3): 161–185.

Sowa, J.F. (1999). *Knowledge Representation: Logical, Philosophical and Computational Foundations*. Pacific Grove: Brooks/Cole Publishing.

Taleb, N.N. (2012). *Antifragile: Things That Gain from Disorder*. New York, NY: Random House Incorporated.

Tolk, A. and Rainey, L.B. (2015). Towards a research agenda for M&S support for system of systems engineering. In: *Modeling and Simulation Support for System of Systems Engineering Applications* (eds. L.B. Rainey and A. Tolk), 581–592. Hoboken, NJ: Wiley.

Tolk, A., Glazner, C.G., and Pitsko, R. (2017). Simulation-based systems engineering. In: *Guide to Simulation-Based Disciplines* (eds. S. Mittal, U. Durak and T. Ören), 75–102. Cham, Switzerland: Springer International Publishing.

Tolk, A., Wildman, W.J., Shults, F.L., and Diallo, S.Y. (2018). Human simulation as the lingua franca for computational social sciences and humanities: potential and pitfalls. *Journal of Cognition and Culture* 18 (5): 462–482.

Traoré, M.K. (2019). Unified approaches to modeling. In: *Model Engineering for Simulation* (eds. L. Zhang, B.P. Zeigler and Y. LaiLi), 43–56. Academic Press.

Williams, R. (2008). *Autonomous Systems Overview*. Technical report. BAE Systems.

Woods, D.D. and Roth, E.M. (1988). Cognitive systems engineering. In: *Handbook of Human-Computer Interaction*, 1e (ed. M.G. Helander), 3–43. Amsterdam, The Netherlands: North-Holland.

Yanco, H.A. and Dury, J.L. (2002). A taxonomy for human-robot interaction. In: *AAAI Fall Symposium on Human-Robot Interaction*, 111–119). Technical Report FS-02-03. Falmouth, Massachusetts: AAAI.

Yanco, H.A. and Dury, J.L. (2004). Classifying human-robot interaction: an updated taxonomy. In: *International Conference on Systems, Man and Cybernetics*, 2841–2846. IEEE Press.

Index

A Framework of Human Systems Engineering: Applications and Case Studies, First Edition.
Edited by Holly A. H. Handley and Andreas Tolk.
© 2021 The Institute of Electrical and Electronics Engineers, Inc. Published 2021 by John Wiley & Sons, Inc.

Printed and bound by CPI Group (UK) Ltd, Croydon, CR0 4YY